Epistemology

SUNY series in Philosophy
George R. Lucas Jr., editor

Epistemology

An Introduction to the Theory of Knowledge

Nicholas Rescher

State University of New York

Published by
State University of New York Press, Albany

© 2003 State University of New York

All rights reserved

Printed in the United States of America

No part of this book may be used or reproduced in any manner whatsoever without written permission. No part of this book may be stored in a retrieval system or transmitted in any form or by any means including electronic, electrostatic, magnetic tape, mechanical, photocopying, recording, or otherwise without the prior permission in writing of the publisher.

For information, address State University of New York Press,
90 State Street, Suite 700, Albany, NY 12207

Production by Michael Haggett
Marketing by Anne M. Valentine

Library of Congress Cataloging-in-Publication Data

Rescher, Nicholas.
 Epistemology : an introduction to the theory of knowledge / Nicholas Rescher.
   p. cm. — (SUNY series in philosophy)
 Includes bibliographical references and index.
 ISBN 0-7914-5811-3 (alk. paper) — ISBN 0-7914-5812-1 (pbk. : alk. paper)
  1. Knowledge, Theory of. I. Title. II. Series

BD161R477   2003
121—dc21

2003057270

10 9 8 7 6 5 4 3 2 1

Contents

Preface ... xi

Introduction ... xiii

KNOWLEDGE AND ITS PROBLEMS

Chapter 1: Modes of Knowledge ... 3
- Is Knowledge True Justified Belief? ... 3
- Modes of (Propositional) Knowledge ... 7
- Other Basic Principles ... 10

Chapter 2: Fallibilism and Truth Estimation ... 15
- Problems of Metaknowledge ... 15
- The Preface Paradox ... 19
- The Diallelus ... 22
- An Apory and Its Reconciliation: K-Destabilization ... 23
- Costs and Benefits ... 26
- More on Fallibilism ... 27
- The Comparative Fragility of Science: Scientific Claims as Mere Estimates ... 29
- Fallibilism and the Distinction Between Our (Putative) Truth and the Real Truth ... 34

Chapter 3: Skepticism and Its Deficits ... 37
- The Skeptic's "No Certainty" Argument ... 37
- The Role of Certainty ... 39
- The Certainty of Logic Versus the Certainty of Life ... 41

	PRAGMATIC INCONSISTENCY	42
	SKEPTICISM AND RISK	45
	RATIONALITY AND COGNITIVE RISK	49
	THE ECONOMIC DIMENSION: COSTS AND BENEFITS	53
	THE DEFICIENCY OF SKEPTICISM	56
Chapter 4:	Epistemic Justification in a Functionalistic and Naturalistic Perspective	61
	EXPERIENCE AND FACT	61
	PROBLEMS OF COMMON-CAUSE EPISTEMOLOGY	62
	MODES OF JUSTIFICATION	64
	THE EVOLUTIONARY ASPECT OF SENSORY EPISTEMOLOGY	68
	RATIONAL VERSUS NATURAL SELECTION	69
	AGAINST "PURE" INTELLECTUALISM	74
	THE PROBLEM OF ERROR	76
	CONCLUSION	78
Chapter 5:	Plausibility and Presumption	81
	THE NEED FOR PRESUMPTIONS	81
	THE ROLE OF PRESUMPTION	85
	PLAUSIBILITY AND PRESUMPTION	87
	PRESUMPTION AND PROBABILITY	90
	PRESUMPTION AND SKEPTICISM	92
	HOW PRESUMPTION WORKS: WHAT JUSTIFIES PRESUMPTIONS	96
Chapter 6:	Trust and Cooperation in Pragmatic Perspective	101
	THE COST EFFECTIVENESS OF SHARING AND COOPERATING IN INFORMATION ACQUISITION AND MANAGEMENT	101
	THE ADVANTAGES OF COOPERATION	103
	BUILDING UP TRUST: AN ECONOMIC APPROACH	104
	TRUST AND PRESUMPTION	106
	A COMMUNITY OF INQUIRERS	108

RATIONAL INQUIRY AND THE QUEST FOR TRUTH

Chapter 7:	Foundationalism and Coherentism	113
	HIERARCHICAL SYSTEMIZATION: THE EUCLIDEAN MODEL OF KNOWLEDGE	113
	CYCLIC SYSTEMIZATION: THE NETWORK MODEL— AN ALTERNATIVE TO THE EUCLIDEAN MODEL	118

	THE CONTRAST BETWEEN FOUNDATIONALISM AND COHERENTISM	123
	PROBLEMS OF FOUNDATIONALISM	128
Chapter 8:	The Pursuit of Truth: Coherentist Criteriology	131
	THE COHERENTIST APPROACH TO INQUIRY	131
	THE CENTRAL ROLE OF DATA FOR A COHERENTIST TRUTH-CRITERIOLOGY	135
	ON VALIDATING THE COHERENCE APPROACH	139
	IDEAL COHERENCE	145
	TRUTH AS AN IDEALIZATION	147
Chapter 9:	Cognitive Relativism and Contexualism	151
	COGNITIVE REALISM	151
	WHAT'S WRONG WITH RELATIVISM	153
	THE CIRCUMSTANTIAL CONTEXTUALISM OF REASON	155
	A FOOTHOLD OF ONE'S OWN: THE PRIMACY OF OUR OWN POSITION	159
	THE ARBITRAMENT OF EXPERIENCE	161
	AGAINST RELATIVISM	165
	CONTEXTUALISTIC PLURALISM IS COMPATIBLE WITH COMMITMENT ON PURSUING "THE TRUTH"	168
	THE ACHILLES' HEEL OF RELATIVISM	170
Chapter 10:	The Pragmatic Rationale of Cognitive Objectivity	173
	OBJECTIVITY AND THE CIRCUMSTANTIAL UNIVERSITY OF REASON	173
	THE BASIS OF OBJECTIVITY	175
	THE PROBLEM OF VALIDATING OBJECTIVITY	177
	WHAT IS RIGHT WITH OBJECTIVISM	180
	ABANDONING OBJECTIVITY IS PRAGMATICALLY SELF-DEFEATING	182
Chapter 11:	Rationality	187
	STAGE-SETTING FOR THE PROBLEM	187
	OPTIMUM-INSTABILITY	188
	IDEAL VERSUS PRACTICAL RATIONALITY: THE PREDICAMENT OF REASON	190
	THE PROBLEM OF VALIDATING RATIONALITY	193
	THE PRAGMATIC TURN: EVEN COGNITIVE RATIONALITY HAS A PRAGMATIC RATIONALE	196

| | Alternative Modes of Rationality? | 198 |
| | The Self-Reliance of Rationality is Not Viciously Circular | 203 |

COGNITIVE PROGRESS

Chapter 12: Scientific Progress — 209

- The Exploration Model of Scientific Inquiry — 210
- The Demand for Enhancement — 211
- Technological Escalation: An Arms Race Against Nature — 212
- Theorizing as Inductive Projection — 215
- Later Need Not Be Lesser — 217
- Cognitive Copernicanism — 221
- The Problem of Progress — 223

Chapter 13: The Law of Logarithmic Returns and the Complexification of Natural Science — 229

- The Principle of Least Effort and the Methodological Status of Simplicity-Preference in Science — 230
- Complexification — 234
- The Expansion of Science — 239
- The Law of Logarithmic Returns — 240
- The Rationale and Implications of the Law of Logarithmic Returns — 245
- The Growth of Knowledge — 248
- The Deceleration of Scientific Progress — 251
- Predictive Implications of the Information/Knowledge Relationship — 253
- The Centrality of Quality and Its Implications — 254

Chapter 14: The Imperfectability of Knowledge: Knowledge as Boundless — 257

- Conditions of Perfected Science — 257
- Theoretical Adequacy: Issues of Erotetic Completeness — 259
- Pragmatic Completeness — 262
- Predictive Completeness — 264
- Temporal Finality — 267

	"Perfected Science" as an Idealization that Affords a Useful Contrast Conception	271
	The Dispensability of Perfection	273

COGNITIVE LIMITS AND THE QUEST FOR TRUTH

Chapter 15:	The Rational Intelligibility of Nature	279
	Explaining the Possibility of Natural Science	279
	"Our" Side	282
	Nature's Side	284
	Synthesis	287
	Implications	289
Chapter 16:	Human Science as Characteristically Human	293
	The Potential Diversity of "Science"	293
	The One World, One Science Argument	297
	A Quantitative Perspective	299
	Comparability and Judgments of Relative Advancement or Backwardness	305
	Basic Principles	308
Chapter 17:	On Ignorance, Insolubilia, and the Limits of Knowledge	315
	Concrete versus Generic Knowledge and Ignorance	315
	Erotetic Incapacity	317
	Divine versus Mundane Knowledge	318
	Issues of Temporalized Knowledge	319
	Kant's Principle of Question Exfoliation	321
	Cognitive Incapacity	323
	Insolubilia Then and Now	324
	Cognitive Incapacity	325
	Identifying Insolubilia	327
	Relating Knowledge to Ignorance	329
	Postscript: A Cognitively Indeterminate Universe	330
Chapter 18:	Cognitive Realism	333
	Existence	333
	Is Man the Measure?	334
	Realism and Incapacity	337

THE COGNITIVE OPACITY OF REAL THINGS	339
THE COGNITIVE INEXHAUSTIBILITY OF THINGS	341
THE CORRIGIBILITY OF CONCEPTIONS	343
THE COGNITIVE INEXHAUSTIBILITY OF THINGS	344
COGNITIVE DYNAMICS	345
CONCEPTUAL BASIS OF REALISM AS A POSTULATE	347
HIDDEN DEPTHS: THE IMPETUS TO REALISM	352
THE PRAGMATIC FOUNDATION OF REALISM AS A BASIS FOR COMMUNICATION AND DISCOURSE	355
THE IDEALISTIC ASPECT OF METAPHYSICAL REALISM	360
SCIENCE AND REALITY	361
Notes	369
Index	403

PREFACE

This book is based on work in epistemology extending over several decades. It combines into a systematic whole ideas, arguments, and doctrines evolved in various earlier investigations. The time has at last seemed right to combine these deliberations into a single systematic whole and this book is the result.

Philosophers are sometimes heard to say that the present is a post-epistemological era and that epistemology is dead. But this is rubbish. If the time ever came when people ceased to care for epistemological questions—as illustrated by the topics treated in the present book—it would not be just epistemology that has expired but human curiosity itself.

I am grateful to Estelle Burris for her competence and patience in putting this material into a form where it can meet the printer's needs.

Nicholas Rescher
Pittsburgh
March 2002

Introduction

The mission of epistemology, the theory of knowledge, is to clarify what the conception of knowledge involves, how it is applied, and to explain why it has the features it does. And the idea of knowledge at issue here must, in the first instance at least, be construed in its modest sense to include also belief, conjecture, and the like. For it is misleading to call cognitive theory at large "epistemology" or "the theory of knowledge." Its range of concern includes not only knowledge proper but also rational belief, probability, plausibility, evidentiation and—additionally but not least—*erotetics*, the business of raising and resolving questions. It is this last area—the theory of rational inquiry with its local concern for questions and their management—that constitutes the focus of the present book. Its aim is to maintain and substantiate the utility of approaching epistemological issues from the angle of questions. As Aristotle already indicated, human inquiry is grounded in wonder. When matters are running along in their accustomed way, we generally do not puzzle about it and stop to ask questions. But when things are in any way out of the ordinary we puzzle over the reason why and seek for an explanation. And gradually our horizons expand. With increasing sophistication, we learn to be surprised by virtually *all* of it. We increasingly want to know what makes things tick—the ordinary as well as the extraordinary, so that questions gain an increasing prominence within epistemology in general.

Any profitable discussion of knowledge does well to begin by recognizing some basic linguistic facts about how the verb *to know* and its cognates actually function in the usual range of relevant discourse. For if one neglects these facts one is well en route to "changing the subject" to talk about something different from that very conception that must remain at the center of our concern. It would clearly be self-defeating to turn away from *knowledge* as we in fact conceive and discuss it and deal with some sort of so-called knowledge different from that whose elucidation is the very reason for being so such a theory. If a philosophical analysis is to elucidate a conception that is in actual use, it has no choice but to address itself to that usage and conform to its actual characteristics.

The first essential step is to recognize that "to know" has both a propositional and a procedural sense: there is the intellectual matter of "knowing that something or other is the case" (*that*-knowledge) and the practical matter of knowing how to perform some action and to go about realizing some end (*how-to*-knowledge). This distinction is crucial because only the former, intellectual and propositional mode of knowledge has generally been the focus of attention in traditional philosophical epistemology, rather than the latter, practical and performatory mode. We shall accordingly focus here on specifically propositional knowledge—that sort of knowledge which is at issue in locutions to the effect that someone knows something-or-other to be the case ("*x* knows that *p*").

The terminology at issue generally so operates that in flatly saying that "*X* knows that *p*" one not only gives a report about *X*'s cognitive posture, but also *endorses* the proposition *P* that is at issue. To disengage oneself from this commitment one must say something like, "*X* thinks (or is convinced) that he knows that *p*." In the present discussion, however, it will be just this latter sort of thing—apparent or purported knowledge—that is under consideration.

Knowledge claims can be regarded from two points of view, namely, *internally and committally*, subject to an acceptance thereof as correct and authentic, and *externally and detachedly*, viewed from an "epistemic distance" without the commitment of actual acceptance, and seen as merely representing *purported* knowledge. We shall here adopt this second perspective, viewing knowledge in an externalized way, so that we shall be dealing with *ostensible* knowledge rather than certifiedly *authentic* knowledge. Our concern is with the merely *putative* knowledge of fallible flesh-and-blood humans and not the capital-*K* Knowledge of an omniscient being.

There is a wide variety of cognitive involvements: one can know, believe or accept (disbelieve or reject), conjecture or surmise or suspect, imagine or think about, assume or suppose, deem likely or unlikely, and so on. And there is also a wide variety of cognitive performances: realizing, noticing, remembering, wondering—and sometimes also their negatives: ignoring, forgetting, and so on. All of these cognitive circumstances belong to "the theory of knowledge"—to epistemology broadly speaking, which accordingly extends far beyond the domain of knowledge as such. But *knowledge* lies at the center of the range, and as the very expression indicates, the "theory of knowledge" focuses on knowledge.

The conception of "knowledge" itself represents a flexible and internally diversified idea. In general terms, it relates to the way in which persons can be said to have access to correct information. This can, of course, occur in rather different ways, so that there are various significantly distinguishable sorts of knowledge in terms of the kind of thing that is at issue:

1. *Knowledge-that* something or other is the case (i.e., knowledge of facts). Examples: I know that Paris is the capital of France. I know that 2 plus 2 is 4.
2. *Adverbial knowledge*. Examples: Knowing what, when, how, why, and so forth.

3. *Knowledge by acquaintance* with individuals or things. Examples: I know Jones. I know the owner of that car.
4. *Performatory (or "how-to") knowledge.* Examples: I know how to ice skate. I know how to swim.

Traditionally epistemology, the theory of knowledge, has focused on knowledge of the first type, propositional or factual knowledge of the sort where "I know that bears are mammals" is paradigmatic. However, the present book culminates in prioritizing the fourth sort of knowledge. It portrays the pivotal use in practical terms, pivoting on the question of how we go about the business of inquiry—of securing tenable answers to our questions.

As long as we are concerned merely with what we know, the idea of limits of knowledge lies outside our ken. We cannot be specific about our ignorance in terms of knowledge: It makes no sense to say "*p* is a fact that I do not know" for if we know something in specific to be a fact we can, for that very reason, no longer be in ignorance about it. However, "*Q* is a question that I cannot answer" poses no difficulty. It is thus only when we turn to questions—when we ask whether or not something is so in situations where we simply cannot say—that we come up against the idea of limits to knowledge. Only as we come to realize that there are questions that we cannot answer does the reality of ignorance confront us. Accordingly, questions are epistemologically crucial because it is in their context that matters of unknowing—also come to the fore. After all, we need information to remove ignorance and settle doubt, and "our knowledge" is constituted by the answers that we accept. What people know—or take themselves to know—is simply the sum total of the answers they offer to the questions they can resolve.

Propositional knowledge is coordinate with the capacity to answer questions, above all, in the case of knowledge-that-*p*, being in a position correctly and appropriately to answer the question: "Is *p* true or not?" And it is this sort of knowledge that will in the main, be at the focus of concern in the present discussion because there is good reason for seeing it as basic to all modes of knowledge in general. This sort of knowledge can further be classified

- by *subject matter* (as per mathematical or botanical knowledge)
- by *source* (personal observation, reliable reportage, etc.)
- by *mode of justification or validation* (personal or vicarious experience, scientific investigation, mathematical calculation, etc.)
- by the *cognitive status* of the matters at issue (empirical facts, convention in linguistic matters, formal relationships in logic or mathematics)
- by *mode of formulation* (verbally, by pictograms, by mathematical symbolism, etc.)

But what is propositional knowledge? It is emphatically not an activity or performance. You cannot answer the question "What are you doing?" with the response "I am knowing that Paris is the capital of France," any more than you could say "I am owning this watch" or "I am liking roses." Knowing a fact is not something that one does; it is a condition one has come to occupy in relation to information. It is not the process (buying, being inaugurated) but the end-state (owning, being president) in which a process culminates. To know something, then, is not to be engaged in an activity but to have entered into a certain condition—a cognitive condition. In sum, propositional knowledge is a cognitive affair—a condition of things coordinate with a suitable relationship between people and facts. And it is this aspect of knowledge that will primarily concern us here.

The fundamental features of propositional knowledge are inherent in the *modus operandi* of knowledge discourse—in the very way in which language gets used in this connection. The following three features are salient in this regard:

1. *Truth Commitment*. Only the truth can be known. If someone knows that *p* then *p* must be true. It simply makes no sense to say "I know that *p*, but it might not be true." or "*X* knows that *p* but it might not be true." Only if one accepts *p* as true can one say of someone that they know that *p*. If one is not prepared to accept that *p* then one cannot say that someone knows it. Otherwise one has to withdraw the claim that actual knowledge is at issue and rest content with saying that the individual *thinks* or *believes* that he know that *p*.

2. *Grounding*. Knowledge must be appropriately grounded. A person may *accept* something without a reason but cannot then be said to *know* it. It makes no sense to say "*X* knows that *p*, but has no sufficient grounds for thinking so." Knowledge is not just a matter of belief—or indeed even of correct belief—but of rationally appropriate belief. Guesswork, conjecture, and so on are not a sufficient basis for what deserves to be characterized as *knowledge*.

3. *Reflexivity*. To attribute a specific item of propositional knowledge to someone else is *ipso facto* to claim it for oneself. It makes no sense to say "You know that *p*, but I don't." Of course one can be generic about it: "You know various things that I don't." But one cannot be specific about it and *identify* these items. To characterize such an item as knowledge is to assert one's own entitlement to it.

4. *Coherence*. Since all items of propositional knowledge must be true, they must in consequence be collectively coherent. It cannot be that *x* knows that *p* but that *y* knows that not-*p*. Since we are committed to the principle that the truth is consistent, the truth-commitment of knowledge demands its consistency.

Knowledge development is a practice that we humans pursue because we have a need for its products. Life is full of questions that must be answered.

(Will that bridge hold up? Is that food safe?) The cognitive project is accordingly a deeply *practical* endeavor, irrespective of whatever purely theoretical interest may attach to its products.

For sure, knowledge brings great benefits. The relief of ignorance is foremost among them. We have evolved within nature into the ecological niche of an intelligent being. In consequence, the need for understanding, for "knowing one's way about," is one of the most fundamental demands of the human condition. Man is *Homo quaerens*. The need for knowledge is part and parcel to our nature. A deep-rooted demand for information and understanding presses in on us, and we have little choice but to satisfy it. Once the ball is set rolling it keeps on under its own momentum—far beyond the limits of strictly practical necessity.

Knowledge is a situational imperative for us humans to acquire information about the world. Homo sapiens is a creature that must, by its very nature, feel cognitively at home in the world. The requirement for information, for cognitive orientation within our environment, is as pressing a human need as that for food itself. The basic human urge to make sense of things is a characteristic aspect of our makeup—we cannot live a satisfactory life in an environment we do not understand. For us intelligent creatures, cognitive orientation is itself a practical need: cognitive disorientation is physically stressful and distressing. As William James observed: "It is of the utmost practical importance to an animal that he should have prevision of the qualities of the objects that surround him."[1]

However, not only is knowledge indispensably useful for our practice but the reverse is the case as well. Knowledge development is itself a practice and various practical processes and perspectives are correspondingly useful—or even necessary—to the way in which we go about constituting and validating our knowledge. Expounding such a praxis-oriented approach to knowledge development is one of the prime tasks of this book. Its principal thesis is that we have not only the (trivial) circumstance that knowledge is required for effective practice, but also the reverse, that practical and pragmatic considerations are crucially at work in the way in which human knowledge comes to be secured.

Part I

Knowledge and Its Problems

Chapter 1

Modes of Knowledge

Synopsis

- *The traditional idea that knowledge is true justified belief requires both clarification and qualification.*
- *Propositional knowledge is subject to various distinctions, among which that between explicit and inferential knowledge and that between occurrent and dispositional knowledge are particularly prominent.*
- *When "knowledge" is construed as inferentially accessible knowledge, it becomes possible to construct what can plausibly be seen as a "logic" of knowledge.*

Is Knowledge True Justified Belief?

It is something of an oversimplification to say the knowledge involves belief. For one thing, believing is sometimes contrasted with knowing as a somewhat weaker cousin. ("I don't just *believe* that, I *know* it.") And there are other contrast locutions as well. ("I know we won the lottery, but still can't quite get myself to believe it.") Still in one of the prime senses of belief—that of acceptance, of commitment to the idea that something or other is so—the knowledge of matters of fact does require an acceptance that is either actual and overt or at least a matter of tacit implicit commitment to accept.

Various epistemologists have sought to characterize knowledge as *true justified belief*.[1] In his widely discussed 1963 article, Edmund Gettier followed up on suggestions of Bertrand Russell by offering two sorts of counterexamples against this view of knowledge as consisting of beliefs that are both true and justified.

Counterexample 1

Let it be that:

1. X believes p
2. p is true
3. X has justification for believing p, for example, because it follows logically from something—say q—that he also believes, although in fact
4. q is false.

Here X clearly has justification for believing p, since by hypothesis thus follows logically from something that he believes. Accordingly, p is a true, justified belief. Nevertheless, we would certainly not want to say that X *knows* that p, seeing that his (only) grounds for believing it are false.

To concretize this schematic situation let it be that:

1. X believes that Smith is in London (which is false since Smith is actually in Manchester).
2. Smith's being in London entails that Smith is in England (which conclusion is indeed true since Smith is in Manchester).
3. X believes that Smith is in England (because he believes him to be in London).

That Smith is in England is accordingly a belief of X's that is both true and for which X has justification. Nevertheless we would clearly not want to say that X *knows* that Smith is in England, since his (only) reason for accepting this is something quite false.

The lesson that emerges here is that knowledge is not simply a matter of having a true belief that is *somehow* justified, but rather that knowledge calls for having a true belief that is *appropriately* justified. For the problem that the counterexample clearly indicates is that in this case the grounds that lead the individuals to adopt the belief just do not suffice to assure that which is believed. Its derivation from a false belief is emphatically *not* an appropriate justification for a belief.

Counterexample 2

Let it be that:

1. X believes p-or-q.
2. q is true (and consequently p-or-q is also true).

3. *X* disbelieves *q*.
4. *X* believes *p*-or-*q*, but does so (only) because he believes *p*.
5. *p* is false.

Here *p*-or-*q* is true. And *X* has justification for believing *p*-or-*q* since it follows from *p* which he believes. And since *p*-or-*q* is true—albeit in virtue of *q*'s being true (when *X* actually disbelieves)—it follows that *p*-or-*q* is a true, justified belief of *X*'s. Nevertheless, in the circumstances we would certainly not say that *X* knows that *p*-or-*q*, seeing that his sole grounds for believing it is once more something that is false.

The difficulty here is that *X* holds the belief *p*-or-*q* which is justified for *X* because it follows from *X*'s (false) belief that *p*, but is true just because *q* is true (which *X* altogether rejects).

To concretize this situation let it be that:

1. *X* believes that Jefferson succeeded America's first president, George Washington, as president
2. *X* accordingly believes that Jefferson or Adams was the second American president, although he thinks that Adams was the third president.
3. Since Adams was in fact the second American president, *X*'s belief that Jefferson or Adams was the second American president is indeed true

So *X*'s (2)-belief is indeed both true and justified. Nevertheless we would certainly not say that *X* *knows* this since his grounds for holding this belief are simply false. However, maintaining that knowledge is constituted by true and *appropriately* justified belief would once again resolve the problem, seeing that a belief held on the basis of falsehoods can clearly not count as appropriate. The difficulty is that the grounds on which the belief is held will in various cases prove insufficient to establish the belief's truth. And this blocks a merely *conjunctive* conception that knowledge is a matter of belief that is both true and justifiably adopted. But this problem is precluded by the *adjectival* conception of knowledge as belief that is at once correct and appropriately seen to be so, so that truth and justification are not separable but blended and conflated.

What is critical for knowledge attribution is that the believer's grounds for the particular belief at issue endorsed by the attributor as well.[2] When one says "A truth is known when it is a justified belief" one is not (or should not) take the line that it is believed *and* (somehow) justified, but rather that it is *justifiedly believed* in that the belief's rationale is flawless. The basic idea is that there can be no problem in crediting *x* with knowledge of *p* if:

X believes *p* on grounds sufficient to guarantee its truth and realizes this to be the case.

And so, the crucial point is that when knowledge is characterized as being true justified belief one has to construe justification in a complex, two-sided way because that belief must be accepted by its believer

- on grounds that *he* deems adequate

and moreover

- these grounds must be that *we* (the attributers of the belief) endorse by way of deeming them adequate as well.

The "subjective" justification of the attributee must be satisfied by the "objective" justification of the attributor if an attribution of knowledge is to be viable.

This line of consideration brings to light the inadequacy of cognitive conclusion: the view that knowledge is belief that is caused in the proper way. Because here "proper way" means proper as we (the attributors) see it, which may fail to be how the holder of that true belief sees it. (Those belief-engendering causes may fail to correspond to his actual cognitive grounds or *reasons* for holding the belief.)

And much the same is true of reliabilism: the view that knowledge is belief produced by a reliable process. For this reliability again holds for that attributer's view of the matter and its issuing from that reliable process may not in fact be the believer's own ground for holding the belief. (Causes can only provide *reasons* when their confirmatory operation is *correctly* recognized.)

So much for what is at issue with someone's actually knowing a fact. But of course here, as elsewhere, there is a distinction between (1) something actually *being* so, and (2) having *adequate grounds for claiming* that it is so. And the former (actually being) always goes beyond the latter (having adequate grounds). We can have good reason for seeing our belief grounds as flawless even when this is actually not the case. In cognitive matters as elsewhere we must reckon with the prospect of unpleasant surprises. The prospect of error is pervasive in human affairs—cognition included.

Consider the issue from another angle. It is part of the *truth conditions* for the claim that something is an apple—a necessary condition for its being so—that was grown on an apple tree, that it contains seeds, and that it not turn into a frog if immersed in a bowl of water for 100 days. And yet many is the time we call something an apple without checking up in these things. The *use condition* for establishment to call something an apple are vastly more lenient. If it looks like an apple, feels like an apple should, and smells like an apple, then that is quite good enough.

And the same sort of thing also holds for knowledge (or for certainty). The truth conditions here are very demanding. But the use conditions that authorize responsible employment of the term in normal discourse are a great deal more relaxed.

Actually, to have knowledge is one sort of thing, something that goes well beyond is required for its being the case that you or I have adequate grounds for claiming that it is the case. With claims appropriate, assurance of all sorts stops well short of *guaranteeing* actual truth. And this is so with our subjectively justified knowledge claims as well.

It is important in this regard to note that one of the basic ground rules of (defeasible) presumption in the cognitive domain inheres in the rule of thumb that *people's conventionally justified beliefs are true*—where "conventional justification" is constituted by the usual sorts of ground ("taking oneself to see it to be so," etc.) on which people standardly base their knowledge claims. And so whenever this groundrule comes into play the preceding formula that true justified beliefs constitute knowledge becomes redundant and knowledge comes to be viewed simply as appropriately justified belief where appropriateness may stop short of full-fledged theoretical adequacy. However what is at issue here is accordingly not a truth of general principle: it is no more than *a principle of practical procedure* which—as we fully recognize—may well fail to work out in particular cases.

Use conditions are geared to the world's operational realities. They bear not on what must invariably be in some necessitarian manner, but on what is usually and normally the case. And here—in the realm of the general rule, the ordinary course of things—it is perfectly acceptable to say that "knowledge is true justified belief." For ordinarily subjective and objective warrant stand in alignment. Those cases where knowledge fails to accompany true and (subjectively) justified belief all represent unusual (abnormal, nonstandard) situations.

The *practical* justification of the principle at issue is nevertheless substantial. For in our own case, at any rate, we have no choice but to presume our conscientiously held beliefs to be true. The injunction "Tell me what *is* true in the matter independently of what you genuinely believe to be true" is one that we cannot but regard as absurd. And the privilege we claim for ourselves here is one that we are pragmatically well advised to extent to others as well. For unless we are prepared to presume that their beliefs too generally represent the truth of the matter one cannot derive information from them. An important principle of practical procedure is at issue. In failing to extend the credit of credence to others we would deny ourselves the prospect of extending our knowledge by drawing on theirs.[3]

MODES OF (PROPOSITIONAL) KNOWLEDGE

When one speaks of "belief" in matters of knowledge one has in view what a person stands committed to accepting by way of propositional ("it is the case that") contentions. Knowledge is not an activity—mental or otherwise. Nor is knowing a psychological process. What is at issue here is not a matter of a mode of action or activity. The question "What are you doing?" cannot be answered by saying "I am knowing that Paris is the capital of France." This is something you may be thinking about or wondering about, but it is not something you can be knowing. The verb "to know" admits of no present continuous: one cannot be engaged in knowing. We can ascribe knowledge without knowing what goes on in people's heads (let alone in their brains). To know something is a matter not of process but of product.

To realize that Columbus knew that wood can float I do not have to probe into the complex of thought processes he conducted in Italian or Spanish or whatever—let alone into his brain processes. Knowledge in the sense a issue here is a matter of a relation being held to obtain between a person and a proposition, and will in general have the format Kxp ("x knows that p"). Like owning or owing, knowing is a state: a condition into which one has entered. It is not an action one does or an activity in which one engages. (Those philosophers who do or have spoken of "the act of knowing" talk gibberish.) Instead what is at issue is a state or condition, namely the sate or condition of a person who stands in a certain sort of relationship—a *cognitive* relationship—to a fact.

What is this relationship like? Here it is both useful and important to distinguish between explicit, dispositional, and inferential knowledge.

Explicit knowledge is a matter of what we can adduce on demand, so to speak. It takes two principal forms:

- *Occurrent knowledge.* This is a matter of actively paying heed or attention to accepted information. A person can say: "I am (at this very moment) considering or attending to or otherwise taking note of the fact that hydrogen is the lightest element." The present evidence of our senses—"I am looking at the cat on the mat"—is also an example of this sort of thing where what is claimed as knowledge is geared to circumvent cognitive activity.
- *Dispositional knowledge.* This is a matter of what people would say or think if the occasion arose—of what, for example, they would say if asked. Even when X is reading Hamlet or, for that matter, sleeping, we would say that an individual knows (in the presently relevant dispositional manner) that Tokyo is the capital of Japan. Here those items of knowledge can automatically be rendered occurrent by suitable stimuli.

Inferential knowledge, by contrast, is (potentially) something more remote. It is something deeply latent and tacit, not a matter of what one *can* produce on demand, but of what one *would* produce if only one were clever enough about exploiting one's occurrent and dispositional knowledge. We understand unproblematically what that means, namely—he must think, reflect, try to remember, exert some intellectual effort and patience. Inferential knowledge is a matter of exploiting one's noninferential (occurent or dispositional) knowledge. It is the sort of extracted knowledge that heeds materials to work out—grist to its mill. Since it is transformative rather than generative, there can be no inferential outputs in the absence of noninferential inputs. (This, of course, is not to say that it cannot be new—or even surprising in bringing previously unrecognizable relationships to light.)

What an individual himself is in a position to infer from known facts are facts he also knows

$$\text{if } Kxp \text{ and } Kx(p \vdash q), \text{ then } Kxq.$$

Here Kxp abbreviates "x knows that p.") This sort of inferential knowledge is a matter of what the individual does—or by rights should—derive from what is known. It is this feature of knowledge—its *accessibility* that provides for the inferential construal of the idea that is at work in these deliberations. For our concern in epistemology is less with the impersonal question of what people do accept than with the normative question of what, in the circumstances, it is both appropriate and practicable for them to accept. Thus in attributing knowledge we look not only to the information that people have in a more or less explicit way but also to what they are bound to be able to infer from this.

This points toward the still more liberal conception of what might be called *available knowledge,* which turns on not what *the individual themselves* can derive from their knowledge but on what *we* can derive from it, so that something is "known" in this sense whenever $(\exists q)(Kxq \ \& \ q \vdash p)$. We arrive here at a disjunctive specification of fundamentally recursive nature:

X knows that p iff (i) X knows p occurrently, or (ii) X knows p dispositionally, or (iii) p can be inferentially derived by using only those facts that X knows.

The "logical omniscience" of the consequences of what one knows is a characterizing feature of this particualr (and overly generous) conception of knowledge. K^* is thus radically different from K.

Since one of our main concerns will be with the limits of knowledge, it will be of some interest to see if there are things not known even in this particularly generous sense of the term that is at issue with *available* knowledge. In general, however, it will be the more standard and received sense of *accessible* knowledge that will concern us here.

OTHER BASIC PRINCIPLES

Accessible knowledge involves its holder in acceptance (endorsement, subscription, credence, belief, etc.). To accept a contention is to espouse and endorse it, to give it credence, to view it as an established fact, to take it to be able to serve as a (true) premiss in one's thinking and as a suitable basis for one's actions.[4] And a person cannot be said to *know* that something is the case when this individual is not prepared to "accept" it in this sort of way. And accordingly, the claim "*x* knows that *p*" is only tenable when *x* holds *p* to be the case. It is senseless to say "*x* knows that *p*," but he does not really believe it or "*x* knows that *p*, but does not stand committed to accepting it."[5]

Accordingly, one cannot be said to know something if this is not true.[6] Let "*Kxp*" abbreviate "*x* knows that *p*." It than transpires that we have:

- *The Veracity Principle*

    If *Kxp*, then *p*.

This relation between "*x* knows that *p*" and "*p* is true" is a necessary link that obtains *ex vi terminorum*. Knowledge must be veracious: The truth of *p* is a presupposition of its knowability: if *p* were not true, we would (*ex hypothesi*) have no alternative (as a matter of the "logic" of the conceptual situation) to withdraw the claim that somebody *know p*.

Some writers see the linkage between knowledge and truth as a merely contingent one.[7] But such a view inflicts violence on the concept of knowledge as it actually operates in our discourse. The locution "*x* knows that *p*, but it is not true that *p*" is senseless. One would have to say "*x thinks* he knows that *p*, but . . ." When even *the mere possibility* of the falsity of something that one accepts comes to light, the knowledge claim must be withdrawn; it cannot be asserted flatly, but must be qualified in some such qualified way as "While I don't actually *know* that *p*, I am virtually certain that it is so."

Knowledge veracity straightaway assures knowledge consistency. If someone knows something, then no one knows anything inconsistent with it.

- *Knowledge-Coherence Principle*

    If *Kxp* and *Kyp*, then *p* compat *q*.

Proof. If *Kxp*, then *p*, and if *Kxq*, then *q*. Hence we have *p* & *q*, and so, since their conjunction is true, *p* and *q* cannot be incompatible.

Again, certain other "perfectly obvious" deductions from what is known must be assumed to be at the disposal of every (rational) knower. We thus have the principles that separably known items are cognitively conjunctive:

- *The Conjunctivity Principle:*

Someone who knows both *p* and *q* separately, thereby also knows their conjunction:

If *Kxp* and *Kxq*, then *Kx*(*p* & *q*)—and conversely.

A person who is unable to exploit his information by "putting two and two together" does not really have *knowledge* in the way that is at issue with specifically *inferential* knowledge. It seems only natural to suppose that any rational knower could put two and two together in this way.

To these principles of *inferential* knowledge we can also adjoin the following:

- *Knowledge-Reflexivity Principle*

Actually to know that someone knows something requires knowing this fact oneself.

If *KxKyp*, then *Kxp*; and indeed: If *Kx*(∃*y*)*Kyp*, then *Kxp*.

Letting *i* be oneself so that *Kip* comes to "I know that *p*" we may note that the Knowledge-Reflexivity Principle entails

If *KiKyp*, then *Kip*; and more generally: if *Kx*(∃*y*)*Kyp*, then *Kxp*.

To claim to know that someone else knows something (i.e., some specific fact) is to assert this item for oneself. Put differently, one can only know with respect to the propositional knowledge of others that which one knows oneself.

However, it lies in the nature of things that there are—or can be—facts that *X* can know about *Y* but *Y* cannot. Thus *X* can know that *Y* has only opinions but no knowledge, but *Y* cannot.

Knowledge entails justification (warrant, grounding, evidence, or the like). One can maintain that someone knows something only if one is prepared to maintain that he has an adequate rational basis for accepting it. It is senseless to say things like "*x* knows that *p*, but has no adequate basis for its acceptance," or again "*x* knows that *p*, but has no sufficient grounding for it." To say that "*x* knows that *p*" is to say (*inter alia*) that *x* has *conclusive* warrant for claiming *p*, and, moreover, that *x* accepts *p* on the basis of the conclusive warrant he has for it (rather than on some other, evidentially insufficient basis). It is senseless to say things like "*x* knows that *p*, but there is some room for doubt" or "*x* knows that *p*, but his grounds in the matter leave something to be desired."[8]

G. E. Moore pointed out long ago that it is anomalous to say "*p* but I don't believe it" or indeed even "*p* but I don't know it."[9] That this circumstance of what might be called "Moore's Thesis" be so is readily shown. For in maintaining a claim one standardly purports to know it. Accordingly, one would not—should not—say p & $\sim Kip$ unless one were prepared to subscribe to

$$Ki(p \ \& \ \sim Kip),$$

where i = oneself. But in view of Veracity and Conjunctivity this straightaway yields

$$Kip \ \& \ \sim Kip,$$

which is self-contradictory. (Observe, however, that "I suspect p but do not fully believe it" [or "really know it"] is in a different boat.)

It is also important to recognize that which is known must be compatible with whatever else is actually known. No part of knowledge can constitute decisive counterevidence against some other part. The whole "body of (genuine) knowledge" must be self-consistent. Accordingly we shall have

If Kxp and Kxq, then compat (p, q)

and consequently:

If Kxp & Kxq, then $\sim(p \vdash \sim q)$.[10]

Given this circumstance, logic alone suffices to assure the implication:

If Kxp and $p \vdash q$, then $\sim Kx \sim q$.

Accordingly, one decisive way of defeating a claim to knowledge is by establishing that its denial follows from something one knows.[11] But of course not knowing something to be false (i.e., $\sim Kx \sim q$) is very different from—and much weaker than—knowing this item to be true (Kxq). And so—as was just noted in the preceding dismissal of "logical omniscience"—the just-indicated principle must *not* be strengthened to the objectionable

If Kxp and $p \vdash q$, then Kxq,

which has already been rejected above.

A further, particularly interesting facet of knowledge-discourse relates to *the automatic self-assumption of particularized knowledge attributions*. It makes no sense to say "You know that p, but I don't" or "x knows that p, but not I." In *conceding* an item of knowledge, one automatically *claims* it for oneself as well. To be sure, this holds only for that-knowledge, and not how-to-knowledge (even how to do something "purely intellectual"—like "answering a certain question correctly"). It makes perfectly good sense to say that someone else knows how to do something one cannot do oneself. Again, abstract (i.e., unidentified) knowledge attributions will not be self-assumptive. One can quite appropriately say "*x* knows everything (or 'something interesting') about automobile engines, though I certainly do not." But particularized and identified claims to factual knowledge are different in this regard. One cannot say "*x* knows that automobile engines use gasoline as fuel, but I myself do not know this." To be sure, we certainly do *not* have the omniscience thesis:

If *Kxp*, then *Kip*   (with i = I myself).

One's being entitled to claim *Ksp* follows from one's being entitled to claim *Kxp*, but the content of the former claim does not follow from the content of the latter.

Moreover, the ground rules of language use being what they are, *mere assertion is in itself inherently knowledge-claiming*. One cannot say "*p* but I don't know that *p*." To be sure, one can introduce various qualifications like "I accept *p* although I don't actually know it to be true," But to affirm a thesis flatly (without qualification) is *eo ipso* to claim knowledge of it. (Again, what is at issue is certainly *not* captured by the—clearly unacceptable—thesis: If *p*, then *Kip*.)

Such "logical principles of epistemology" will of course hinge crucially on the exact construction that is to be placed on the conception of "knowledge." In particular, if it were not for the inferential availability character of this concept, the situation would be very different—and radically impoverished—in this regard.

One further point. The statement "possibly somebody knows *p*" $\Diamond(\exists x)Kxp$ says something quite different from (and weaker than) "somebody possibly knows *p*" $(\exists x)\Diamond Kxp$. It is accordingly necessary to distinguish

1. Only if *p* is true will it be possible that somebody knows it:

    If $\Diamond(\exists x)Kxp$, then *p*

from

2. Only if *p* is true will there be someone who possibly knows it:

    If $(\exists x)\Diamond Kxp$, then *p*.

Because the modality in (1)'s antecedent precedes that qualified the statement carries us beyond the bounds of the actual world into the realm of merely possible existence. But with (2)'s antecedent we remain within it: we discuss only what is possible for the membership of *this* world. Hence (2) is a plausible thesis, while (1) is not.

Be this as it may, the cardinal point is that of clarifying the nature of knowledge. And in this regard *putative* knowledge is a matter of someone's staking a claim to truth for which that individual (subjectively) deems himself to have adequate grounds, while *actual* knowledge by contrast is a matter of someone's correctly staking a claim to truth when that individual has (objectively) adequate grounds.

The various principles specified here all represent more or less straightforward facts about how the concept at issue in talk about actual knowledge actually functions. Any philosophical theory of knowledge must—to the peril of its own adequacy—be prepared to accommodate them.

Chapter 2

Fallibilism and Truth Estimation

SYNOPSIS

- *Metaknowledge, the development higher-order knowledge about our knowledge itself, is one of the principal tasks of epistemology. And one of its key lessons is that of fallibility, the almost inevitable liability of our knowledge to the discovery of error.*
- *The acknowledgment of error with its denial the overall conjunction of one's affirmations seems paradoxical. (Indeed it is characterized as "The Preface Paradox.") Nevertheless, such fallibility represents a fact of life with which we must come to terms.*
- *The ancient problem of the* diallelus *posed by the absence of any cognitive-external standard of cognitive adequacy—also has to be reckoned with.*
- *There is a wide variety of possible responses to this view of the cognitive situation.*
- *But in balancing costs and benefits it emerges that a fallibilism that views our knowledge-claims in the light of best-available estimates is itself our best-available option.*
- *Such a* fallibilism *means that our "scientific knowledge" is no more (but also no less) than our best estimate of the truth.*
- *Accordingly, we have to see our knowledge claims in scientific matters as representing merely putative truth, that is, as truth-estimates.*

Problems of Metaknowledge

The reflexive aspect of human cognition is one of its most characteristic and significant features. Nothing is more significant for and characteristic of our human cognitive situation than our ability to step back from what we deem ourselves to know and take a critically evaluative attitude toward it.

The development of metaknowledge—of information about our knowledge itself—is a crucial component of the cognitive enterprise at large. Metaknowledge is higher-order knowledge regarding the facts that we know (or believe ourselves to know); the object of its concern is our own knowledge (or putative knowledge). The prospect of metaknowledge roots in the reflexivity of thought—the circumstance we can have doubts about our doubts, beliefs about our beliefs, knowledge about our knowledge. The development of metaknowledge is a crucial component of epistemology, and in its pursuit we encounter some very interesting but also disconcerting results, seeing that attention to the actual nature of our knowledge yields some rather paradoxical facts.

Of course, knowledge as such must be certain. After all, we are very emphatically fair-weather friends to our knowledge. When the least problem arises with regard to a belief we would not, could not call it *knowledge*. There is no such thing as defeasible knowledge. Once the prospect of defeat is explicitly acknowledged, we have to characterize the item as merely putative knowledge.

Let K^* be the manifold of propositions that we take ourselves to know: the collection of our appropriately (though not necessarily correctly) staked knowledge claims, specifically including the theoretical claims that we endorse in the domain of the sciences. Thus K^* is the body of our *putative* knowledge, including all of the claims which, in our considered judgment, represent something that a reasonable person is entitled—in the prevailing epistemic circumstances—to claim as an item of knowledge. It includes the sum total of presently available (scientific) knowledge as well as the information we manage to acquire in ordinary life.

Given this conception of knowledge, it is clear that we can appropriately endorse the rule:

(R) For any p: If p belongs to K^*, then p is true.

In effect, (R) represents the determination to equate K^* with K, the body of our *actual* knowledge. We accept this precisely equation because its membership in K^* represents acceptance as true; a claim would not be a K^*-member did we not see it as true. However, $K^* = K$ represents not a theoretical truth but a practical principle—a matter of procedural policy. Unfortunately, we know full well that we are sometimes mistaken in what we accept as knowledge, even when every practicable safeguard is supplied. And this is true not only for the ordinary knowledge of everyday life but for our scientific knowledge as well. Accordingly, the rule (R) is not a true generalization but rather represents a rule of thumb; it is a practical pinnacle of procedure rather than a flat out truth. It reflects our determination to treat our putative knowledge as actual.

The following considerations confront us with some of the inescapable facts of (cognitive) life:

Cognitive Imperfection

We are (or at any rate should be) clearly aware of our own liability to error. We cannot avoid recognizing that any human attempt to state the truth of things will include misstatements as well. Our "knowledge" is involved not just in errors of omission but in errors of commission as well. We do—and should—recognize that no matter how carefully we propose to implement the idea of membership in our "body of accepted knowledge," in practice some goats will slip in along with the sheep. (The skeptic's route of cognitive nihilism—of accepting absolutely nothing—is the only totally secure way to avoid mistakes in acceptance.)

An Epistemic Gap

The epistemic gap that inevitably arises between our objective factual claims and the comparatively meager evidence on which we base them also means that such claims are always at risk. In epistemic as in other regards, we live in a world without guarantees.

Scientific Fallibilism

We do (or at any rate should) realize full well that our scientific knowledge of the day contains a great deal of plausible error. We are (or should be) prepared to acknowledge that the scientists of the year 3000 will think the scientific knowledge of today to be every bit as imperfect and extensively correction-requiring as we ourselves think to be so with respect to the science of 300 years ago. To be sure, we can safely and unproblematically make the conditional prediction that *if* a generalization states a genuine law of nature, *then* the next century's phenomena will conform to it every bit as much as those of the last. But we can never—in the prevailing condition of our information—predict with unalloyed confidence that people will still continue to regard that generalization as a law of nature in the future. Scientific progress brings in its wake not only new facts but also change of mind regarding the old ones; what we have in hand are not really certified "laws of nature" as such, but mere *theories*—that is, *laws as we currently conceive them to be in the prevailing state of the scientific art.*

After all, if any induction whatsoever can safely be drawn from the history of science it is this: that much of what we currently accept as the established knowledge of the day is wrong, and that what we see as our body of *knowledge* encompasses a variety of errors. In fact, there are few inductions *within* science that are more secure than this induction *about* science. Despite the occasional overenthusiastic pretentions of some scientists to having it exactly right in understanding nature's phenomena—to be in a position to say "the last word"—the fact remains that the realities of the scientific situation pose virtually insurmountable obstacles in this regard.[1]

Blind Spots

There are not only errors of commission but errors of omission as well—blind spots as it were. We know *that* there are facts we do not know, though we cannot say *what* these items of ignorance are. We know *that* there are answers we cannot give to some questions, but do not know what those missing answers are. We know there are various specific truths we do not know, but cannot identify any of them. We realize the incompleteness of our knowledge—and can even often localize where this incompleteness exists—but of course cannot characterize those missing items of information positively by identifying them as an individuated item.[2]

Predictive Biases

People's best-made predictions—like their best-made plans—"gang aft agley." We can safely predict that the cost of any major construction project will exceed our most carefully constructed cost-prediction; and we can safely predict that its actual completion will postdate our most carefully contrived prediction of the date of its accomplishment. In this way, we can come to recognize that our most carefully contrived predictions will in certain areas go *systematically* amiss. This sort of thing also happens in other prediction situations. For example, a statistical analysis of weather forecasts shows that present-day meteorologists systematically underestimate the *change* in temperatures from one day to the next by approximately 10 percent. Now it is clear that metaknowledge of *this* sort can be eminently productive. For we can indeed make effective use of it to *correct* our errors, proceeding to replace our first-order predictions by second-order predictions designed to correct them in the light of the systematic biases revealed by experience at the first-order level.

Estimation Biases

Predictive biases of the just-indicated sort are reflections of the more general situation of estimation biases. For we can come to recognize that in various situations of estimation there is a general tendency of a displacement of people's best estimates of various factors on the direction of too big, too little, too long, too short, and so on. And then, of course, we can make straightforward use of such metaknowledge regarding the error-tendencies in people's base level knowledge to effect second-order improvements. For example, the combination of estimates drawn from different sources or methods (say the *average* for the sense of simplicity) will generally be more accurate than the best estimate of *most* of the individual contributions.[3]

There emerge certain rather paradoxical facts about our knowledge. For much of what we see and accept as knowledge—and in fact knowledge of the highest quality, namely, scientific knowledge—has to be acknowledged as being little more than yet unmasked error. All too clearly, the cognitive resources at our disposal never enable us to represent reality with full adequacy in the present condition of things—irrespective of what the date on the calendar happens to be.

THE PREFACE PARADOX

It is important to recognize that operating with the distinction between real and putative knowledge—though unavoidable—is a tricky business. We certainly can apply this distinction retrospectively. ("Yesterday I took myself to know that p but I was quite wrong about it.") But we cannot apply it to what is presently before us. ("I know that p but I really don't" is in deep semantical trouble on grounds of simple inconsistency.) There is, clearly, something very frustrating about this situation. The self-critical aspect of our metaknowledge—the circumstance that it involves highly general claims about the imperfections of our knowledge—endows it with a paradoxical aspect.

This circumstance is readily brought to view by considering some further examples. One of the most vivid of these is the so-called Preface Paradox.[4] The conscientious author of a fact-laden book apologizes in the Preface for the error the book contains. "Several friends have read the MS and helped me to eliminate various errors. But the responsibility for the errors that yet remain is entirely mine." But why not simply correct these mistakes? Alas, one cannot. (If only one knew what they were one would of course correct them.) They are lost in a fog of unknowing.

Yet there is a straightforward logical conflict here. One cannot consistently accept a collection of contentions distributively and yet also maintain that they are not all true collectively. From the standpoint of logic, the things we claim as true severally and individually must also be claimed as true conjointly. And yet this is not realistic. It just does not reflect—always and everywhere—the realities as we must actually acknowledge them.

The Preface Paradox situation is indicative of a larger predicament. We know or must presume that (at the synoptic level) our science contains various errors of omission and commission—though we certainly cannot say where and how they arise. And it does not matter how the calendar reads. This state of affairs hold just as much for the science of the future as for that of our own day. Natural science is not only imperfect but imperfectable.[5] And this fact has profound implications for the nature of our "scientific knowledge." Our knowledge as an aggregate of accepted claims is one thing. But our metacognitive recognition—indeed knowledge—of the presence of error in the whole collection is another of no lesser significance. Both are facts. And our knowledge here is to all intents and purposes consistent with our metaknowledge. The paradox is that our metaknowledge *conflict* with our knowledge, seeing that one of the things we cannot avoid adding to our knowledge is the item of metaknowledge that some of our knowledge claims are mistaken.

The Preface Paradox analogy accordingly indicates that we cannot use present knowledge to *correct* itself—even where we recognize and acknowledge that it requires correction. And our deliberations here indicate that we cannot use present knowledge to *complete* itself even though we know it requires completion. Either way, we have the predicament of remaining powerless in the face of acknowledged shortcomings.

The point of such considerations emerges in the context of the following challenge:

> You are clever enough to realize that what you purport as knowledge contains errors of various sorts. So why not just refrain from such purportings and tell us the truth instead.

This plausible challenge is in fact absurd. We have no viable option here. We have no access way to the truth save via what we *think* to be true—there simply is no such thing. "Tell us what *is* true independently of what you *think* to be so" is an absurd challenge. Despite our recognition that it contains errors—that all too often what we think to be true just is not—we have no alternative but to accept our best estimate of the truth as a viable surrogate for the real thing. In matters of truth estimation as elsewhere, we have no alternative but to do the best we can.

To be sure, actual knowledge as such has to be absolutely certain. It makes no sense to say "X knows that p but it is uncertain whether or not p is true." If that were the case, we would have to qualify our knowledge claim by saying something like "*X thinks* he knows that p."

But absolute certainty is all too often difficult to obtain. We know full well that our knowledge claims are often wrong. We realize full well that much of our "knowledge" is not more than *purported* knowledge—that our knowledge is defeasible—that our claims to knowledge must often be retracted in the course of cognitive progress.

But while all of this is true, and our knowledge claims gang aft agley and may well prove to be wrong, nevertheless we cannot say which ones. Although we are—or should be—cognitive fallibilists the fact remains that we are error-blind. While we can say with confidence that some—perhaps many—of our knowledge claims are wrong, we can never say in advance with respect to which ones this possibly will be realized. We are—or should be—reasonably confident *that* our knowledge contains error, but we can never say with comparable reasonable confidence pinpoint just where it is that this prospect will be realized.

All the same, while we cannot *pinpoint* potential error in our knowledge by way of specific identification we can indeed say where it is at the generic level of types or regions of knowledge that error is more or less likely to occur. Specifically the following observations are in order in this regard.

1. Other things equal, objective claims to knowledge in the language of what is are more vulnerable than subjective ones in the language of what seems to be. It is more risky to claim to know that the cat *is* black than to claim to know that it looks black. A cognitive claim that something *is* so is always more vulnerable than one that claim than it *appears* so.
2. Other things being equal, general claims to knowledge are more vulnerable than particular ones. It is more risky to claim to know that that *this* weather report underestimates rainfall than it is to claim that *some* forecast underestimates rainfall.
3. Vague and indefinite knowledge claims are always more secure than detailed and definite ones. It is less risky to claim to know that someone was born early in the twentieth century than to claim that person was born on 5 January 1904.
4. Guarded or qualified claims are always securer than their unguarded and unqualified counterparts. If I claim to know that there is good reason to think such-and-such I am on firmer ground than in claiming to know that such-and-such is actually the case.

And there are doubtless other general principles along these lines. But the overall lesson is perfectly clear. By revising our knowledge claims so as to make them more qualified, less general, more indefinite and more grounded we can always ensure their certainty. This enhancement of certainty will, all too clearly, be achieved at the price of informativeness. But at least there is always the option of protecting our knowledge claims in this sort of way whenever any sort of reason for doing this comes to view. All the same, we have to face some harsh realities where the issue is one of securing significant knowledge of factual matters. Here we must be prepared to come to terms with the fact of life that our knowledge is something of a Swiss cheese replete with holes of unrecognized error.

THE DIALLELUS

We have a set of standards for deciding what to accept as true—let us call it *the Criterion*. But how can we possibly test such a standard—how can we subject it to quality control. Clearly we cannot do so by means of the usual and straightforward process of comparing its deliverances with the real truth. Given that the Criterion is itself our standard of deciding what is true, we have no way of effecting such a comparison.

An important negative line of reasoning regarding man's prospects of attaining knowledge about the world has been known from the days of the sceptics of antiquity under the title of the "*diallelus*" (*ho diallêlos tropos*)—or "the wheel"—which presents a particular sort of vicious-circle argumentation (*circulus vitiosus in probandi*). Montaigne presented this Wheel Argument, as we may term it, as follows:

> To adjudicate [between the true and the false] among the appearances of things we need to have a distinguishing method (*un instrument judicatoire*); to validate this method we need to have a justifying argument; but to validate this justifying argument we need the very method at issue. And there we are, going round on the wheel.[6]

But the argument is far older than Montaigne, and goes back to the skeptics of classical antiquity. The classical formulation of the argument comes from Sextus Empiricus:

> [I]n order to decide the dispute which has arisen about the criterion we must possess an accepted criterion by which we shall be able to judge the dispute; and in order to possess an accepted criterion, the dispute about the criterion must first be decided. And when the argument thus reduces itself to a form of circular reasoning (*diallēlus*), the discovery of the criterion becomes impracticable, since we do not allow them to adopt a criterion by assumption, while if they offer to judge the criterion by a criterion we force them to a regress *ad infinitum*. And furthermore, since demonstration requires an approved demonstration, they are forced into circular reasoning.[7]

It is difficult to exaggerate the significance of this extremely simple line of reasoning, which present the skeptic's "No Criterion" Argument to the effect that an adequate standard of knowledge can never be secured. One of the prime aims of the subsequent deliberations is to determine just how serious the implications of this situation are.

AN APORY AND ITS RECONCILIATION: K-DESTABILIZATION

It is helpful to look once more at some individually plausible but collectively inconsistent beliefs regarding the nature of our knowledge—at any rate our knowledge at the level of scientific generality and precision rather than that of everyday discourse imprecision. Let us return to the situation already sketched above regarding our commitment—or seeming commitment—to the (putative) facts constituting our body of knowledge K:

(I) K includes only truths: If $p \in K$, then p.

After all, if we did not see p as being true, then we would not have included it in K in the first place. The point is that the claims that committed our body of putative knowledge are ipso facto what we take or accept as being generic knowledge.

But as we have seen, it is pretty well inescapable that we also stand committed to:

(II) K includes some falsehoods: $(\exists p)(p \in K \ \& \ \sim p)$.

And these two theses are clearly contradictories: they stand in patent conflict with one another. How are we to respond to this situation?

Clearly there is just no way of getting around (II). No induction from the history of science stands more secure than this idea that some of the theories we currently accept as true are just plain false.

But (II) is an item of metaknowledge that destabilizes (I). And so we have no real choice but to abandon (I)—or rather, to weaken it. After all, we here confront a fundamental axiom of philosophical method: "Never simply abandon. In the face of insuperable difficulties, introduce distinctions and qualifications that enable you to save what you can of your commitments."

Let us inventory the available ways of effecting the unavoidable weakening of thesis (I). The object of the enterprise here is to survey the avenues open to us for obtaining an at least plausible weaker variant of this thesis. Those that are available here are mapped out in Table 2.1. Let us briefly examine the tenor and tendency of each of these positions.

Probabilism is a historic position which was variously espoused in the middle ages as a response to radical skepticism regarding human knowledge in general. (It was already anticipated in antiquity by some of the moderate

TABLE 2.1
A Survey of Possibilities

If p ∈ K, then p is true.

POSSIBLE QUALIFICATIONS OF THIS THESIS: Change the ending from "is true" to:

A. is (only) *probably* true. PROBABILISM: whatever we accept as true is no more than probably so. We never achieve secure truths but only probability.

B. is (only) *putatively* true. SOPHISTRY: K consists of mere opinion. There is no point in speaking of what *is* so but only of what people *think* to be so.

C. is (only) *putatively* BUT NOT *actually* true. RADICAL SKEPTICISM: Our so-called knowledge is no more than currently accepted falsehood.

D. is (only) *approximately* true. APPROXIMATIONISM: Our so-called knowledge never achieves more than rough correctness—"being in the right ball-park").

E. is (only) *presumptively* true. PRESUMPTIONISM or FALLIBILISM: Our so-called knowledge only consists of tentative estimate—claims we are only provisionally to endorse as truths, subject to future defeat.

F. is *generally* (or *almost always*) true. STATISTICISM or REGULISM: We change that basic thesis from a universal generalization to a general—but exception-admitting—rule by replacing "is" by "is generally."

G. is true *provided* that p also belongs to $K^+ \subset K$. DISCRIMINATIONISM: The idea that K consists of a secure core (K^+) and a more problematic remainder (K^-).

H. is *without any claim or pretentions to truth as such*: its acceptance is no more than merely a practical expedient to guide our conduct in certain regards. (INSTRUMENTALISM)

Plan Z. Accept inconsistencies and take them in stride. (INCONSISTENCY TOLERANCE).

---

skeptics of the Middle Academy.) More recently, it has been adopted with respect to scientific theories by the mainstream logical positivists and logical empiricists of the prewar era, in particular Rudolf Carnap and Hans Reichenbach.

Sophistry is also an historic position that goes back to classical antiquity. It would have us see the members of *K* as representing no more than plausible opinion. As Nietzsche put it, human "knowledge" is fiction masquerading as truth. Present-day "postmodernism," for all intents and purposes returns to this position—as does the sociology of knowledge school of the so-called strong program.

Skepticism represents perhaps the most drastic step—that of accepting . . . *nothing at all* and remaining *adoxastos* (doctrinally uncommitted), as the Greek skeptics put it. It calls for simply suspending judgment in matters of scientific theorizing. This position has a long and depressing history for classical antiquity to our own day with exponents ranging from Sextus Empiricus to Peter

Unger. On this approach, scientific knowledge is seen as effectively amounting to currently accepted error that has not been unmasked as such yet. This, in essence, was Nietzsche's view: what we see as our "scientific" knowledge as merely a matter of useful illusion.

Approximationism or something pretty close to the instinctive views of most scientists, who hesitate to claim that they have got us exactly right, but like to think they have at least got it pretty much right—that our favored scientific theories are correct "in the essentials" as it were. This was—roughly speaking—the view of C. S. Peirce.

Presumptionism or Fallibilism. On this view, a scientific theory is a temporary expedient to be replaced when and if something better comes along. One commitment to it is merely provisional and tentative—though, to be sure, we can only *displace* something with something else. One could characterize the falsificationist position of Karl Popper as a version of this approach.

Regulism or Statisticism. This position calls for replacing (I) by "K includes *mainly* truths," and being prepared to tolerate exceptions. Here (I) is read as stating merely a general (but not strictly universal) rule—one which, as such, admits of exceptions, though of course we cannot identify them in advance. In the context of scientific theories, this approach has no historical precedents that I know of.[8]

Discriminationism. The strategy of this approach is to divide K into two components: the unproblematic K^+ (whose members must, of course, be mutually consistent), and the problematic (K^-). And then it proposes to adopt:

(I') If $p \in K^+$, then p. [Note that the membership of K^* must accordingly be seen individually certain and collectively consistent.]

(II") Not- (If $p \in K^-$, then p) or equivalently $(\exists p) (p \in K^-$ & not-$p)$

There does not seem to be on record any theoretician who has adopted this discriminationist position, though there is (as we shall see) much to be said for it.

Instrumentalism. This view of the matter takes science out of the business of claims to truth. Its work is seen not as providing answers to substantive questions but to providing practical guidance for the conduct of affairs in matters of application and of inquiry. Our scientific theories are thus not actual truth-purporting beliefs but belief-neutral instruments for predicting and controlling the phenomenon. This position—or something substantially like it—has been held by some pragmatists (e.g., John Dewey)—and by such twentieth-century philosophers of science as P. W. Bridgman and Bas van Fraassen.

Inconsistency Tolerance. This rather drastic approach of simply taking consistencies in stride view is in favor among some logicians[9] and was to some extent favored by Ludwig Wittgenstein as regards pure mathematics. Some physicists have entertained the idea in with quantum mechanics. However, philosophers of science have been distinctly disinclined to resort to so radical a recourse.

It must be noted—and conceded—that these several positions are probably not exhaustive and that for sure they are in many cases not mutually exclusive. Thus, for example, one can combine instrumentalism and discriminationism by relegating all scientific *theories* to the "problematic remainder," while positioning particular observation reports (these are creatures of the moon") within the secure core. Here one would take the stance that we do not actually *believe* these problematic theories but see them as mere instruments for predicting true core statements.

Costs and Benefits

The crucial fact about these various alternative approaches is that none of the options that they afford us are cost-free: each has its characteristic balance of advantages and disadvantages, of costs and benefits. Let us examine somewhat more closely the issue of the costs of adopting the various plans.

Plan A: *Probabilism.* The domain of the probable is not inferentially closed. So there is now no justification for simply combining a variety of accepted propositions as collective premises for inference. We arrive at a situation of deductive disfunctionality. Moreover, we cannot now make our way to any definite (unprobabilistic) assertion. We secure no answers to our questions about how things work in the world. (To distribute probabilities across a spectrum of alternatives is emphatically not to provide an answer.)

Plan B: *Sophistry.* In its substance, this approach represents a know-nothing position, that destroys any prospect of a workable distinction between knowledge and opinion.

Plan C: *Radical Skepticism.* This position is simply a cognitive nihilism. By seeing all of our scientific knowledge as categorically mistaken it throws out the baby with the bathwater. It is an all-annihilating H-bomb that allows us no answers, no guidance, no nothing.

Plan D: *Approximationism.* This approach confronts us with the notorious difficulties of defining approximation. If we claim to have it right "in its essentials," it would be nice to know (as we do not and cannot) what those essentials are. Then too there are all those problems of confirming approximations. (In cognitive contexts, is somewhere between difficult and impossible to preestablish how matters will evolve over the long run.)

Plan E: *Presumptionism, Fallibilism.* On this approach we can only ever stake only guarded claims to truth. We insist that whatever we maintain as true need hold only *pro tem*, to be replaced when and if something better comes along. (Note that we must here differentiate between "is presumably true" and "is presumably-true.")

Plan F: *Regularism/Statisticism*. This approach indulges a certain overoptimism. For it does not do justice to the extent to which K has problems, seeing that K is simply not plausible to see the bulk our present scientific theories as exact and definitive truths.

Plan G: *Discriminationism*. This approach confronts us with the burden of delineating that unproblematic case $K^+ \wp K$. It saddles us with the problem of differentiating K^- from K^+—of identifying (at least some of) unproblematic knowledge.[10]

Plan H: *Instrumentalism*. This approach has the disadvantage of cognitive vacuity. Like skepticism (to which it is closely akin) it leaves us without answers to our questions.

Plan Z: *Inconsistency Tolerance*. This approach would have us simply take the inconsistency between principles (I) and (II) in stride, and to accept at face value the incoherence of what we know. This involves abandoning the traditional idea that it is a nonnegotiable demand of rationality to insist on the consistency of one's knowledge claims. The modern semanticological theory of inconsistency tolerance ("passconsistency") has created mechanisms for implementing this sceemingly absurd strategy in a more or less workable way.[11] This approach represents so radical a recourse, however—one that cuts so deep a swath across so large a terrain—that it cannot count as more than a plan of last resort, a desperate expedient to be adopted if and when all else fails.

More on Fallibilism

It would be an error to think that a conclusion based on fuller information is necessarily an improvement, presenting us with a result that cannot be false if its "inferior" predecessor was already true. For if this were so, then as more information is added, an outcome's probability would increasingly move in the same direction, and could not oscillate between increases and decreases. But this is clearly not the case.

Consider the following example, based on the question: "What will John do to occupy himself on the trip?" Suppose we require an answer to this question. But suppose further that the following data becomes successively available:

1. He loves doing crosswords.
2. He loves reading mysteries even more.
3. He didn't take any books along.

It is clear that we are led to and fro. Initially (at stage (1) of information access) we incline to the answer that he will be working crosswords. At the next stage, when item (2) arrives we change our mind and incline to the answer that he will be reading. At stage (3) we abandon this idea and go back to our initial view. And of course a subsequent stage, say one where we have

4. One of his fellow passengers lends him a book

can nevertheless reverse the situation and return matters to step (2). And who know what step (5) will bring? The crucial point is that additional information need not serve to settle matters by bringing us closer to the truth.

When we are dealing with *assured* truths, additional information cannot unravel or destabilize what we have. For truths have to be compatible with one another and new truths cannot come into logical conflict with old ones. Truths, once we have them, are money in the bank. But *likely* truths—plausible or probable truths—are something else again. Here additional evidence can unravel old conclusions. P can be highly probable relative to Q and yet improbable relative to $Q \& R$. Inductive conclusions are never money in the bank. It is an ever-present prospect that we will have to revisit and reassess them in the light of new information.

The fact is that throughout the realm of inductive or plausible reasoning, a larger, conjunctive fact $F \& G$ can always point to a conclusion at odds with that indicated by F alone: the circumstance that fact F renders X highly probable is wholly compatible with the existence of another fact G of such a sort that $F \& G$ renders X highly improbable. And yet what G undermines here can always be restored by some yet further additional fact.

The reality of it is that inductive "inference" is what logicians call *nonmonotonic* through learning the future that even if premiss p supports conclusion q, premiss $p \& r$ need not support q. For since an inductive linkage is not airtight something can always creep into the gap to soil things. Since owls can blink, you will confer when told that an owl is in the cage it too can blink. And this seems sensible enough. But if told additionally that a *stuffed* owl is at issue, this conclusion is no longer forthcoming.

Exactly because additional information always has the potential of constraining a change of mind—rather than merely providing additional substantiation for a fixed result—we have no assurance that further information produces "a closer approximation to the truth." Conclusions based on additional information may in some sense be comparatively "better" or "securer" but they need certainly not be "truer" or "more accurate." (There are no degrees here, true is true.)

Whenever what we accept is deductively derived from truths the results be realize are bond to be secure. Here what we do not know can do no damage to the knowledge we have. But when what we accept is inductively derived from probabilities and plausibilities it is vulnerable to unraveling. Here what we do not know can do great damage. With deductions from truths, ignorance may be bliss, but with inductions from plausibilities and probabilities it poses potential threats. This said, the fact remains that here as elsewhere we can only do the best we can.

On the basis of these considerations, it emerges that what is to all appearances the best available course of resolution is that afforded by the mixed strategy of combining a presumptivistic Fallibilism (Plan E) with a judicious Discriminationism (Plan G). In effect, this strategy calls for endorsing two positions:

1. That we see the scientific theories of the day as representing—in the main—not categorical truths, but rather merely *best estimates* that we can accept provisionally and *pro tem*, as viable truth-surrogates adopted with due tentativity until such time as something better comes along.
2. That while this position is something which we adopt on a "by and large" and "in general" basis (symbolized by that qualification "in the main"), we nevertheless stand prepared to endorse as categorically true certain particular categories of contentions, so that we do not see *everything* as fallible. On the one hand this secure core (K^+) of categorically endorsed contentions will include such predominantly *observational* information as "There are craters on the moon." On the other hand, as regards predominantly *theoretical* information, this unproblematic core will also contain (among others) two further groups:

    a. *Vague knowledge* (schoolbook science). [For example, "The structure of matter is granular—that is to say, atomic."][12]
    b. *Metaknowledge*. [For example, "The bulk of our scientific theories will eventually have to be replaced or emended."][13]

Knowledge on such a view is, for the most part, our cognitively optimal *estimate* of the truth.

This position may be motivated on the basis of a cousin of the epistemological "Sherlock Holmes Principle": "When you have eliminated the impossible, then whatever remains, however improbable, must be the truth."[14] However, in the present context, this precept should be so transmuted so as to take variant form of an *axiological* principle: "When you have eliminated the even less appealing alternatives, then whatever remains, however unappealing, must be accepted." That is—when all the other alternatives are even less appealing, then a given alternative, however unappealing, must be endorsed. (In effect, this represents an axiologically geared policy of rational choice in the context of a fixed range of alternatives.)

It is on this basis of comparative cost-effectiveness (rather than any claims to inherent and unconditional appropriateness) that the present deliberations will unfold so as to support the claims of a Fallibilistic Discriminationism—or, perhaps better, a Discriminating Fallibilism.[15]

THE COMPARATIVE FRAGILITY OF SCIENCE: SCIENTIFIC CLAIMS AS MERE ESTIMATES

Scientific "knowledge" at the level of deep theory is always *purported* knowledge: knowledge as we see it today. In our heart of hearts, we realize that we may see it differently tomorrow—or the day after. We must stand ready to acknowledge the fragility of our scientific theorizing. All we are ever able to do in natural science is to select the optimal answer to the questions we manage to formulate within the realm of alternatives specifiable by means of the conceptual machinery of the day. And we have no reason to doubt—nay, we have every reason to believe—that the day will come when this conceptual basis will be abandoned, in the light of yet unrealizable developments, as altogether inadequate.

Give that S_n is (*ex hypothesi*) our own current body of (putative) scientific knowledge—the science of the day, S_t with $t = n$ for "now"—we ourselves unquestionably stand committed to the inference schema: If $p \in S$, then p. At the level of specific claims, we have little alternative but to look on *our* knowledge as *real* knowledge—a thesis p would not be part of "*our*" truth" if we did not *take* it to form part of "*the* truth." The answers we give to our questions are literally the best we can provide. The "knowledge" at issue is to be knowledge according to our own lights—the only ones we have got. But despite our resort to this adequacy principle at the level of particular claims, we nevertheless should not have the hubris to think that our science has "got it right."

The relationship of "our (putative) scientific knowledge" to "the (real) truth" has to be conceived of in terms of *estimation*. At the frontiers of generality and precision, "our truth" in matters of scientific theorizing is not—and may well never actually be—the real truth. Science does not *secure* the truth (deliver it into our hands in its definitive finality). We have no alternative to acknowledging that our science, as it stands here and now, does not present the real to provide us with a tentative and provisional *estimate* of it. However confidently it may affirm its conclusions, the realization must be maintained that the declarations of natural science are provisional and tentative—subject to revision and even to outright rejection. The most we can ever do is to take our science (S_n) as the imperfect best we can do here and now to conjecture "the real truth."

To be sure, there is no question that we can *improve* our science. The history of science cries out for the Whig interpretation. Every applicable standard, from systemic sophistication to practical applicability, yields reason to think that later science is *better* science. But of course this does not mean that later science is any the *truer*—for true it cannot be if it, too, is destined for eventual rejection. The standards of scientific acceptability do not and cannot assure actual or indeed even probable or approximate truth. As Larry Laudan has argued with substantial evidence and eloquence: "No one has been able even to say what it would mean to be 'closer to the truth,' let alone to offer criteria for determining how we could assess such proximity."[16] All claims that emerge from scientific theorizing are vulnerable—subject to improvement and replacement. We can claim that later, "superior" science affords a *better warranted estimate* of the truth, but we cannot claim that it manages somehow to capture more of the truth or to approximate to it more closely.[17]

Interestingly enough, in point of vulnerability science compares unfavorably with the commonplace knowledge of matters of prescientific fact—our knowledge regarding sticks and stones and sealing wax and the other paraphernalia of everyday life. And there is good reason for this.

Increased security can always be purchased for our estimates at the price of decreased accuracy. We estimate the height of a tree at around 25 feet. We are *quite sure* that the tree is 25 ± 5 feet high. We are *virtually certain* that its height is 25 ± 10 feet. But we are *completely and absolutely sure* that its height is between 1 inch and 100 yards. Of this we are "completely sure" in the sense that we are "absolutely certain," "certain beyond the shadow of a doubt," "a certain as we can be of anything in the world," "so sure that we would be willing to stake our life on it," and the like. For any sort of estimate whatsoever, there is always a characteristic trade-off between the evidential security or reliability of the estimate, on the one hand (as determinable on the basis of its probability or degree of acceptability), and, on the other hand, its contentual *definiteness* (exactness, detail, precision, etc.). A situation of the sort depicted by the concave curve of Figure 2.1 obtains.

Now, the crucial point is that natural science eschews the security of indefiniteness. In science we operate at the right-hand side of the diagram: we always strive for the maximal achievable universality, precision, exactness, and so on. The law-claims of science are *strict*: precise, wholly explicit, exceptionless, and unshaded. They involve no hedging, no fuzziness, no incompleteness, and no exceptions. In stating that "the melting point of lead is 327.545°C at standard pressure," the physicist asserts that *all* pieces of (pure) lead will unfailingly melt at *exactly* this temperature; he certainly does not mean to assert that most pieces of (pure) lead will *probably* melt at *somewhere around* this temperature. By contrast, when we assert in ordinary life that "peaches are delicious," we mean something like "most people will find the eating of suitably grown and duly matured peaches a relatively pleasurable experience." Such statements have all sorts of built-in safeguards like "more or less," "in ordinary circumstances," "by and large," "normally," "if other things are equal," and the like. They are not laws but rules of thumb, a matter of practical lore rather than scientific rigor. In natural science, however, we deliberately accept risk by aiming at maximal definiteness—and thus at maximal informativeness and testability. The theories of natural science take no notice of what happens ordinarily or normally; they seek

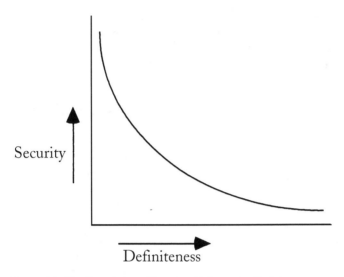

FIGURE 2.1. The Degradation of Security with Increasing Definiteness.

to transact their explanatory business in terms of high generality and strict universality—in terms of what happens always and everywhere and in all circumstances. In consequence, we must recognize the vulnerability of our scientific statements. The fact that the theoretical claims of science are "mere estimates" that are always cognitively at risk and enjoy only a modest life span has its roots in science's inherent commitment to the pursuit of maximal definiteness. Its cultivation of informativeness (of definiteness of information) everywhere forces science to risk error

A tenable fallibilism must take this inverse relationship between definiteness and security into careful account. Consider the "atomic theory," for example. It has an ancient and distinguished history in the annals of science, stretching from the speculations of Democritus in antiquity, through the work of Dalton, Rutherford, and Bohr, to the baroque complexities of the present day. "It is surely unlikely that science will ever give up on atoms!" you say. Quite true! But what we are dealing with here is clearly not *a* scientific theory at all, but *a vast family* of scientific theories, a great bundle loosely held together by threads of historical influence and family resemblances. There is not much that Rutherford's atoms, and Bohr's, and our contemporary quantum theorists' have in common. As such, the "atomic theory" is no more than a rough generic schema based on the more or less metaphorical intuition that "matter is granular in the small, composed of tiny structures separated in space," This is surely incomplete and indeterminate—a large box into which a vast number of particular theories can be fitted. It claims *that* there are atoms, but leaves open an almost endless range of possibilities as to what they are like. This sort of contention may well be safe enough; at this level of schematic indiscriminateness and open-endedness, scientific claims can of course achieve security. But they do so only at the expense of definiteness—of that generality and precision that reflect what science is all about.

The quest for enhanced definiteness is unquestionably a prime mover of scientific inquiry. The ever-continuing pursuit of increasing accuracy, greater generality, widened comprehensiveness, and improved systematicity for its assertions is the motive force behind scientific research. And this innovative process—impelled by the quest for enhanced definiteness—drives the conceptual scheme of science to regions ever more distant from the familiar conceptual scheme of our everyday life. For the ground rules of ordinary life discourse are altogether different. Ordinary-life communication is a practically oriented endeavor carried on in a social context: it stresses such maxims as "Aim for security, even at the price of definiteness"; "Protect your credibility"; "Avoid misleading people, or—even worse—lying by asserting outright falsehoods"; "Do not take a risk and 'cry wolf.'"

The aims of ordinary-life discourse are primarily *practical,* largely geared to social interaction and the coordination of human effort. In this context, it is crucial that we aim at credibility and acceptance—that we establish and maintain a good reputation for reliability and trustworthiness. In the framework of common-life discourse, we thus take our stance at a point far removed from that of science. Very different probative orientations prevail in the two areas. In everyday contexts, our approach is one of situational satisficing: we stop at the first level of sophistication and complexity that suffices for our present needs. In science, however, our objectives are primarily theoretical and governed by the aims of disinterested inquiry. Here our approach is one of systemic maximizing: we press on toward the ideals of systemic completeness and comprehensiveness. In science we put ourselves at greater risk because we ask much more of the project.

A view along the following lines is very tempting: "Science is the best, most thoroughly tested knowledge we have; the 'knowledge' of everyday life pales by comparison. The theses of science are really secure; those of everyday life, casual and fragile." But the very reverse is the case: our scientific theories are vulnerable and have a short life span; it is our claims at the looser level of ordinary life that are relatively secure and stable.

But, of course, the claims of everyday life have sacrificed definiteness in order to gain this security and reliability. One recent writer has quite correctly written that:

> we sometimes forget . . . how completely our own knowledge has absorbed what the cavemen knew, and what Ptolemy, Copernicus, Galileo and Newton knew. True, some of the things our predecessors thought they knew turned out not to be knowledge at all: and these counterfeit facts have been rejected along with the theories that "explained" them. But consider some of the facts and low-level regularities our ancestors knew, which we know, and which any astronomical

theory is obligated to explain: the sky looks about the same every night; it is darker at night than during the day; there is a moon; the moon changes its appearance regularity; there is a cycle of seasons.[18]

To be sure, the sort of "knowledge" at issue here does not go very far in affording us a detailed understanding of the ways of the world. It lacks detailed scientific substance. But just exactly that is the basis of its comparative security.

We must recognize, without lapsing into skepticism, that the cognitive stance characteristic of science requires the acceptance of fallibility and corrigibility, and so requires a certain tentativeness engendering the presumption of error. Because the aims of the enterprises are characteristically different, our inquiries in everyday life and in science have a wholly different aspect; the former achieves stability and security at the price of sacrificing definiteness, a price that the latter scorns to pay.

Fallibilism and the Distinction Between Our (Putative) Truth and the Real Truth

While we would like to think of our science as "money in the bank"—as something safe, solid, and reliable—history unfortunately militates against this comfortable view of our scientific theorizing. In science, new knowledge does not just supplement but generally upsets our knowledge-in-hand. We must come to terms with the fact that—at any rate, at the scientific level of generality and precision—*each* of our accepted beliefs *may* turn out to be false, and many of our accepted beliefs *will* turn out to be false.

Let us explore the ramifications of a cognitive fallibilism that holds that science—our science, the body of our scientific beliefs as a whole consists largely, and even predominantly, of false beliefs, embracing various theses that we will ultimately come to see (with the wisdom of hindsight) as quite untenable. It is useful to distinguish between potential error at the *particularistic* and *collectivistic* levels—between *thesis defeasibility* and *systemic fallibilism*. The former position makes the claim that *each* of our scientific beliefs *may* be false, while the latter claims that *many* or *most* of our scientific beliefs *are* false. The two theses are independent. Of the pair, "Each might be false" and "Most are false," neither entails the other. (Each of the entrants in the contest might be its winner, but it could not turn out that most of them indeed are. Conversely, most of the integers of the sequence 1, 2, 3, ... come after 3, but it cannot be that all might do so.)

Though the theses are independent, both are true. Thesis defeasibility follows from the fact that it is perfectly possible—and not only theoretically but *realistically* possible—that one could play the game of scientific inquiry correctly by all the accepted rules and still come to a result that fails to be true. For the truth of our objective claims hinges on "how matters really stand in the world," and not on the (inevitably incomplete) grounding or evidence of justification that we ourselves have in hand. We cannot cross the epistemic gap between the apparent and the real by any *logically* secure means; and so we must always be careful to distinguish between "our (putative or ostensible) truth," *what we think to be true*, on the one hand, and, on the other, "the real truth," *what actually is true*—the authentic and unqualified truth of the matter. In scientific theorizing, we are never in a position to take this epistemic gap as closed.

Systemic fallibilism is the more serious matter, however, for it means that one must presume that our whole scientific picture of the world is seriously flawed and will ultimately come to be recognized as such in that many of our present-day beliefs at the scientific frontier will eventually have to be abandoned. We learn by empirical inquiry about empirical inquiry, and one of the key things we learn is that at no actual stage does science yield a final and unchanging result. All the experience we can muster indicates that there is no justification for regarding our science as more than an inherently imperfect stage within an ongoing development. And we have no responsible alternative to supposing the imperfection of what we take ourselves to know. We occupy the predicament of the "Preface Paradox" exemplified by the author who apologizes in his preface for those errors that have doubtless made their way into his work, and yet blithely remains committed to all those assertions he makes in the body of the work itself. We know or must presume that (at the synoptic level) there are errors, though we certainly cannot say where and how they arise. There is no realistic alternative to the supposition that science is wrong—in various ways and that much of our supposed "knowledge" of the world is no more than a tissue of plausible error. We are thus ill advised to view the science of our day—or of *any* day—as affording "the final truth of the matter."

It goes without saying that we can never be in a position to claim justifiedly that our current corpus of scientific knowledge has managed to capture "the whole truth." But we are also not even in a position justifiably to claim that we have got "nothing but the truth." Our scientific picture of nature must always be held provisionally and tentatively, however deeply we may be attached to some of its details.

All attempts to equate the real truth with apparent truth of some sort are condemned to failure: having the same generic structure, they have the same basic defect. They all view "the real truth" as "the putative truth when arrived at under conditions C" (be it in favorable circumstances or under careful procedures). And in each case the equation is destroyed by the ineliminable possibility of a separation between "the real truth" and the *putative* truth that we ourselves actually reach (by *whatever* route we have arrived there). None of our scientific beliefs are so "cataleptic" or so "clear and distinct" that they cannot run awry. Nobody's putative "scientific truth" can be authenticated as genuine—be it that of the experts or of those who prudently implement established epistemic methods. With claims to knowledge at the level of scientific generality and precision, there is no way in which we can effect an inferentially secure transition to the real truth from anybody's *ostensible* truth (our own included).

The transition from appearance to reality can perhaps be salvaged by taking God or the community of perfect scientists or some comparable idealization to be the "privileged individual" whose putative truth is at issue. The doctrine can thus be saved by resorting to the Myth of the God's Eye View: With S^* as the cognitive corpus of "perfected science," we can indeed establish an equivalence between S^* membership and "the real truth." But this desperate remedy is of little practical avail. For only God knows what "perfected science" is like; we ourselves certainly do not. And the Schoolmen were right: in these matters we have no right to call on God to bail us out. (*Non in scientia recurrere est ad deum.*) While the idea at issue may provide a useful contrast-conception, it is something that we cannot put to concrete implementation.

Of course, we stand committed to our truths; imperfect they may be, but we have nowhere else to go. We have no alternative to proceeding on the "working hypothesis" that in scientific matters *our* truth is *the* truth. But we must also recognize that is simply not so—that the working hypothesis in question is no more than just that, a convenient fiction.

We recognize that the workaday identification of our truth with *the* truth is little more than a makeshift—albeit one that stands above criticism because no alternative course is open to us. Fallibilism is our destiny. Given our nature as a creature that makes its way in the world by use of information we have to do the best we can. The recognition that this may well not be good enough is simply an ineliminable part of the human condition. No deity has made a covenant with us to assure that what we take to be "adequately established" theories must indeed be true in these matters of scientific inquiry.

The proper lesson here is not the skeptic's conclusion that our "science" counts for naught, but simply, and less devastatingly, the realization that our scientific knowledge—or putative and purported knowledge—regarding how things work in the world is flawed; that the science of the day is replete with inexactitudes and errors that we ourselves are impotent to distinguish from the rest. With respect to the claims of science, we realize that our herd is full of goats that nevertheless appear altogether sheeplike to us.

Chapter 3

Skepticism and Its Deficits

Synopsis

- *The fact that knowledge must be certain seems to open the door to skepticism.*
- *However, what is at issue is not* categorical *or* desolate *certainty but* practical *or* effective *certainty.*
- *What is called for is not the theoretical centrality of logic but the practical certainty of life.*
- *Skepticism's case rests on the impossibilities engendered by unrealistic and unachievable standards.*
- *The statement "I know that p but it could possibly be the case the not-p" is not so much a logical as a pragmatic contradiction.*
- *Skepticism is fostered by an unrealistic unwillingness to run cognitive risks.*
- *This itself runs counter to the demands of reason.*
- *It is at variance not with cognitive reason alone but with practical reason as well.*
- *For the sckptic loses sight of the very reason for being of our cognitive endeavors.*
- *The fact of it is that skepticism is* economically irrational: *The price we would pay in endorsing the skeptic's position is so high as to outweigh whatever benefit could possibly accrue from it.*

THE SKEPTICS "NO CERTAINTY" AGREEMENT

As stressed above, attributions of authentic knowledge do not admit of any element of doubt or qualification, so that authentic knowledge must be unqualifiedly certain and undeniably true. On the other hand, philosophers since Plato's day[1] have stressed the unattainability of absolutes in our knowledge of this world. And the plausibility of such a view is readily established.

A straightforward and plausible argument is at work here. If a contention is to be absolutely, secure relative to the grounds by which it is supported, then its content must not go beyond the content of those contentions that serve as grounds for the claim. But with factual statements there is always an "evidential gap," because the data at our disposal were exhaust the content of our objective claims. "The apple *looks red* to me" is an autobiographical statement about me while "The apple *is red*" is an objective claim that has involvements (e.g., how it will look to others in different sorts of light) that I have not checked and cannot check in toto. Factual claims are invariably such that there is a wide gap between the evidence we need to have at our disposal to make a claim warrantedly and the content of this claim. The milkman leaves the familiar sort of bottle of white liquid on the doorstep. One does not hesitate to call it "milk." A small cylinder of hard, white, earthen material is lying next to the blackboard. One does not hesitate to call it "chalk." The content of such claims clearly ranges far wider than our meager Evidence and extends to chemical composition, sources of origin, behavior under pressure, and so on and so forth. And this story is a standard one. For the fact is that all of our statements regarding matters of objective fact (i.e., "That is an apple" as opposed to "Something appears to me to be an apple") are such that the content of the claim—its overall set of commitments and implications—moves far beyond the (relatively meager) evidence for it that is actually at our disposal. And this evidential gap between the evidence-in-hand and the substantive contention at issue seemingly foredooms any prospect of attaining certainty.[2] For all intents and purposes this circumstance squarely violates the absolutism of claims to knowledge. What we have called "the facts of cognitive life" are to all appearances such that the definitive conditions for knowledge cannot be met in the factual domain. This, at any rate, is the skeptic's contention.

The skeptic bolsters this claim by inserting into the evidential gap the sharp wedge of a knowledge-defeating possibility—the supposition that life is but a dream, or the hypothesis of the Cartesian arch-deceiver and its latter-day successor, the wicked powerful scientist. The unattainability of any knowledge in matters of objective fact is supported by the skeptic on the basis of the impossibility in principle of ruling out such certainty-defeating possibilities. Do we, for example, know for sure that the person who jumps off the Empire State Building will crash downward? Why should that person not float gently skyward? This would, no doubt, surprise us, but surprises do happen; in such matters of generalization "we may be in no better position than the chicken which unexpectedly has its neck wrung."[3] After all, the skeptic insists:

We are thus caught in the inexorable grip of the skeptic's "No Certainty" Argument:[4]

1. All knowledge-claims are committed to a demand for absolute certainty
2. Objective factual claims are always evidence-transcending: they are never in a position to meet absolutistic demands.

∴ Our objective factual claims can never amount to actual *knowledge*.[5]

The task of these present deliberations is to examine whether—and how—this sort of sceptical argumentation can be defeated.[6]

The Role of Certainty

Skepticism's pet thesis to the contrary notwithstanding, however, something's being absolutely certain—or even *being* true—in fact just is not a necessary precondition for staking a rationally warranted *claim* to knowledge. Consider the two propositions:

- P is justifiably held by X to be true.
- P is true.

Does the first entail or presuppose the second? Surely not. For the evidence-in-hand that suffices to justify someone in holding a thesis to be true need not provide a deductive guarantee of this thesis. For a strictly analogous situation obtains with the pair:

- P is justifiably held by X to be certain.
- P is certain.

Again, the first proposition does not entail or require the second. The standard gap between the epistemic issue of what someone justifiably holds to be and the ontological issue of what is again comes into the picture. One must be willing to admit in general the existence of a gap between warranted assertability and ultimate correctness, holding that on occasion even incorrect theses can be maintained with due warrant. And there is no decisive reason for blocking the application of this general rule to certainty-claims in particular. It would be a fallacy to think that claims to certainty are warranted only if they themselves are certain.

Here, as elsewhere, use-conditions stop short of truth-conditions. The rules of reasoned discourse are such that we are rationally entitled to make our descriptive claims—*notwithstanding the* literally endless implicative ramifications of their content, ramifications over whose obtaining we in fact have no rational control without making specific verificatory checks. And note that even if such checks were made, their results would be incomplete and would pertain to a few specific samples drawn from an infinite range.

And this state of things has an important application when knowledge-discourse is at issue. For here too we must distinguish between evidence and claim, between the *warrant* for a claim and its *content*, its evidentiation and its implications.

The assertion "I know that *p*" has all of those absolutistic facets we have considered (certainty, unqualifiedness, etc.). All are ineliminably parts of the content of the claim. But with this particular factual claim, as with any other, one need not establish full rational control over the whole gamut of its entailments and implications. One can appropriately assert "I know that P" when one has adequate rational warrant for this assertion, and this warrant may well stop at adequately conclusive[7]—rather than comprehensively exhaustive evidence for the claim.

The thesis that knowledge must be certain requires critical scrutiny and analysis in the light of these considerations. For "certainty" here must not be construed to mean "derived by infallible processes from theoretically unassailable premises" since one is surely justified in "being certain" in circumstances that do not logically preclude any possibility of error. The operative mode of "certainty" here is not some absolutistic sense of logical infallibility—it is the realistic concept that underlies our actual, real-life processes of argumentation and reasoning. It is impossible to give too heavy emphasis to the crucial fact that to say of a thesis that it "is certain" is to say no more than it is as certain as in the nature of the case, a thesis of this sort reasonably could be rendered. And this does not—and need not—preclude any possibility of error, but any real or genuine possibility of error.

The cognitivist who rejects skepticism and purports to know some fact need not insist that he has intrinsically irrefutable and logically conclusive evidence that it obtains. It is simply not necessary for him to make an assertion of this sort in support of his contention. It is sufficient that his evidence for his claim *p* is as good as that for anything of *p*'s type can reasonably be asked for. To be sure we must be certain of what we know, but the "certainty" that must attach to knowledge-claims need not be absolutistic in some way that is in principle unrealizable. It must be construed in the sense of as certain as can reasonably be expected in the circumstance. A claim to knowledge does not—as the skeptics charge—transgress by offering in principle infeasible guarantees; it simply means that one is to rest secure in the assurance that everything has indeed been done that one can possibly ask for within the limits of reasonableness to ascertain the fact at issue.

A claim to knowledge extends an assurance that all due care and caution has been exercised to ensure that any *real* possibility of error can be written off: it issues a guarantee that every proper safeguard has been exercised. Exactly this is the reason why the statement "I know *p*, but might be mistaken" is self-inconsistent. For the man who claims to know that *p*, thereby issues a guarantee that the qualification "but I might be mistaken" effectively revokes. What is established by the self-defeating nature of locutions of the type of "I know that *p* but *might* well be wrong" is thus not that knowledge is inherently indefeasible, but simply that knowledge-claims offer guarantees and assurances so strong as to preempt any safeguarding qualifications: they preclude abridgment of the sort at issue in protective clauses like "1 might well be wrong."

THE CERTAINTY OF LOGIC VERSUS THE CERTAINLY OF LIFE

It is not possible to overemphasize that the certainty of knowledge is the certainty of life—realistic certainty and not that of some transcendentally inaccessible realm.[8] It is the certainty that precludes any realistic possibility of error: any possibility of error that is "worth bothering about," the closing of every loophole that one can reasonably ask for.[9] This is, and must be so because knowledge claims are asserted and denied here, in this world—and not in some transcendentally inaccessible one, so that the norms and ground rules governing their use must be appropriately applicable (at least in principle) here and now. Accordingly, there is no contradiction in terms involved in saying that the absolutistic aspect of a knowledge claim is compatible with an element of (claim-externalized) qualification.

As we have seen, the "epistemic gap" between available evidence and asserted content means that the falsity of our objective factual claims is always logically compatible with the evidence at our disposal. But this undoubted fact remains epistemically irrelevant. To be sure, the evidential gap is real and undeniable. And since it exists, it is always possible to insert an hypothesis into it to sever what we ourselves think to be the case from "the real truth of the matter." But there are hypotheses and hypotheses—sensible ones as well as those which cannot but strike us as strained and bizarre. An hypothesis capable of undoing "There are rocks in the world" (to take an ordinary-life example) or that in the present cosmic era $s = \frac{1}{2}gt^2$ (to take a scientific one) would illustrate the latter, far-fetched variety. That either thesis is false is "unthinkable"—any hypothesis capable of undoing the thesis at issue is too peculiar and "unrealistic" to afford a real possibility of error. Admittedly there are such (far-fetched) hypotheses and such (implausible) possibilities. But their very far-fetchedness and implausibility mean that the possibilities of error they pose are not realistic. The upsets at issue are simply too drastic—the whole demonology of deceitful deities, powerful mad scientists, and so on brings our entire view of the world crashing down about our ears. Such possibilities cannot be ruled out from the domain of the imaginable, but we can and do exclude them from the arena of the practical politics of the cognitive situation.

There is no changing the fact that the person who claims to know something also becomes committed thereby to its implications (its logical consequences and its presuppositions). But a claim to knowledge can be made reasonably and defensibly even by one who realizes that it involves commitments and ramifications that may not stand up in the final analysis to the challenges of a difficult and often recalcitrant world. No assurances that extend beyond the limits of the possible can be given—or sensibly asked for. The absolute certainty of our knowledge claims is not and cannot be the sort of thing which one is in principle precluded from realizing. (*Ultra posse nemo obligatur*. To reemphasize: the certainty of knowledge is the certainty of life!)

After all, the "certainty" of knowledge claims can seemingly be understood in two very different perspectives:

1. as an unattainable ideal, a condition at which a knowledge claim aims but which in the very nature of things it cannot attain—to its own decisive detriment.
2. as an assurance, a promise, a guarantee that everything needful has been done for the ascertainment of the knowledge claim, and this must be construed in socially oriented terms as a real-life resource of the operative dynamics of communication.

Various philosophers—and most skeptics—insist on the former interpretation, an insistence which is as unnecessary as it is unrealistic.[10] For it is clearly the second, mundane or realistic interpretation that is operative in the conception of knowledge we actually use within the setting of real life.

It is thus tempting to speak of a contrast between "the hyperbolic certainty of the philosopher" and "the mundane certainty of the plain man" in the setting of the actual transaction of our cognitive business.[11] Philosophers have often felt driven to a conception of knowledge so rigid as to yield the result that there is little if anything left that one ever be said to know. Indeed, skeptical thinkers of this inclination launch on an explication of the "nature of knowledge" which sets the standards of its attainment so high that it becomes in principle impossible for anything to meet such hyperbolic demands. Against this tendency it is proper to insist that while what is known must indeed be true—and certainly true—it is nevertheless in order to insist that the conceptions at issue can and should be so construed that there are realistic and realizable circumstances in which our claims to *certainly* and to *knowledge* are perfectly legitimate and altogether justified. A doctrine which admits the defeasibility of quite appropriate claims to knowledge need involve no contradictions in terms.

Pragmatic Inconsistency

The line between authentic and putative knowledge simply cannot be drawn in our own case. The challenge, "Indicate to me something that you indeed *think* you know but really don't" is a challenge I cannot possibly meet. You might possibly be able to do this for me, but neither you not I can bring it off on our own account.

This line of consideration brings to light the effective basis of the untenability of "I know that p, but it is not certain that p" or "I know that p, but it could possibly be the case that not-p." The crux is not that these are logically inconsistent but that they are incoherent on grounds of being pragmatically self-defeating—that is, self-defeating in the context of the relevant communicative aims and objectives. For it is clear that the former part of these statements extends an assurance that the qualifications of the second part effectively abrogate.

A knowledge-claim requires that the grounding-in-hand is sufficient to preclude any real possibility of error, whereas the but-addendum says that there are still such possibilities worth worrying about. And so the statement as a whole is inconsistent: it takes away with one hand what it gives with the other.[12]

This difference between real and merely conjectural possibilities of error is crucial for rational warrant for claims or concessions to knowledge. A real possibility must be case-specific and not abstractly generic and somehow based on general principles alone. And this, it must be insisted, is not incompatible with the existence of a "purely theoretical" (let alone "purely logical") prospect of error. There is thus no real anomaly in holding, on the one hand, that knowledge "must be certain" (in the *effective* sense of this term) and, on the other hand, that a valid knowledge claim "might possibly be wrong" (with "might" construed in the light of a merely theoretical or "purely logical" mode of possibility).

For example, we need not in the usual course of things exert ourselves in an endeavor to rule out the imaginative skeptic's recourse to the whole demonology of uncannily real dreams, deceitful demons, powerful evil scientists, and so forth. The general principles and presumptions of the domain suffice to put all this aside. To claim knowledge in specific cases, all we need do is eliminate those case-specific considerations that would countervail against the claim at issue.

Some philosophers have endeavored to evade skepticism by emptying knowledge claims of any and all pretentions to certainty. For example, in his interesting book, *Knowledge*, Keith Lehrer writes:

> Thus, our theory of knowledge is a theory of knowledge without certainty. We agree with the sceptic that if a man claims to know for certain, he does not know whereof he speaks. However, when we claim to know, we make no claim to certainty. We conjecture that to speak ~n this way ~s a departure from the most customary use of the word "know." Commonly, when men say they know, they mean they know for certain.[13]

But this approach of extruding certainty from knowledge and so of going over to a construction of knowledge at variance with that of our standard usage—

avoids skepticism too high a price. Claims to certainly cannot simply be eliminated from the conception of knowledge. Our present position, by contrast, emphatically preserves the certainty-involvement of all our knowledge claims. It simply insists that the certainty of knowledge is the certainty of real life—the sort of certainty that is not timeless and untarnishable but can be abrogated by the difficult circumstances of an uncooperative world.

To be sure, a skeptical objection may well be offered:

> If we sometimes fall into error even when doing the best we can, how can we tell that we have not done so in the present case? And if we cannot *guarantee* truth, then how can we speak of *knowledge*?

The answer is simple. If we have done all that can reasonably be asked of us, the best that can realistically be done, then there can be no need for any *further* assurance. The objection suggests that we must be in a position to do better than the best we can. And of course we cannot—and so this must not be asked of us.

Accordingly, the "No Certainty" Argument becomes invalidated: It is not true that knowledge claims are committed to a demand for absolute certainty in any hyperbolically inaccessible way. They are indeed committed to a demand for certainty, but this "certainty" must be construed realistically—in the effective, mundane, and practical sense of the term. The certainty at issue in our knowledge-claims is not inherently unattainable; it is simply that the grounding in hand must be strong enough to indicate that further substantiation is superfluous in the sense of yielding every reasonable assurance that the thing at issue is as certain as something of its sort need appropriately be. To repeat: it suffices to ask for an adequate grounding for these claims; logically exhaustive grounding is not a reasonable requirement, for the simple reason—so eloquently stressed by the skeptic himself—that it is in principle incapable of being satisfied.

The skeptic sets up standards on "knowledge" that are so unrealistic as to move outside the range of considerations at work in the conception as it actually functions in the language. He insists on construing knowledge in such a way as to foist on the cognitivist a Sisyphus-like task by subjecting all knowledge-claims to an effectively undischargeable burden of proof. But this hyperbolic standard separates the skeptic from the concept of knowledge as it functions in common life and in (most) philosophical discussions.

To be sure, the skeptic might well ask: "What gives you the right to impose your probative standards on *me*?

After all, if it is knowledge as the language deals with it that we wish to discuss—and not some artificial construct whose consideration constitutes a change of subject-then we must abide by the ground rules of the conceptual scheme that is at issue. And this simply is nor a negotiable matter of playing the game by X's rules instead of Y's. It is not a conventional game that is defined by the rules we arbitrarily adopt, but an impersonal set of rules established in the public instrumentality of a language. We must not afford the skeptic the luxury of a permission to rewrite at his own whim the warranting standards for the terminology of our cognitive discourse.

The skeptic simply is not free to impose his hyperbolic probative standards on us. If he wishes to dispute about knowledge, he must take the concept as he finds it in the language-based conceptual system that we actually use. (He cannot substitute a more rigoristic standard for counting as knowledge, any more than he can substitute a more rigoristic standard for counting as a dog, say a standard that ruled chihuahuas out as just too small.) In failing to make effective contact with the conceptual scheme in which our actual knowledge claims in fact function, the skeptic assumes an irrational posture in the debate. And the crucial fact is that our actual standards here root in the ground rules for rational controversy and the conditions for making out a conclusive (probatively solid) case, ground rules and conditions in which all of the standard mechanisms of presumptions and burden of proof are embedded.

Skepticism and Risk

The scientific researcher, the inquiring philosopher, and the plain man, all desire and strive for information about the "real" world. The sceptic rejects their ventures as vain and their hopes as foredoomed to disappointment from the very outset. As he sees it, any and all sufficiently trustworthy information about factual matters is simply unavailable as a matter of general principle. To put such a radical scepticism into a sensible perspective, it is useful to consider the issue of cognitive rationality in the light of the situation of risk taking in general.

There are three very different sorts of personal approaches to risk and three very different sorts of personalities corresponding to these approaches, as follows:

Type 1: *Risk avoiders*
Type 2: *Risk calculators*

    2.1: cautious
    2.2: daring

Type 3: *Risk seekers*

The type 1, risk-avoidance, approach calls for risk aversion and evasion. Its adherents have little or no tolerance for risk and gambling. Their approach to risk is altogether negative. Their mottos are Take no chances, Always expect the worst, and Play it safe.

The type 2, risk-calculating, approach to risk is more realistic. It is a guarded middle-of-the-road position, based on due care and calculation. It comes in two varieties. The type 2.1, cautiously calculating, approach sees risk taking as subject to a negative presumption, which can however, be defeated by suitably large benefits. Its line is: Avoid risks unless it is relatively clear that a suitably large gain beckons at sufficiently suspicious odds. If reflects the path of prudence and guarded caution. The type 2.2, daringly calculating, approach sees risk taking as subject to a positive presumption, which can, however, be defeated by suitably large negativities. Its line is: Be prepared to take risks unless it is relatively clear that an unacceptably large loss threatens at sufficiently inauspicious odds. It reflects the path of optimistic hopefulness.

The type 3, risk-seeking, approach sees risk as something to be welcomed and courted. Its adherents close their eyes to danger and take a rosy view of risk situations. The mind of the risk seeker is intent on the delightful situation of a favorable issue of events: the sweet savor of success is already in his nostrils. Risk seekers are chance takers and go-for-broke gamblers. They react to risk the way an old warhorse responds to the sound of the musketry: with eager anticipation and positive relish. Their motto is: Things will work out.

In the conduct of practical affairs, risk avoiders are hypercautious; they have no stomach for uncertainty and insist on playing it absolutely safe. In any potentially unfavorable situation, the mind of the risk avoider is given to imagining the myriad things that could go wrong. Risk seekers, on the other hand, leap first and look later, apparently counting on a benign fate to ensure that all will be well; they dwell in the heady atmosphere of "anything may happen." Risk calculators take a middle-of-the-road approach. Proceeding with care, they take due safeguards but still run risks when the situation looks sufficiently favorable. It is thus clear that people can have very different attitudes toward risk.

So much for risk taking in general. Let us now look more closely at the cognitive case in particular.

The situation with regard to specifically cognitive risks can be approached as simply a special case of the general strategies sketched above. In particular, it is clear that risk avoidance stands coordinate with skepticism. The skeptic's line is: Run no risk of error; take no chances; accept nothing that does not come with ironclad guarantees. And the proviso here is largely academic, seeing that little if anything in this world comes with ironclad guarantees-certainly nothing by way of interesting knowledge. By contrast, the adventuresome syncretist is inclined, along with radical Popperians such as P. K. Feyerabend, to think that anything goes. His cognitive stance is tolerant and open to input from all quarters. He is gullible, as it were, and stands ready to endorse everything and to see good on all sides. The evidentialist, on the other hand, conducts his cognitive business with comparative care and calculation, regarding various sorts of claims as perfectly acceptable, provided that the evidential circumstances are duly favorable.

The skeptic accepts nothing, the evidentialist only the chosen few, the syncretist virtually anything. In effect, the positions at issue in skepticism, syncretism, and evidentialism simply replicate, in the specifically cognitive domain, the various approaches to risks at large.

It must, however, be recognized that in general two fundamentally different kinds of misfortunes are possible in situations where risks are run and chances taken:

1. We reject something that, as it turns out, we should have accepted. We decline to take the chance, we avoid running the risk at issue, but things turn out favorably after all, so that we lose out on the gamble.
2. We accept something that, as it turns out, we should have rejected. We do take the chance and run the risk at issue, but things go wrong, so that we lose the gamble.

On the one hand, if we are risk seekers, we will incur few misfortunes of the first kind, but things being what they are, many of the second kind will befall us. On the other hand, if we are risk avoiders, we shall suffer few misfortunes of the second kind, but shall inevitably incur many of the first. The overall situation has the general structure depicted in Figure 3.1.

Clearly, the reasonable thing to do is to adopt a policy that minimizes misfortunes overall. It is thus evident that both type 1 and type 3 approaches will, in general, fail to be rationally optimal. Both approaches engender too many misfortunes for comfort. The sensible and prudent thing is to adopt the middle-of-the-road policy of risk calculation, striving as best we can to balance the positive risks of outright loss against the negative ones of lost opportunity. Rationality thus counterindicates approaches of type 1 and type 2, taking the line of the counsel, Neither avoid nor court risks, but manage them prudently in the search for an overall minimization of misfortunes. The rule of reason calls for sensible management and a prudent calculation of risks; it standardly enjoins on us the Aristotelian golden mean between the extremes of risk avoidance and risk seeking. Turning now to the specifically cognitive case, it may be observed that the skeptic succeeds splendidly in averting misfortunes of the second kind. He makes no errors of commission; by accepting nothing, he accepts nothing false. But, of course, he loses out on the opportunity to obtain any sort of information. The skeptic thus errs on the side of safety, even as the syncretist errs on that of gullibility. The sensible course is clearly that of a prudent calculation of risks.

Being mistaken is unquestionably a negativity. When we accept something false, we have failed in our endeavors to get a clear view of things—to answer our questions correctly. And moreover, mistakes tend to ramify, to infect environing issues. If I (correctly) realize that p logically entails q but incorrectly believe not-q, then I am constrained to accept not-p, which may well be quite

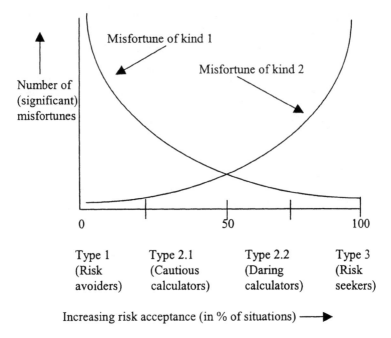

FIGURE 3.1 Risk Acceptance and Misfortunes

wrong. Error is fertile of further error. So quite apart from practical matters (suffering painful practical consequences when things go wrong), there are also the purely cognitive penalties of mistakes—entrapment in an incorrect view of things. All this must be granted and taken into account. But the fact remains that errors of commission are not the only sort of misfortune there are.[14] Ignorance, lack of information, cognitive disconnection from the world's course of things—in short, errors of omission—are also negativities of substantial proportions. This too is something we must work into our reckoning.

In claiming that his position wins out because it makes the fewest mistakes, the skeptic uses a fallacious system of scoring, for while he indeed makes the fewest errors of one kind, he does this at the cost of proliferating those of another. Once we look on this matter of error realistically, the skeptic's vaunted advantage vanishes. The skeptic is simply a risk avoider, who is prepared to take no risks and who stubbornly insists on minimizing errors of the second kind alone, heedless of the errors of the first kind into which he falls at every opportunity.

Ultimately, we face a question of value trade-offs. Are we prepared to run a greater risk of mistakes to secure the potential benefit of an enlarged understanding? In the end, the matter is one of priorities—of safety as against information, of ontological economy as against cognitive advantage, of an epistemological risk aversion as against the impetus to understanding. The ultimate issue is one of values and priorities, weighing the negativity of ignorance and incomprehension against the risk of mistakes and misinformation.

The crucial fact is that inquiry, like virtually all other human endeavors, is not a cost-free enterprise. The process of getting plausible answers to our questions also involves costs and risks. Whether these costs and risks are worth incurring depends on our valuation of the potential benefit to be gained. And unlike the committed skeptic, most of us deem the value of information about the world we live in to be a benefit of immense value something that is well worth substantial risks.

Philosophical epistemologists tend to view knowledge acquisition in purely theoretical terms and to abstract it from the crass business of exerting effort, expending resources, and running risks. But this is simply unrealistic. Even the purest and most theoretical of inquiries has its practical and mundanely economic dimension.

The key point is that our epistemological procedures such as explanation, substantiation, justification are all essentially communicative acts performed in the context of an interpersonal interchange of ideas and information that is subject to communally established ground rules. A fundamental communicative purpose underlies all these probative activities. The end of the road of the process of justification is clearly reached when anything further that could be brought in would be less plausible than what is already at hand. Even as one must not explain by the yet more obscure (*non est explanandum obscurum per obscurius*), so we must not defend by the yet more dubious. In the preceding dialogue, a stage is reached when the existence of the pen in question is (ex hypothesi) something that is as sure under the epistemic circumstances at issue as one could ever be of the further supporting considerations that could be adduced on its behalf. And when this stage is reached, there is no point in going on.[15]

RATIONALITY AND COGNITIVE RISK

Skepticism not only runs us into practical difficulties by impeding the cognitive direction of action, but it has great theoretical disadvantages as well. The trouble is that it treats the avoidance of mistakes as a paramount good, one worth purchasing even at a considerable cost in ignorance and lack of understanding. For the radical sceptic's seemingly high-minded insistence on definitive truth, in contradistinction to merely having reasonable warrant for acceptance—duly followed by the mock-tragic recognition that this is of course unachievable—is totally counterproductive. It blocks from the very outset any prospect of staking reasonable claims to information about the ways of the world. To be sure, the averting of errors of commission is a very good thing when it comes free of charge, or at any rate cheap. But it may well be bought at too high a cost when it requires us to accept massive sacrifices in for going the intellectual satisfaction of explanation and understanding. It would of course be nice if we could separate errors of commission from errors of omission and deploy a method free from both. But the realities do not permit this. Any method of inquiry that is operable in real life is caught up in the sort of trade-off illustrated in Figure 3.1.

From such a standpoint, it becomes clear that skepticism purchases the avoidance of mistakes at an unacceptable price. After all, no method of inquiry, no cognitive process or procedure that we can operate in this imperfect world, can be altogether failure free and totally secure against error of every description. Any workable screening process will let some goats in among the sheep. With our cognitive mechanisms, as with machines of nay sort, perfection is unattainable; the prospect of malfunction can never be eliminated, and certainly not at any acceptable price. Of course, we could always add more elaborate safeguarding devices. (We could make automobiles so laden with safety devices that they would become as large, expensive, and cumbersome as busses.) But that defeats the balance of our purposes. A further series of checks and balances prolonging our inquiries by a week (or a decade) might avert certain mistakes. But for each mistake avoided, we would lose much information. Safety engineering in inquiry is like safety engineering in life. There must be proper balance between costs and benefits. If accident avoidance were all that mattered, we could take our mechanical technology back to the stone age, and our cognitive technology as well.

The skeptics' insistence on safety at any price is simply unrealistic, and it is so on the essentially economic basis of a sensible balance of costs and benefits. Risk of error is worth running because it is unavoidable in the context of the cognitive project of rational inquiry. Here as elsewhere, the situation is simply one of nothing ventured, nothing gained. Since Greek antiquity, various philosophers have answered our present question, Why accept anything at all?, by taking the line that man is a rational animal. Qua animal, he must act, since his very survival depends on action. But qua rational being, he cannot act availingly, save insofar as his actions are guided by his beliefs, by what he accepts. This argument has been revived in modern times by a succession of pragmatically minded thinkers, from David Hume to William James.

The contrast line of reasoning to this position is not, If you want to act effectively, then you must accept something. Rather, its line is, If you want to enter into the cognitive enterprise, that is, if you wish to be in a position to secure information about the world and to achieve a cognitive orientation within it, then you must be prepared to accept something. Both approaches take a stance that is not categorical and unconditional, but rather hypothetical and conditional. But in the classically pragmatic case, the focus is on the requisites for effective action, while our present, cognitively oriented approach focuses on the requisites for rational inquiry. The one approach is purely practical, the other also theoretical. On the present perspective, then, it is the negativism of automatically frustrating our basic cognitive aims (no matter how much the sceptic himself may be willing to turn his back on them) that constitutes the salient theoretical impediment to skepticism in the eyes of most sensible people.

Now here it must be conceded that life without knowledge, reason, or belief is certainly not in principle impossible: animals, for example, manage very well. Or again, a somewhat less radical strategy is available, one that countenances acceptance (and belief), but only on a wholly unreasoned basis (e.g., instinct, constraint by the appearances, etc). The skeptic can accordingly hold—and act on—all those beliefs that people ordinarily adopt, with only this difference: that he regards them as reflecting mere appearances and denies that the holding of these beliefs is justified. But such a view simply heads rationality underfoot.

Ludwig Wittgenstein wrote: "The squirrel does not infer by induction that it is going to need stores next winter as well. And no more do we need a law of induction to justify (*rechtfertigen*) our actions or our predictions." This gets the matter exactly wrong. Had Wittgenstein written *perform* (or carry out) in place of *justify,* all would have been well. But once justification (*Rechfertigung*) is brought upon the stage, induction or some functional equivalent for establishing the need is just exactly what does become necessary. Saying that someone is being rational commits us to holding that this person is in a position to rationalize what he says and what he does—to justify it and exhibit it as the sensible thing to do. The rational use of a technique inevitably requires a great deal of factual backing from our knowledge—our purported knowledge—regarding how things work in the world. The rational man as such requires the availability of cogent reasons for action, and only with the abandonment of a rigorous all-out skepticism can such reasons possibly be obtained.

The *traditional* pragmatic argument against skeptical agnosticism goes roughly as follows:

> On the plane of abstract, theoretical reasoning the skeptical position is, to be sure, secure and irrefutable. But scepticism founders on the structure of the human condition—that man finds himself emplaced *in medias res* within a world where his very survival demands action. And the action of a rational being requires the guidance of belief. Not the inferences of theory and cognition but the demands of practice and action make manifest the untenability of the sceptic's position.

Conceding that skepticism cannot be defeated on its own ground, that of pure theory, it is held to be invalidated on practical grounds by an incapacity to support the requisites of human action. Essentially this argument is advanced by such diverse thinkers as the Academic Skeptics of classical antiquity, David Hume and William James.[16]

Unfortunately however, this position leaves it open for the sceptic to take to the high ground of a partisan of rigorous rationality. For the skeptic may well take the following line:

> This charge of stultifying practice is really beneath my notice. Theoretical reason and abstract rationality are what concerns the true philosopher. The issue of what is *merely practical* does not concern me. As far as "mere practice" goes, I am perfectly prepared to conform my actions to the pattern that men in general see fit to follow. But one should recognize that the demands of theoretical rigor point in another—and altogether sceptical—direction.[17]

The present line of argument does not afford the skeptic this comfortable option. Its fulcrum is not the issue of *practice* as such, but the issue of pragmatic rationality included—since it is our specifically cognitive practice of rational inquiry and argumentation that is at issue. In affecting to disdain this the skeptic must now turn his back not simply on the practice of ordinary life, but rationality itself. His "victory" is futile because he conveniently ignores the fact that the whole enterprise of reason-giving is aimed at rationale construction and is thus pointless save in the presence of a route to adequacy in this regard—the standard machinery for assessing probative propriety. The skeptic in effect emerges as unwilling to abide by the evidential ground rules that govern the management of rational deliberation along the established lines.

In the final analysis, the skeptic thus runs afoul of the demands of that very rationality in whose name he so high-mindedly claims to speak. Rationality, after all, is not a matter of *logic* alone—of commitment to the logical principles of consistency (i.e., not to accept what contradicts accepted premises) and completeness (i.e., to accept what is entailed by accepted premises), which are, after all, purely hypothetical in nature ("If you accept . . . , then—") Cognition is not just a *hypothetical* issue of making proper inferences from given premises; it involves also the *categorical* issue of giving their proper evidential weight to the premises themselves. Thus rationality indispensably requires a categorical and material constraint inherent in the conception of evidence—namely, to abide by the established evidential ground rules of various domains of discussion in terms of the locus of presumption and the allocation of benefit of doubt.

The skeptic is not embarked on a *defense* of reason, but on a self-imposed *exile* from the enterprise of cogent discussion and the community of rational inquirers. And at this juncture he is no longer left in possession of the high ground. In refusing to give to the standard evidential considerations the presumptive and *prima farie* weight that is their established value on the market of rational interchange, the skeptic, rather than being the defender of rigid reason, is in fact profoundly irrational. The sceptic *seemingly* moves within the orbit of rationality. But by his refusal to acknowledge the ordinary probative rules of plausibility, presumption, evidence, and so on, he effectively opts out of the rational enterprise in the interests of a private, idiosyncratic rigorism that results from attachment to an inappropriately hyperbolic standard.

The line of thought being developed here does not involve any immediate resort to practical considerations. Since the days of the Academic Skeptics of Greek antiquity, philosophers have often answered our present question "Why accept anything at all" by taking roughly the following line: Man is a rational animal. *Qua* animal he must act—his very survival depends on action. But *qua* rational being he cannot act availingly save insofar as his actions are guided by what he accepts. The *practical* circumstances of the human condition preclude the systematic suspension of belief as a viable policy.

THE ECONOMIC DIMENSION: COSTS AND BENEFITS

Knowledge has a significant economic dimension because of its substantial involvement with costs and benefits. Many aspects of the way we acquire, maintain, and use our knowledge can be properly understood and explained only from an economic point of view. Attention to economic considerations regarding the costs and benefits of the acquisition and management of information can help us both to account for how people proceed in cognitive matters and to provide normative guidance toward better serving the aims of the enterprise. Any theory of knowledge that ignores this economic aspect does so at the risk of its own adequacy.

Homo sapiens has evolved within nature to fill the ecological niche of an intelligent being. With us, the imperative to understanding is something altogether basic: things being as they are, we cannot function, let alone thrive, without knowledge of what goes on about us. The knowledge that orients our activities in this world is itself the most practical of things—a rational animal cannot feel at ease in situations of which it can make no cognitive sense. We have questions and want (nay, *need*), to have answers to them. And not just answers, but answers that cohere and fit together in an orderly way can alone satisfy a rational creature. This basic practical impetus to (coherent) information provides a fundamental imperative to cognitive intelligence.

We humans want and need our cognitive commitments to comprise an intelligible story, to give a comprehensive and coherent account of things. For us, cognitive satisfaction is unattainable on any other basis—the need for information, for knowledge to nourish the mind, is every bit as critical as the need for food to nourish the body. Cognitive vacuity or dissonance is as distressing to us as physical pain. Bafflement and ignorance—to give suspensions of judgment the somewhat harsher name that is their due—exact a substantial price from us. The quest for cognitive orientation in a difficult world represents a deeply practical requisite for us. That basic demand for information and understanding presses in upon us and we must do (and are pragmatically justified in doing) what is needed for its satisfaction. For us, cognition is the most practical of matters: Knowledge itself fulfills an acute practical need.

It has come to be increasingly apparent in recent years that knowledge is cognitive capital, and that its development involves the creation of intellectual assets, in which both producers and users have a very real interest. Knowledge, in short, is a good of sorts—a commodity on which one can put a price tag and which can be bought and sold much like any other—save that the price of its acquisition often involves not just money alone but other resources, such as time, effort, and ingenuity. Man is a finite being who has only limited time and energy at his disposal. And even the development of knowledge, important though it is, is nevertheless of limited value it is not worth the expenditure of every minute of every day at our disposal.

Charles Sanders Peirce proposed to construe the "economy of research" at issue in knowledge development in terms of the sort of balance of assets and liabilities that we today would call cost-benefit analysis.[18] On the side of benefits of scientific claims, he was prepared to consider a wide variety of factors: closeness of fit to data, explanatory value, novelty, simplicity, accuracy of detail, precision, parsimony, concordance with other accepted theories, even antecedent likelihood and intuitive appeal. And in the liability column, he placed those demanding factors of "the dismal science": the expenditure of time, effort, energy, and money needed to substantiate our claims.

The introduction of such an economic perspective does not of course detract from the value of the quest for knowledge as an intrinsically worthy venture with a perfectly valid *l'art pour l'art* aspect. But as Peirce emphasized, one must recognize the inevitably economic aspect of any rational human enterprise—inquiry included. It is clear that staking claims to knowledge is the risk of error. But, of course, knowledge brings critical benefits. Our species has evolved within nature into the ecological niche of an intelligent being. In consequence, the need for understanding, for "knowing one's way about," is one of the most fundamental demands of the human condition. *Homo sapiens* is *Homo quaerens*. The requirement for information, for cognitive orientation within our environment, is as pressing a human need as that for food itself. We are rational animals and must feed our minds even as we must feed our bodies.

The need for knowledge is part and parcel of our nature. A deep-rooted demand for information and understanding presses in on us, and we have little choice but to satisfy it. Once the ball is set rolling it keeps on under its own momentum—far beyond the limits of strictly practical necessity. The great Norwegian polar explorer Fridtjof Nansen put it well. What drives men to the polar regions, he said, is

> The power to the unknown over the human spirit. As ideas have cleared with the ages, so has this power extended its might, and driven Man willy-nilly onwards along the path of progress. It drives us in to Nature's hidden powers and secrets, down to the immeasurably little world of the microscopic, and out into the unprobed expanses of the

Universe.... it gives us no peace until we know this planet on which we live, from the greatest depth of the ocean to the highest layers of the atmosphere. This Power runs like a strand through the whole history of polar exploration. In spite of all declarations of possible profit in one way or another, it was that which, in our hearts, has always driven us back there again, despite all setbacks and suffering.[19]

The discomfort of unknowing is a natural component of human sensibility. To be ignorant of what goes on about us is almost physically painful for us—no doubt because it is so dangerous from an evolutionary point of view. As William James observed: "The utility of this emotional effect of [security of] expectation is perfectly obvious "natural selection," in fact, was bound to bring it about sooner or later. It is of the utmost practical importance to an animal that he should have prevision of the qualities of the objects that surround him."[20]

It is a situational imperative for us humans to acquire information about the world. We have questions and we need answers. Homo sapiens is a creature that must, by nature, feel cognitively at home in the world. Relief from ignorance, puzzlement, and cognitive dissonance is one of cognition's most important benefits. These benefits are both positive (pleasures of understanding) and negative (reducing intellectual discomfort through the removal of unknowing and ignorance and the diminution of cognitive dissonance). The basic human urge to make sense of things is a characteristic aspect of our makeup—we cannot live a satisfactory life in an environment we do not understand. For us, cognitive orientation is itself a practical need: cognitive disorientation is actually stressful and distressing.

The benefits of knowledge are twofold: theoretical (or purely cognitive) and practical (or applied). The theoretical/cognitive benefits of knowledge, on the one hand, relate to its satisfactions in and for itself, for understanding is an end unto itself and, as such, is the bearer of important and substantial benefits—benefits that are purely cognitive. The practical benefits of knowledge, on the other hand, relate to its role in guiding the processes by which we satisfy our (noncognitive) needs and wants. The satisfaction of our needs for food, shelter, protection against the elements, and security against natural and human hazards all require information. And the satisfaction of mere desiderata comes into it as well. We can, do, and must put knowledge to work to facilitate the attainment of our goals, guiding our actions and activities in this world into productive and rewarding lines. And this is where the practical payoff of knowledge comes into play.

The impetus to inquiry to investigation, research, and acquisition of information—can thus be validated in strictly economic terms with a view to potential benefits of both theoretical and practical sorts. We humans need to achieve both an intellectual and a physical accommodation to our environs. Efforts to secure and enlarge knowledge are worthwhile insofar as they are cost effective, in that the resources we expend for these purposes are more than compensated for through benefits obtained. And in this regard skepticism simply proves uneconomic.

The Deficiency of Skepticism

The skeptic too readily loses sight of the very reason of being of our cognitive endeavors. The object of rational inquiry is not just to avoid error but to answer our questions, to secure *information* about the world. And here, as elsewhere, "Nothing ventured, nothing gained" is the operative principle. Granted, a systematic abstention from cognitive involvement is a surefire safeguard against one kind of error. But it affords this security at too steep a price. The shortcoming of that "no-risk" option is that it guarantees failure from the very outset.

It is self-defeating to follow the radical skeptic into letting discretion be the whole of epistemic valor by systematically avoiding accepting anything whatsoever in the domain of empirical fact. To be sure, when we set out to acquire information we may well discover in the end that, try as we will, success in reaching our goal is beyond our means. But we shall *certainly* get nowhere at all if we do not even set out on the journey—which is exactly what the skeptic's blanket proscription of acceptance amounts to.

In "playing the game" of making assertions and laying claims to credence, we may well lose: our contentions may well turn out to be mistaken. But, in a refusal to play this game at all we face not just the possibility but the certainty of losing the prize—we abandon any chance to realize our cognitive objectives. A skeptical policy of systematic avoidance of acceptance is fundamentally *irrational,* because it blocks from the very outset any prospect of realizing the inherent goals of the enterprise of factual inquiry. In cognition, as in other sectors of life, there are no guarantees, no ways of averting risk altogether, no option that is totally safe and secure. The best and most we can do is to make optimal use of the resources at our disposal to "manage" risks as best we can. To decline to do this by refusing to accept any sort of risk is to become immobilized. The skeptic thus pays a great price for the comfort of safety and security. If we want information—if we deem ignorance no less a negativity than error—then we must be prepared to "take the gamble" of answering our questions in ways that risk some possibility of error. A middle-of-the-road evidentialism emerges as the most sensible approach.

Perhaps, no other objection to radical skepticism in the factual domain is as impressive as the fact that, for the all-out skeptic, any and all assertions about the world's objective facts must lie on the same cognitive plane. No contention—no matter how bizarre—is any better off than any other in point of its legitimate credentials. For the thoroughgoing skeptic there just is no rationality-relevant difference between "More than three people are currently living in China" and "There are at present fewer than three automobiles in North

America." As far as the cognitive venture goes, it stands committed to the view that there is "nothing to choose" in point of warrant between one factual claim and another. Radical skepticism is an H-bomb that levels everything in the cognitive domain.

The all-out skeptic writes off at the very outset a prospect whose abandonment would only be rationally defensible at the very end. As Charles Sanders Peirce never tired of maintaining, inquiry only has a point if we accept from the outset that there is some prospect that it may terminate in a satisfactory answer to our questions. He indicated the appropriate stance with trenchant cogency: "The first question, then, which I have to ask is: Supposing such a thing to be true, what is the kind of proof which I ought to demand to satisfy me of its truth?"[21] A general epistemic policy which would as a matter of principle make it impossible for us to discover *something which is* ex hypothesi *the case* is clearly irrational. And the skeptical proscription of all acceptance is obviously such a policy—one that abrogates the project of inquiry at the very outset, without according it the benefit of a fair trial. A presumption in favor of rationality—cognitive rationality included—is rationally inescapable. It could, to be sure, eventuate at the end of the day that satisfactory knowledge of physical reality is unachievable. But, until the end of the proverbial day arrives, we can and should proceed on the idea that this possibility is not in prospect. "Never bar the path of inquiry," Peirce rightly insisted. And radical skepticism's fatal flaw is that it aborts inquiry at the start.

Reason's commitment to the cognitive enterprise of inquiry is absolute and establishes an insatiable demand for extending and deepening the range of our information. As Aristotle observed, "Man by nature desires to know." In this cognitive sphere, reason cannot leave well enough alone, but insists on a continual enhancement in the range and depth of our understanding of ourselves and of the world about us. Cognitive rationality is a matter of using cogent reasons to govern one's acceptance of beliefs—of answering one's questions in the best feasible way. But is this project realizable at all? We must consider the skeptic's longstanding challenge that it is not. For, skeptics—in their more radical moments, at any rate—insists that there is never a satisfactory justification for accepting anything whatsoever. By rejecting the very possibility of securing trustworthy information in factual matters, skepticism sets up a purportedly decisive obstacle to implementing these aims of reason.

The skeptical challenge to the project of empirical inquiry based on principles of cognitive rationality has a very plausible look about it. Our means for the acquisition of factual knowledge are unquestionably imperfect. Where, for example, are the "scientific truths" of yesteryear—those earth-shaking syntheses of Aristotle and Ptolemy, of Newton and Maxwell? Virtually no part of them has survived wholly unscathed. And given this past course of bitter experience, how can we possibly validate our *present* acceptance of factual contentions in a rationally convincing way? This alone indicates the need of coming to terms with a radical skepticism, which maintains that rational cognition is unattainable—that the quest for knowledge is in principle a vain and vacuous pursuit.

The skeptic *seemingly* moves within the orbit of rationality, but only seemingly so. For, in fact skepticism runs afoul of the only promising epistemological instrumentalities that we have. Philosophical sceptics generally set up some abstract standard of absolutistic certainty and then try to show that no knowledge claims in a certain area (sense, memory, scientific theory, and the like) can possibly meet the conditions of this standard. From this circumstance, the impossibility of such a category of "knowledge" is accordingly inferred. But this inference is totally misguided. For, what follows is rather the inappropriateness or incorrectness of the standard at issue. If the vaunted standard is such that knowledge claims cannot possibly meet it, the moral is not "too bad for knowledge claims," but "too bad for the standard." Any position that precludes in principle the possibility of valid knowledge claims thereby effectively manifests its own unacceptability.

The skeptic's argument is a double-edged sword that cuts both ways and inflicts the more serious damage on itself. It is senseless to impose on something qualification conditions which it cannot in the very nature of things meet. An analogue of the old Roman legal precept is operative here—one is never obliged beyond the limits of the possible (*ultra posse nemo obligatur*). It cannot rationally be required of us to do more than the best that is possible in any situation—cognition included. But rationality also conveys the comforting realization that more than this cannot be required of us: we are clearly entitled to see the best that can be done as good enough.

Historically, to be sure, committed skeptics like Sextus Empiricus occasionally been prepared to grant that they are destroyers of *logos* (reason and discourse), insisting that they use reason and discourse simply as instruments for their own destruction. They have been prepared to grasp the nettle and admit—perhaps even welcome—the consequence that their position does not only gainsay knowledge, but actually rejects the whole project of cognitive rationality (at any rate at the level of our factual beliefs). But whatever satisfaction this posture may afford the skeptic, it has little appeal to those who do not already share his position since (on its own telling) no cogent *reason* can be given for its adoption. The collapse of the prospect of rational inquiry and communication is the ultimate sanction barring the way to any rational espousal of radical skepticism. It is a price that a fanatically dedicated devotee of the skeptical position may be willing to meet, but it is clearly one that people who are not so precommitted cannot possibly pay. Sensible people require cogent reasons for what they do, and only with the abandonment of a rigoristic all-out skepticism can such reasons be obtained.

The only sort of critique of skepticism that makes sense to ask for is a *rational* critique. And, viewed from this standpoint, the decisive flaw of skepticism is that it makes rationality itself impossible.

In a last-ditch defense of a skeptical position, the radical sceptic may still take the following line: "Departing from *your* view of 'rationality' and falling short by *your* standards is not something that *I* need regard as a genuine failing. Indeed, my very thesis is that your "rationality" has no suitable credentials." To meet this desperate but profound tactic we must shift the ground of argumentation. It now becomes advantageous to approach the whole issue from a new point of departure, namely that of the prospects of communication. For, it turns out that not only does his refusal to accept claims preclude the radical skeptic from participating in the enterprise of *inquiry*, he is blocked from the enterprise of communication as well.

The process of transmitting information through communication is predicated on the fundamental convention that ordinarily and normally what one declares to be so is something that (1) one accepts as true, and (2) one claims to be rationally warranted in accepting.[22] In rejecting the ground rules of our *reasoning* as inappropriate, the skeptic also abandons the ground rules of our *communication*. In denying the prospect of any sort of rational warrant—however tentative—the all-out skeptic embarks upon a self-imposed exile from the community of communicators, seeing that the communicative use of language is predicated on conceding the warranting presuppositions of language use. To enter into a discussion at all, one must acquiesce in the underlying rules of meaning and information-transmission that make discussion in general possible. But, if *nothing* can appropriately be accepted, then no rules can be established—and thus no statements made, since meaningful discourse requires an agreement on the informative conventions. If the skeptic denies us (and himself) knowledge—or at least plausible beliefs—about the meanings of words and symbols, then there is just no way he can communicate with us (and perhaps even with himself). A radical skepticism, in the final analysis, engenders the collapse of communication and deliberation, thereby leading to a withdrawal from the human community.

Skepticism defeats from the very start any prospect of realizing our cognitive purposes and aspirations. It runs counter to the teleological enterprise to which we humans stand committed in virtue of being the sort of intelligent creatures we are. It is ultimately this collision between skepticism and our need for the products of rational inquiry that makes the rejection of skepticism a rational imperative. A radical skepticism conflicts not only with the realities and practicalities of the human situation but also no less decidedly with its equally important idealities.[23]

Chapter 4

Epistemic Justification in a Functionalistic and Naturalistic Perspective

Synopsis

- *Experience is always personal—a biographical episode in the life of an individual. The move from such subjectivity to objective, person-independent fact is never automatic.*
- *Some suggest that a coordination can be effected through common-cause considerations. But this embarking is on a sea of difficulties.*
- *The epistemologically appropriate justification of objective claims calls for a recourse to pragmatic argumentation.*
- *The validation at issue emerges from evolutionary considerations, although it is* rational *rather than natural selection that is crucial here.*
- *There, of course, remains an ineliminable prospect of error in relation to our experientially based knowledge claims. But this need not stand in the way of claiming the appropriateness—or even the truth—of our claims to objective knowledge.*
- *As such considerations indicate, the concept of presumption is a crucial involvement of cognitive validation.*

Experience and Fact

Once the idea of sensory certainty is called into question, the way is cleared to seeing sense-based knowledge in a practicalistic light. To gain a firm grip on the issue, it helps to draw a crucial distinction between cognitive *experiences* (which are always personal and subjective) and objective *situations* (which are not). "I take myself to be seeing a cat on that mat" is one sort of thing and "There actually is a cat on that mat" is something quite different. The former is purely person-coordinated and thereby merely autobiographical while the latter objective claim is decidedly impersonal.[1] (Only in the special case of automatically self-appertaining issues such as feeling aches and pains will subjective experiences authorize claims in objective territory: to feel a headache is to have a headache.)

Two facts are crucial in this connection:

- Experience as such is always personal: it is always a matter of what occurs within the thought-realm of some individual. Thus all that is ever secured directly and totally by experience are claims in the order of subjectivity.
- Such subjective claims are always autobiographical: they can never assure objective claims about facts regarding the "external" world without some further ado.

These facts separate experience from objectivity. With objectively world-regarding facts we always transcend experience as such. The assertoric context of an objectively factual statement ("There is a cat there") is something that outruns any merely experiential fact ("I take myself to be seeing a cat there"). There is always an epistemic gap between subjectivel experience and objective fact.

But how can one manage to cross this epistemic gap between subjective experience and subject-transcending reality? Some theorists think that *causality* provides a natural and effective bridge across the gap.[2] Let us examine this prospect.

Problems of Common-Cause Epistemology

The basic idea of the causal approach is epistemic justification roots in the circumstance.

Sosa's endorsement of causal epistemology pivots on that experience causes our beliefs in some appropriate or standard way. However, this formula makes it transparently clear that epistemic justification will turn not just on causality alone but its suitability and thereby on the way in which this causality operates. And this little qualification of "standard appropriateness" has to carry a big burden of weight. Of course we cannot do without it: mere causality as such does not engender epistemic justification. (I was not epistemically justified in expecting the villain's death after being shot on stage even when the wicked prop man had substituted a real loaded pistol—unbeknownst to one and all.) And the fly in the ointment is that things become unraveled when we proceed to take this idea of *appropriate* causation seriously.

Suppose that I espouse the objective claim "That there is a cat on that mat" because I have a seeing-a-cat-on-the-mat experience, that is, because I take myself to be looking at a cat on the mat. What is now required for that experience to "be caused in the appropriate way"—that is, being appropriately produced by a cat on that mat? Clearly the only correct/appropriate/standard mode of experience production here is one that proceeds via an actual cat on that mat. Unless one construes "experience" in a question-begging way that is not just psychologically phenomenological but objectively authentic, experiences will not suffice for justifactory authentications. That is, something like the following causal story must obtain:

> There really is a cat on the mat. And in the circumstances it is this situation—duly elaborated in its causal complexity by way of involving a light source, eyes looking in the right direction, photons emanating from the light source impacting on the retina, etc.—that stimulates my cerebral cortex in such a way as to engender this cat-on-the-mat seeing experience.

But note the difficulty here. Unless a real cat on that real mat is appropriately the causal source of my taking myself to be seeing a cat on the mat, that cat-on-the-mat belief of mine is simply not "caused in the appropriate way." (If Pavlovian conditioning leads me to experience a cat-conviction when confronted by any domestic quadruped—cats included—then my "Cat on the mat belief," even when indeed caused by a cat's presence on the mat, is not caused in the right sort of way for knowledge.) But on any such account *the APPROPRIATE production of that belief not merely assures but actually PRESUPPOSES its truth.* Only when one assumes an account that already assures the truth of the belief at issue can one tell that the belief is epistemically justified. On this basis, determining the epistemic justification of a belief does not just substantiate the truth of that belief but requires it. On such a construction of the matter, epistemic justification is available only when it is too late because it is no longer needed.

This condition of things creates big problems when it indeed is "the epistemic being of sensory experience upon knowledge" that we have in view. For given that epistemic justification is our only cognitive pathway to establishing the truth of beliefs, there is no point to a concept of justification that requires an appropriateness that is available only after this truth has already been established. If appropriateness cannot be determined independently of establishing the actual truth, then it is question-beggingly self-defeating to include it among the criteria of our truth-claims.

Sosa's endorsement of the idea that "it is enough [for the epistemic justification of claims to knowledge] that the experience cause the belief in some appropriate or standard way" (p. 284) places him squarely into the increasingly popular tradition of causal epistemology. This approach looks to the following plausible-looking thesis generally favored by causal epistemologists:

> A person X *knows* that p is the case iff the (causal) processes that actualize p as a state-of-affairs (in nature) and the (causal) processes that engender X's belief that this state of affairs obtains are suitably coordinated.

Implementing this idea requires the following three elements:

1. A set [p] of causal factors engendering P's actualization via a causal relationship: [p] $\Rightarrow$ p.
2. A set [X believes p] of causal factors engendering X's belief in P via a causal relationship: [X believes p] $\Rightarrow$ x believes p.
3. A (duly specified) relationship of "harmonious coordination" between the two opposite sets of causal factors [p] and [X believes p].

Those three elements must all be in proper alignment with one another for a knowledge claim to qualify as appropriate.

But a rocky and winding road now lies before us. For one thing, it now becomes extremely difficult ever to attribute knowledge to others. If I know anything about history, I know that Edward Gibbon knew that Julius Caesar was assassinated. But I certainly do not have much of a clue about the specific causal factors that produced this belief in Gibbon. (Indeed I am not even in a position to say anything much about the causal sources of my own metabelief.) All those now requisite causal issues are lost in a fog of unknowing.[3] I can plausibly conjecture *that* a cogent causal story can be told but I do not have much of clue as to *what* it is and would thus be hard put to evidentiate it. Moreover—and more damagingly yet—the sorts of claims that I could ever make about this causal story are in epistemically far more problematic than is my belief that Gibbon knew that Caesar was assassinated. The problem is that the causal approach to knowledge explains the obscure by what is yet far or more so. To establish simple knowledge claims on its basis we would need to go through a rigamarole so cumbersome in its demands for generally unavailable information as to lead sensible people to despair of the whole project. If that is what is required to substantiate claims about other people's knowledge, then skepticism on the subject offers the best option. If any plausible alternatives to common-cause epistemology are available, they surely deserve our most sympathetic consideration.

Modes of Justification

To see how a cogent alternative approach to the appropriate epistemic justification of objective claims might be developed, it is necessary to go back to square one and begin with a closer look at the question of just what it is to "justify" a belief.

Belief justification is a complex idea subject to a considerable variety of distinctions and elaborations. But for present purposes the crucial distinction is that between:

- *strong epistemic justification*, that is, justification for accepting the belief as definitively true, for seeing it as meriting outright acceptance.
- *weak epistemic justification*, that is, justification for according the belief some credit, for seeing it as a plausible prospect.

With strong justification we regard the issue of acceptability as settled, with weak justification we regard it as yet substantially open, viewing whatever commitment we have to the belief as tentative, provisional, and defeasible.

Now the critical considerations in the light of this distinction are (1) that while we must grant foundationalists that perceptual experience provides epistemic justification, nevertheless where objective factual claims are concerned we have to acknowledge this can only be justification of the weak sort, and (2) that while we must grant the coherentists that strong justification is in general not available on the basis of experience alone but is something that is only to be had on a larger contextual and systemic basis.

Thus consider again that cat-on-the-mat experience where "I take myself to be seeing a cat on the mat." On its basis I would arrive quite unproblematically at the following contentions:

- It seems plausible to suppose that there is a cat on the mat
- There is presumably a cat on the mat

But to claim unqualified assurance that there indeed is a cat on the mat in such circumstances would be stretching matters too far. The Sceptic is right in this, that there is a possibility of illusion or delusion—that something controversial might be being done with mirrors or puppets, and so on. But the indicated pro-inclination toward the theses at issue is certainly warranted. Conclusiveness may be absent but plausibility is certainly there.

Yet how is one to get beyond such tentativity? To step from that visual experience to an objective factual claim on the order of

- There actually is a cat there
- There actually is a mat there
- The cat is actually emplaced on the mat

is a move that can be made—but not without further ado. Let us consider what sort of "further ado" is required here.

The position at issue is a "direct realism" of sorts. The step from a sensory experience ("I take myself to be seeing a cat") to an objective factual claim ("There is a cat over there and I am looking at it" is operationally direct but epistemically mediated. And it is mediated not by an *inference* but by a *policy*, namely the policy of trusting one's own senses. This policy itself is based neither on wishful thinking not on arbitrary decisions: it emerges in the school of praxis from the consideration that a long course of experience has taught us that our senses generally guide us aright—that the indications of visual experience, unlike, say, those of dream experience, generally provide reliable information that can be implemented in practice.

Consider the prospect of devising an inference of the format:

—I take myself to be seeing a COTM (cat on the mat)

—Additional Premise X

Therefore: There is a COTM

Clearly if that Additional Premise X were "Whatever I take myself to be seeing is actually there" we would be home free. But it is also clear that no such factual premiss able to effect the transit from subjective appearances to objective reality is going to be available.

We thus seem to have two prime choices. One is to weaken the conclusion to: "There is probably a COTM." But, of course, probabilities do not fix objective facts: we can count cats alright, but not probable cats. And so we now lose our grip on the actual objective constitution of things.

The other prime alternative is to shift that Additional Premise from the reaches of available fact to that of epistemic policy. On this basis, we would move to a more complex course of reasoning that would run somewhat as follows:

1. [*Experiential fact*] By all presently available indications there is a COTM ($\cong$ I take myself to be seeing a COTM).
2. [*Experiential fact*—encompassed in (1)] There are no concrete counterindications in hand: no case-specific evidence indicating that it is not the case that there is a COTM.
3. [*Principle of rational procedure*] It is rational to adopt the policy of accepting noncounterindicated visual indications as veridical.
4. [*Derivative contention*—from (1)-(3)] It is rational to adopt the claim "There is a COTM" as veridical.
5. [*Principle of rational procedure*] To implement whatever course is rationally warranted in the circumstances.
6. [*Policy Implementation*—of (5) in the light of (1)-(4)] Accepting the claim: "There is a COTM"

The transit from the premise of our rational argument to its contention is now not effected by a process of deductive or inductive *interference* at all. Instead, it is now the product of a prudent policy of procedure geared to the implementation of a rationality principle. The long and short of it is that the epistemic gap between subjective experience and subject-transcending reality is crossed neither by a quasi-cogical inference nor by a conceptually grounded general principle but by a *practical policy*: a procedural modus operandi grounded in the functional teleology of the communicative enterprise.

To be sure, the line of thought at issue here leads to the question of how a principle of procedure is to be validated. This turns the problem in a new direction. And here the crucial point is that we need not necessarily proceed on a factual basis With factual claims validation is a matter of securing evidence of truth. But with procedural process validation can take many forms. Preeminently there are such alternatives as:

1. *This-will-work argumentation* to the effect that the method at issue is efficient and effective in delivering the envisioned results.

But this is not our only option. For there is also the prospect of:

2. *This-is-bound-to-work argumentation* to the effect that it *any* method or standardization procedure will deliver the envisioned results, then the method at issue will do so as well.

And going beyond this we also have:

3. *This-or-nothing argumentation* to the effect that the method at issue is our only hope for realizing the envisioned result: that if it does not work, then nothing will.

and

4. *This-or-nothing-better argumentation* to the effect that the method at issue is bound to do at least as well as any available alternative.

The principal point is now clear. In order to validate those practical policies of procedure at issue in the revised course of reasoning at issue in our argument we need not proceed on the factual/evidential order of justification at all but can shift to the practical/methodological arena of deliberation where altogether different ground rules of justification and validation apply.

But how would this emergence of policy validation from a body of experience work in practice? The idea of presumption provides the key that unlocks the issue.

The Evolutionary Aspect of Sensory Epistemology

A presumption-based practicalistic theory of knowledge development clearly differs from a causal physicalisitic account. And it enjoys substantial advantages.

Recent philosophy of knowledge has been pervaded by a "revolt against dualism" that insists on continuities and overlaps where earlier theorists had seen clear divisions and sharp separations. William James stressed the continuity of thought and inveighed against division and compartmentalization. Whitehead condemned bifurcation. John Dewey inveighed against all dualisms. And their successors have been enthusiastic in continuing their tendency in matters of detail. W. V. Quine strives to consign the analytic/synthetic distinction to oblivion. Donald Davidson denigrates the conceptual/substantive distinction. And innumerable contemporary philosophy of science object to the fact/theory distinction because they regard all factual claims as "theory laden."

But throughout all his turning away from division and dualism, one traditional distinction has remained sacrosanct: that between cause and reason, between efficient causation and final causation. Where the medievals tended to construe *causa* broadly and to group paternal and volitional "causation" together as reflecting differences in mode but not in kind, the moderns from Descartes onward have erected a Chinese wall between the two. Immanual Kant stressed the distinction: "There are two and only two kinds of causality conceivable by us: causality is either according to nature [and is thus efficient causality] or according to free will and choices [and is thus final causality]."[4] And so even Davidson, in the very midst of his polemic against the conceptual/substantive distinction, argues that:

> [we must] give up the idea that meaning or knowledge is grounded in something [external to cognition proper] . . . No doubt meaning and knowledge depend on experience and experience ultimately on sensation. But this is the "depend" of causality, not that of evidence or justification.[5]

And so as Davidson sees it, causation is one sort of thing and justification another, and never the twain shall meet in the cognitive realm. For while beliefs may well have causes, these do not matter in the context of justification. Causes are not reasons since "nothing can count as a *reason* for holding a belief except another belief."[6] All belief validation (justification) is discursive and proceeds by means of other beliefs. "Causes are rationally inert" is the motto here: they may provide the occasion for the provision of reasons but they themselves can never provide reasons. Or so Davidson has it.

But not so with causal epistemologists. For they propose to join the two hemispheres together. And with good reason. For having abandoned such other Kantian dichotomies as synthetic/analytic, empirical/conceptual, a priori/a posteriori we would do well to abandon this cause/reason division also. The fact is that causes and reasons can and should be seen as systemically coordinated not by a Leibnizian preestablished harmony but by a Darwinian process of evolution. With intelligent beings whose modus operandi is suitably shaped by their evolutionary heritage, the step from a suitable experience to a belief is at once *causal* AND *rational:* we hold the belief *because* of the experience both in the order of efficient (causal) *and* in the order of final (rational) causation. With creatures such as ourselves, experiences of certain sorts are dual-purpose; owing to their—and our—evolutionary nature their occurrence both causally engenders and rationally justifies the holding of certain beliefs through a disposition that evolution has engendered exactly because it is objectively justified to some substantial extent. If this is evolutionary reliabilism, then we are well-advised to make the most of it.[7] Let us examine a bit more closely what is at issue here.

RATIONAL VERSUS NATURAL SELECTION

Scientifically minded epistemologists nowadays incline to consider how the workings of the "mind" can be explained in terms of the operations of the "brain."[8] But this approach has its limits. Biological evolution is doubtless what accounts for the cognitive machinery whose functioning provides for our *possession* of intelligence, but explaining the ways in which we *use* it largely calls for a rather different sort of evolutionary approach, one that addresses the development of thought-procedures rather than that of thought-mechanisms—of "software" rather than "hardware." What is at issue here is a matter of cultural-teleological evolution through a process of *rational* rather than Darwinian *natural* selection. Very different processes are accordingly at work, the one as it were blind, the other purposive. (In particular, biological evolution reacts only to *actually realized* changes in environing conditions: cultural evolution in its advanced stages can react also to *merely potential* changes in condition through people's capacity to think hypothetically and thereby to envision "what could happen if" certain changes occurred.) Once intelligence appears on the scene to any extent, no matter how small, it sets up pressures toward the enlargement of its scope, powerfully conditioning any and all future cultural evolution through the rational selection of processes and procedures on the basis of purposive efficacy.

Rationality thus emerges as a key element in the evolutionary development of *methods* as distinguished from *faculties*. The "selective" survival of effective methods is no blind and mechanical process produced by some inexorable agency of nature; rational agents place their bets in theory *and* practice in line with methods that prove themselves successful, tending to follow the guidance of those that succeed and to abandon—or readjust—those that fail. Once we posit a method-using community that functions under the guidance of intelligence—

itself a factor of biologically evolutionary advantage—only a short step separates the pragmatic issue of the applicative success of its methods (of *any* sort) from the evolutionary issue of their historical survival. As long as these intelligent rational agents have a prudent concern for their own interests, the survival of relatively successful methods as against relatively unsuccessful ones is a foregone conclusion.

Rational selection is a complex process that transpires not in a "population" but in a culture. It pivots on the tendency of a community of rational agents to adopt and perpetuate, through example and teaching, practices and modes of operation that are relatively more effective for the attainment of given ends than their available alternatives. Accordingly, the historical development of methods and modes of operation within a society of rational agents is likely to reflect a course of actual improvement. Rational agents involved in a course of trial-and-error experimentation with different processes and procedures are unlikely to prefer (for adoption by themselves and transmission to their successors) practices and procedures which are ineffective or inefficient.

This line of consideration does not envision a direct causal linkage between the historical survival of method users and the functional effectiveness of their methods. The relationship is one of common causation. The intelligence that proves itself normal conducive also forms functional efficacy. In consequence, survival in actual use of a method within a community of (realistic, normal) *rational* agents through this very fact affords evidence for its being successful in realizing its correlative purposes.[9]

These deliberations regarding rational selection have to this point been altogether general in their abstract bearing on methodologies of any shape or description. They apply to methods across the board, and hold for methods for peeling apples as much as of methods for substantiating knowledge-claims. But let us now focus more restrictedly on specifically *cognitive* methods, and consider the development of the cognitive and material technology of intellectual production.

There is certainly no need to exempt cognitive methodology from the range of rational selection in the evolution of methodologies. Quite to the contrary: there is every reason to think that the cognitive methods and information-engendering procedures that we deploy in forming our view of reality evolve selectively by an historic, evolutionary process of trial and error—analogous in role though different in character from the biological mutations affecting the bodily mechanisms by which we comport ourselves in the physical world. Accordingly, cognitive methods develop subject to revision in response to the element of "success and failure" in terms of the teleology of the practice of rational inquiry. An inquiry procedure is an *instrument* for organizing our experience into a systematized view of reality. And as with any tool or method or instrument, the paramount question takes the instrumentalistic form: Does it work? To be more specific, does it produce the desired result? Is it successful in practice? Legitimation along these lines is found in substantial part on the fact of survival through historical vicissitudes in the context of this pivotal issue of "working out best." This sort of legitimation has at the basis of the cultural development of our cognitive resources via the varieties and selective retention of our epistemically oriented intellectual products.[10]

It is clear that there are various alternative approaches to the problem of determining "how things work in the world." The examples of such occult cognitive frameworks as those of numerology (with its benign ratios), astrology (with its astral influences), and black magic (with its mystic forces) indicate that alternative explanatory frameworks exist, and that these can have very diverse degrees of merit. Now in the Western tradition the governing standards of human rationality are implicit in the goals of *explanation*, *prediction*, and preeminently *control*. (And thus the crucial factor is not, for example, sentimental "at-oneness with nature"—think of the magician vs. the mystic vs. the sage as cultural ideals.) These standards revolve about considerations of *practice* and are implicit in the use of our conceptual resources in the management of our affairs.

Given the reasonable agent's well-advised predilection for *success* in one's ventures, the fact that the cognitive methods we employ have a good record of demonstrated effectiveness in regard to explanation, prediction, and control is not surprising but only to be expected: the community of rational inquirers would have given them up long ago were they not comparatively successful. The effectiveness of our cognitive methodology is thus readily accounted for on an evolutionary perspective based on rational selection and the requirements for survival through adoption and transmission.

Yet, people are surely not all that rational—they have their moments of aberration and self-indulgence. Might not such tendencies selectively favor the survival of the ineffective over the effective—of the fallacious rather than the true—and slant the process of cognitive evolution in inappropriate directions? C. S. Peirce clearly recognized this prospect:

> Logicality in regard to practical matters . . . is the most useful quality an animal can possess, and might, therefore, result from the action of natural selection; but outside of these it is probably of more advantage to the animal to have his mind filled with pleasing and encouraging visions, independently of their truth; and thus, upon unpractical subjects, natural selection might occasion a fallacious tendency of thought.[11]

However, the methodological orientation of our approach provides a safeguard against an unwarranted penchant for such fallacious tendencies. At the level of individual beliefs "pleasing and encouraging visions" might indeed receive a survival-favoring impetus. But this unhappy prospect is effectively removed where a *systematic* method of inquiry is concerned—a method that must by its very synoptic nature lie in the sphere of the pragmatically effective.

The objection looms: But is it true that evolutionary survival among cognitive methods is inherently rational? Hasn't astrology survived to the present day—as its continuing presence in newspaper columns attests? The response runs: Astrology has indeed survived. But *not* in the scientific community, that is, among people dedicated in a serious way to the understanding, explanation, and control of nature. In the Western, Faustian[12] intellectual tradition of science, the ultimate arbiter of rationality is represented by the factor of knowledge-wed-to-action, and the ultimate validation of our beliefs lies in the combination of theoretical and practical success—with "practice" construed primarily in its pragmatic sense. All these "occult" procedures may have survived in some ecological niche in Western culture. But in *science* they are long extinct.

It is accordingly not difficult to give examples of the operation of evolutionary processes in the cognitive domain. The intellectual landscape of human history is littered with the skeletal remains of the extinct dinosaurs of this sphere. Examples of such defunct methods for the acquisition and explanatory utilization of information include astrology, numerology, oracles, dream-interpretation, the reading of tea leaves or the entrails of birds, animism, the teleological physics of the Pre-Socratics, and so on. No doubt, such processes continue in issue in some human communities to this very day; but clearly not among those dedicated to serious inquiry into nature's ways—that is, scientists. There is nothing intrinsically absurd or inherently contemptible about such unorthodox cognitive programs—even the most occult of them have a long and not wholly unsuccessful history. (Think, for example, of the prominent role of numerological explanation from Pythagoreanism, through Platonism, to the medieval Arabs, down to Kepler in the Renaissance.) Distinctly different scientific methodologies and programs have been mooted: Ptolemaic "saving the phenomena" versus the hypothetico-deductive method, or again, Baconian collectionism versus the post-Newtonian theory of experimental science, and so forth. The emergence, development, and ultimate triumph of scientific method of inquiry and explanation invite an evolutionary account—though clearly one that involves rational rather than natural selection.

The scientific approach to factual inquiry is simply one alternative among others, and it does not have an unshakable basis in the very constitution of the human intellect. Rather, the basis of our historically developed and entrenched cognitive tools lies in their (presumably) having established themselves in open competition with their rivals. It has come to be shown before the tribunal of bitter experience—through the historical vagaries of an evolutionary process of selection—that the accepted methods work out most effectively in actual practical vis-à-vis other tried alternatives. Such a legitimation is not absolute, but only presumptive. It does, however, manage to give justificatory weight to the historical factor of being in de facto possession of the field. The emergence of the principles of scientific understanding (simplicity, uniformity, and the like) is thus a matter of *cultural* rather than *biological* evolution subject to *rational* rather than *natural* selection.

To be sure cultural evolution is shaped and canalized by constraints that themselves are the products of biological evolution. For our instincts, inclinations, and natural dispositions are all programmed into us by evolution. The transition from a biologically advantageous economy of effective physical effort to a cognitively advantageous economy of effective intellectual efforts is a short and easy step.

An individual's heritage comes from two main sources: a biological heritage derived from the parents and a cultural heritage derived from the society. However, in the development of our knowledge, this second factor becomes critical. To establish and perpetuate itself in any community of *rational* agents, a practice or method of procedure must prove itself in the course of experience. Not only must it be to some extent effective in realizing the pertinent aims and ends, but it must prove itself to be more efficient than comparably available alternatives. With societies composed of rational agents, the pressure of means-ends efficacy is ever at work in forging a process of cultural (rather than natural) selection for replacing less by more cost-effective ways of achieving the group's committed ends—its cognitive ends emphatically included. Our cognitive faculties are doubtless the product of biological evolution, but the processes and procedures by which we put them to work are the results of a *cultural* evolution, which proceeds through rationally guided trial and error subject to a pragmatic preference for retaining those processes and procedures that prove theorists efficient and effective. Rational people have a strong bias for what works.[6] And progress is swift because once rationality gains an inch, it wants a mile.

We know that various highly "convenient" principles of knowledge-production are simply false:

- What seems to be, is.
- What people say is true.
- The simplest patterns that fit the data are actually correct.

We realize full well that such generalizations do not hold—however nice it would be if they did. Nevertheless throughout the conduct of inquiry we accept them as principles of *presumption*. We follow the higher-level metarule: "In the absence of concrete indications to the contrary, proceed as though such principles were true—that is, accept what seems to be (what people say, etc.) as true." The justification of this step as a measure of practical procedure is not the factual consideration that, "In proceeding in this way, you will come at correct information—you will not fall into error." Rather it is the methodological justification: "In proceeding in this way you will efficiently foster the interests of the cognitive enterprise: the benefits will—on the whole—outweigh the costs."

Against "Pure" Intellectualism

The process of presumption accordingly provides us with a nondiscursive route to knowledge. After all, it is or should be—clear that beliefs can be justified not just by other beliefs but also by experiences. My belief that the cat is on the mat need not rest on the yet different belief that I am under the impression of seeing it there; it can rest directly and immediately on my visual experience. My reason for holding that belief of mine is not yet another belief but an experience—which both occasions and at the same time considered from another point of view validates and justifies that belief.[13] But in point of validation the linkage of belief to experience is not direct but systemic.

Contemporary theorists often see the cognitive domain as confined to the realm of verbalized belief. There is no *hors de texte* says Derrida. The only reason for a belief is yet another belief says Davidson. But this sort of postmodern cognitivistic "wisdom" is folly. The world of thought is not self-contained; it is integral to the wider world of nature, part of a realm in which events happen and experiences occur. A perfectly good reason for believing that the cat is on the mat is just that we experience (i.e., observe) it to be there. The acceptability of beliefs frequently roots not in other beliefs but in experience—and experience must here be understood in rather general and broadly inclusive terms.

For while the productive order of causes and the explanatory order of reasons are fundamentally distinct: they melt together into a seamless whole with rational agents whose thoughts and acts reflect the impetus of reasons. In the case of rational beings, these two factors of causality and justification can come together in a conjoint fusion because here informatively meaningful perceptions and physical stimuli run together in coordinated unison. Since intelligent agents operate both in the realm of the causality of nature and the causality of reason it transpires that for them experiences such as a "cat-on-mat vision" have a double aspect, able at once to engender and (in view of imprinted practical policies) to justify suitable beliefs.

Accordingly, such intelligent agents have dual-function experiences that conjointly both cause their beliefs and provide them with reasons for holding them. For the justifactory links of experience to belief validation is not directly causal. We can—of course—construct a causal account linking experience to belief. But what plays a *justifactory* role in the rational validation of belief is not the *existence* of such a causal account—it is, rather, the contextual *belief* (itself the product of an ample body of experience) that such an account can be constituted.

Yet what sort of rationale is there for taking this line? The answer lies in the Darwinian revolution. For in its wake we can contemplate the evolutionary development of intelligent beings—the emergence of intelligent creatures for which the realization of reasons can be causally effective and, conversely, suitably operative causes can assume the form of reasons. We should, in fact, proceed to see intelligence itself as a capacity that renders it possible for certain modes of experience-causation concurrently to provide reasons for beliefs.

It is exactly here—in explaining the modus operandi of evolutionarily emergent rational beings—that causes and reasons must *not* be separated. The "experience of having a cat perception of a certain sort"—exactly because it is a cognitively significant experience—at once and concurrently constitutes the *cause* of someone's disposition to assent to "The cat is on the mat" and affords them with a *reason* for making this claim. In the cognitive experience of intelligent beings the regions of causes and of reasons are not disjoint but rather coordinated: one and the same of experience can at once provide for the ground and for the reason of a belief.

The fact is that intelligent agency brings something new on nature's scene. Certain sorts of eventuations are now *amphibious* because they are able to function at once and concurrently *both* in the realm of natural causes *and* in the realm of reasons. My perceptual experience of "seeing the cat on the mat" is at once the cause of my belief and affords my reason for holding that belief. Q: "What causally produced his belief that the cat is on the mat?" A: "He saw it there." "Why—with what reason—does he claim that the cat is on the mat?" A: "He saw it there." His seeing experience is a matter of dual action. With intelligent agents, such as ourselves, *experiences* can do double-duty as eventuations in nature and as reasons for belief. For we have an evolution-imprinted "rational disposition" to effect the transition from a subjective experience ("I take myself to be seeing a cat on the mat") to the endorsement of an duly coordinated objective claim ("There is a cat on the mat.")

In a way, the difference between causes and reasons in the realm of cognitive lies in the angle of vision, in whether those productive experiences are viewed from an agent's first-person or an observer's third-person vantage point. What we have here is one uniform sort of process experienced in two different modes: internally to the experiencing subject as psychic; externally to the third party observer as physical. And since a single underlying process is at issue, psychophysical coordination is assured. The perspective in question provides for an automatic response to Mark Twain's question "When the body gets drunk, does the mind stay sober?"—and also to the reverse question: "When the mind decides to get up and leave, does the body remain behind?" What we have here is not a supranaturally preestablished harmony but an evolution-programmed coordination: two different ways of processing one uniform basic sort of material. The question of causal effect versus rational response is not one of primacy but rather one of coordination. (That of an analogy: the same occurrence—sugar ingestion—engenders one result in relation to the tongue (taste) and another in relation to the nose (odor). The upshot is a cognitive theory close to the ontological "*neutral monism*" of Bertrand Russell or the "*radical empiricism*" of William James.

To be sure, we must distinguish the mental (conceptual, rational) order of the physical (causal) order. But where human knowledge is concerned evolutionary processes can and do coordinate these two orders into a parallel alignment. An intelligent being is by natural design one for which certain transactions in the causal order appear (from the internal experiential standpoint) as processes in the mental order capable of rationally engendering (i.e., in the modus operandi natural to the mental order) the acceptance of verbalized responses of the type we characterize as beliefs. In sum, the experiential route to belief validation and the causal sorts of belief production come into coordinative fusion for such a creature in virtue of its evolution-designed modus operandi. Only on this basis can we validate seeing our experience-provided "reasons" as authentic *reasons*. (After all, if beliefs alone could ever justify beliefs it is hard to see how the process of belief-justification could ever get under way?)

The Problem of Error

But are there not deceptive experiences—experiences that both engender and "justify" erroneous beliefs? Of course there are! But the threat is not as serious as it looks. Even in cognitive matters we can—strange to say—manage to extract truth from error. Accordingly, one fundamental feature of inquiry is represented by the following observation:

> *Insofar as our thinking is vague, truth is accessible even in the face of error.*

Consider the situation where you correctly accept P-or-Q. But—so let it be supposed—the truth of this disjunction roots entirely in that of P while Q is quite false. However, you accept P-or-Q only because you are convinced of the truth of Q; it so happens that P is something you actually disbelieve. Yet despite your error, your belief is entirely true.[14] Consider a concrete instance. You believe that Smith bought some furniture because he bought a table. However it was, in fact, a chair that he bought, something you would flatly reject. Nevertheless your belief that he bought some furniture is unquestionably correct. The error in which you are involved, although real, is not so grave as to destabilize the truth of your belief.

Ignorance is reflected in an inability to answer questions. But one has to be careful in this regard. Answering a question informatively is not just a matter of offering a *correct* answer but also a matter of offering an *exact* answer.

Thus consider the question "What is the population of Shanghai?" If I respond "More than ten and less than ten billion" I have provided a *correct* answer, but one that is not particularly helpful. This example illustrates a more far-reaching point.

> *There is in general an inverse relationship between the precision or definiteness of a judgment and its security: detail and probability stand in a competing interconnection.*

As we saw in chapter 2, increased confidence in the correctness of our estimates can always be purchased at the price of decreased accuracy. For any sort of estimate whatsoever there is always a characteristic trade-off relationship between the evidential *security* of the estimate, on the one hand (as determinable on the basis of its probability or degree of acceptability), and on the other hand its contentual *definitiveness* by way of exactness, detail, precision, and the like. A *complementarity* relationship of sorts thus obtains here as between definiteness and security. The moral of this story is that, insofar as our ignorance of relevant matters leads us to be vague in our judgments, we may well manage to enhance the likelihood of being right. The fact of the matter is that we have:

By constraining us to make vaguer judgments, ignorance can actually enhance our access to correct information (albeit at the cost of less detail and precision).

Thus if I have forgotten that Seattle is in Washington State then if "forced to guess" I might well erroneously locate it in Oregon. Nevertheless, my vague judgment that "Seattle is located in the Northwestern United States" is quite correct. This state of affairs means that when the truth of our claims is critical we generally "play it safe" and make our commitments less definite and detailed.

Consider, for example, so simple and colloquial a statement as "The servant declared that he could no longer do his master's bidding." This statement is pervaded by a magisterial vagueness. It conveys very little about what went on in the exchange between servant and master. We are told virtually nothing about what either of them actually said. What the object of their discussion was, what form of words they used, the manner of their discourse (did the master order or request? was the servant speaking from rueful incapacity or from belligerent defiance?), all these are questions we cannot begin to answer. Even the relationship at issue, whether owner/slave or employer/employee is left in total obscurity. In sum, there is a vast range of indeterminacy here—a great multitude of very different scenarios would fit perfectly well to the description of events which that individual statement puts before us. And this vagueness clearly provides a protective shell to guard that statement against a charge of falsity. Irrespective of how matters might actually stand within a vast range of alternative circumstances and conditions, the statement remains secure, its truth unaffected by which possibility is realized. And in practical matters in particular, such rough guidance is often altogether enough. We need not know just how much rain there will be to make it sensible for us to take an umbrella. And so their potential fragility does not—or need not—stand in the way of the utility—or even of the truth—of our claims to objective knowledge.

CONCLUSION

A whole generation of epistemologists have turned against the idea of "the experientially given" by supplementing the perfectly sound idea that *Nothing is given in experience by way of categorical truth regarding matters of objective fact* with the very mistaken conception that *In epistemic contexts* GIVEN *must be construed as* GIVEN AS TRUE. When these theses are combined, it indeed follows that there are no experiential givens. But of course the situation is drastically transformed when one acknowledges that beyond the given *as true*, there is also the given *as plausible*, or *as probable*, or *as presumption-deserving*. Once this step is taken, Sosa's question of "the epistemic bearing or sensory experience upon our knowledge" acquires a very different aspect. For consider the salient question: Does the fact that I take myself to be seeing a cat on a mat—that I have a "seeing a cat on a mat" experience—"entitle" or "suffice to justify" me to adopt the belief that there actually is a cat on the mat there? This question, once raised, this leads back to the underlying issues: What is at issue with belief *entitlement* in such a context? Does it require guaranteed correctness beyond the reach of any prospect of error—however far-fetched and remote? It is clear that the answer is: surely not! Entitlement and justification here is no more than the sort of rational assurance that it makes sense to ask for in the context at issue: a matter of reasonable evidentiation rather than categorical proof. The fundamental question, after all, is not "Does subjective experience *unfailingly guarantee* its objective proportions?" but rather "Does it *appropriately evidentiate* it?"

The shift from a proof-oriented "given as true" to a presumption-oriented "given as plausible or credible" also has the advantage that in epistemology we are not driven outside the epistemic realm of reasons-for into that of ontologically problematic causes-of. A sensible epistemology can without difficulty remain within the epistemic realm of reasons by drawing a due distinction between reasons-for-unblinking-acceptance-as-true and reasons-for-endorsement-as-plausible.

Accordingly, the principal theses of these deliberations regarding the epistemic bearing of sensory experience on our knowledge can be set out at follows:

1. There indeed are experiential givens. But these "givens" are actually "takens." They are not products of inference (hence "givens"), but of an epistemic endorsement policy or practice (hence "takens").
2. Most critically, those experiential givens are not "givens as categorically and infallibly true," but rather merely as plausible. What is at issue is not something *categorically certified* but merely something *presumptive*.
3. The move from plausibility to warranted acceptance ("justified belief") is automatic in those cases where *nihil obstat*—that is, whenever there are no case-specific counterindications.

Epistemic Justification

4. Although this way of proceeding does not deliver categorical guarantees or infallible certitude, such things are just not required for the rational validation (or "epistemic justification") of belief.
5. On this basis, subjective experience can—and does—validate our claims to objective factual knowledge. But the step at issue in moving across the subjectivity/objectivity divide is indeed a step—a mode of praxis. And the modus operandi at issue in this practice or policy is at once validated by experience and established through a complex process in which rational and natural selection come into concurrent operation.[15]

From the angle of this perspective it appears that the causal theory of epistemic justification leaves one very crucial validating factor out of sight—the role of presumption as filtered through the process of evolution in its relation to our cognitive hardware and software. What this theory offers us is, in a way, Hamlet without the ghost.

The critical role that presumptions play in this regard make it desirable to take a closer look at what is at issue here.

Chapter 5

Plausibility and Presumption

Synopsis

- *Presumption is a cognitive resource borrowed from its origins in the setting of legal deliberations.*
- *The concept of presumption plays a crucial role in epistemology.*
- *It is intimately intertwined with the conception of plausibility,*
- *and also—but more distantly—with that of probability.*
- *Presumption provides an important instrument in the critiques of skepticism.*
- *The use of presumptions is initially validated on grounds of practical rationality, but ultimately achieves retrovalidation on factual grounds. (The case for presumption is fundamentally analogous to that from trust in general.)*

The Need for Presumptions

A *presumption* is something distinctive that is characteristically its own within the cognitive domain. What is at issue here is not knowledge, nor probability, nor postulation, nor assumption, but something quite different and destructive, namely a provisional gap-filler for an informational void. The key idea of presumption thus roots in an analogy with the legal principle: innocent until proven guilty. A presumption is a thesis that is *provisionally* appropriate—on which can be maintained *pro tem*, viewed as acceptable until or unless sufficiently weighty counterconsiderations arise to displace it. On this basis, a presumption is a contention that remains in place until something better comes along.

For example we generally presume that what our senses tell us is correct. We know full well that our senses sometimes deceive us: the held at an angle under water looks bent even though touch tells us it is straight. Nevertheless we are entitled to trust sight and maintain what we see to be so until something else (like touch) comes along to disagree. Again, we generally presume that what uniformity tells us deserves the benefit of doubt. We know full well that generally reliable informants can provide misinformation, but nevertheless deem ourselves entitled to treat what otherwise reliable people tell us as correct unless and until some counterindication comes to view. We know full well that what our expert advisers in medical, or economic, or legal matters tell us is often incorrect but nevertheless are well advised to treat it as true until good reason to do otherwise comes to view. All of these instances—sensory indications, informal reports, expert judgments—illustrate matters of presumption. Much of what we take ourselves to know is actually presumptive knowledge: claims to knowledge that may in the end have to be withdrawn.

It must be stressed from the very start that the acceptance or assertion of claims and contentions is not something that is uniform and monolithic but rather something that can take markedly different types or forms. For we can "accept" a thesis as necessary and certain on logicoconceptual grounds ("Forks have tynes"), or as unquestionably true, although only contingently so ("The Louvre Museum is in Paris"), or in a much more tentative and hesitant frame of mind as merely plausible or presumably true ("Dinosaurs became extinct owing to the earth's cooling when the sun was blocked by commetary debris"). When a proposition is so guardedly endorsed we do not have it that, à la Tarski, asserting p is tantamount to claiming that "p is true." What is at issue with the guided endorsement of a claim p as plausible is simply the contention that "p is *presumably* true." And what this means is that we will endorse and employ p insofar as we can do so without encountering problems but are prepared to abandon it should problems arise. In effect we are not dedicated partisans of such a claim but merely its fair-weather friends.

The sort of tentative plausibility at issue with presumptions is preevidential in its bearing. As far as evidentiation goes, plausibility awaits further developments. Plausibilities are accordingly something of a practical epistemic device. We use them where this can render effective service for purposes of inquiry. But we are careful to refrain from committing ourselves to them unqualifiedly and come what may. And we would, in particular, refrain from using them where this leads to contradiction. In sum, our commitment to them is not absolute but situational: whether or not we endorse them will depend on the context. To reemphasize: the "acceptance" that is at issue here represents no more than a merely tentative or provisional endorsement.

Plausibilities are thus one thing and established truths another. We "accept" plausible statements only tentatively and provisionally, subject to their proving unproblematic in our deliberations. Of course, problems do often arise. *X* says twenty-five people were present, *Y* says fifteen. Sight tells us that the stick held at an angle under water is bent, touch tells us that it is straight. The hand held in cold water indicates that the tepid liquid is warm but held in hot water indicates that it is cold. In such cases we cannot have it both ways. Where our sources of information conflict—where they point to aporetic and paradoxical conclusions—we can no longer accept their deliverances at face value, but must somehow intervene to straighten things out. And here plausibility has to be our guide, subject to the idea that the most plausible prospect has a favorable presumption on its side.

Plausibility is in principle a comparative matter of more or less. Here it is not a question of yes or no, of definitive acceptability, but one of a cooperative assessment of differentially eligible alternatives of comparative advantages and disadvantages. We are attached to the claims that we regard as plausible, but this attachment will vary in strength in line with the epistemic circumstances. And this fact has significant ramifications. For the idea of plausibility functions in such a way that *presumption always favors the most plausible of rival alternatives.*

Presumption represents a way of filling in—at least *pro tem*—the gaps that may otherwise confront us at any stage of information. The French *Code civil* defines "presumptions" as: "*des conséquences qui la loi ou le magistrat tire d'un fait cunnu à un fait incunnu.*"[1] A presumption indicates that in the absence of specific counterindications we are to accept how things "as a rule" are taken as standing. For example, there is, in most probative contexts, a standing presumption in favor of the usual, normal, customary course of things. The conception of burden of proof is correlative with that of a presumption. The "presumption of innocence" can serve as a paradigm example here.

In its legal setting, the conception of presumption has been explicated in the following terms:

> A presumption in the ordinary sense is an inference. . . . The subject of presumptions so far as they are mere inferences or arguments, belongs not to the law of evidence, or to law at all, but to rules of reasoning. But a legal presumption, or, as it is sometimes called, a presumption of law, as distinguished from presumptions of fact, is something more. It may be described, in (Sir James) Stephen's language, as "a rule of law that courts and judges shall draw a particular inference from a particular fact, or from particular evidence, unless and until the truth" (perhaps it would be better to say "soundness") "of the inference is disproved."[2]

Such a *presumptio juris* is a supposition relative to the known facts which, by legal prescription, is to stand until refuted.

Accordingly, presumptions, though possessed of significant probative weight, will in general be defeasible, subject to defeat in being overthrown by sufficiently weighty countervailing considerations. In its legal aspect, the matter has been expounded as follows:

> [A] presumption of validity . . . retains its force in general even if subject to exceptions in particular cases. It may not by itself state all the relevant considerations, but it says enough that the party charged should be made to explain the allegation or avoid responsibility; the plaintiff has given a reason why the defendant should be held liable, and thereby invites the defendant to provide a reason why, in this case, the presumption should not be made absolute. The presumption lends structure to the argument, but it does not foreclose its further development.[3]

Presumptions are, as it were, in tentative and provisional possession of the cognitive terrain, holding their place until displaced by something more evidentially substantial.[4] In the law, such presumptions are governed by strict and specific rules (the accused is presumed innocent until proven guilty; the individual missing for seven years is presumed dead). In this way, the standing of a presumption is usually tentative and provisional, not absolute and final. A presumption only stands until the crucial issues that "remain to be seen" have been clarified, so that it is actually seen whether the presumptive truth will in fact stand up once everything is said and done."[5] Absent competition, it stays in place, on analogy with the politician's maxim: "You can't beat somebody with nobody." A presumptive answer is one that enables us to make do until that point (if ever) when something better comes along. This is a resource we constantly employ in communication, allowing matters to go unsaid that—in the circumstances—are sufficiently probable. (If I tell you that I met a man in town, it can go unsaid that he wore clothes.)

Specifically defeasible presumptions are closely interconnected with the conception of burden of proof:

> The effect of a presumption is to impute to certain facts or groups of facts *prima facie* significance or operation, and thus, in legal proceedings, to throw upon the party against whom it works the duty of bringing forward evidence to meet it. Accordingly, the subject of presumption is intimately connected with the subject of burden of proof, and the same legal rule may be expressed in different forms, either as throwing the advantage of a presumption on one side, or as throwing the burden of proof on the other.[6]

In effect, a defeasible presumption is just the reverse of a burden of proof (of the "burden of further reply" variety). Whenever there is a "burden of proof" for establishing that *P* is so, the correlative defeasible presumption that not-*P* stands until the burden has been discharged definitively. Archbishop Whately has formulated the relationship at issue in the following terms:

> According to the most current use of the term, a "presumption" in favour of any supposition means, not (as has been sometimes erroneously imagined) a preponderance of probability in its favour, but such a *preoccupation* of the ground as implies that it must stand good till some sufficient reason is adduced against it; in short, that *burden of proof* lies on the side of him who would dispute it.[7]

It is in just this sense, for example, that the "presumption of innocence" in favor of the accused is correlative with the burden of proof carried by the state in establishing his guilt.

For a proposition to count as a presumption is something altogether different from its counting as an accepted truth. A presumption is a plausible present-day to truth whose credentials may well prove insufficient, a runner in a race it may not win. The "acceptance" of a proposition as a presumptive truth is not acceptance at all but a highly provisional and conditional epistemic inclination toward it, an inclination that falls far short of outright commitment.

Our stance towards presumptions is unashamedly chat of fair-weather friends: we adhere to them when this involves no problems whatsoever, but abandon them at the onset of difficulties. But it is quite clear that such *loose* attachment to a presumption is by no means tantamount to no attachment at all.[8] However, a presumption stands secure only "until further notice," and it is perfectly possible that such notice might be forthcoming.

The Role of Presumption

The classical theories of perception from Descartes to the sense-datum theorists of the first half of the twentieth century all involve a common difficulty. For all of them saw a real and deep problem to be rooted in the question:

> Under what circumstances are our actual experiences genuinely veridical? In particular: which facts about the perceptual situation validate the move from "I (take myself to) see a cat on the mat" to "There is a cat on the mat"? How are we to monitor the appropriateness of the step from "perceptual experiences" to actual perceptions of real things-in-the-world, seeing that experience is by its very nature something personal and subjective.

The traditional theories of perception—all face the roadblock of the problem: How do we get from here to there, from subjective experience to warranted claims of objective fact?

However, what all these theories ignore is the fact that in actual practice we operate within the setting of a concept-scheme that reverses the burden of proof here: that our perceptions (and conceptions) are standardly treated as innocent until proven guilty. The whole course of relevant experience is such that the standing presumption is on their side. The indications of experience are taken as true provisionally—allowed to stand until such time (if ever) when concrete evidential counterindications come to view. Barring indications to the contrary, we can and do move immediately and unproblematically from "I take myself to be seeing a cat on the mat" to "There really is a cat on the mat and I actually see it there." But what is at issue here is not an *inference* (or a deriving) from determinable facts but a mere *presumption* (or a taking). The transition from subjectivity to objectivity is automatic, though, to be sure, it is always provisional, that is, subject to the proviso that all goes as it ought. For unless and until something goes amiss—that is, unless there is a mishap of some sort—those "subjective percepts" are standardly allowed to count as "objective facts."

To be sure, there is no prospect of making an inventory of the necessary conditions here. Life is too complex: neither in making assertions nor in driving an automobile can one provide a comprehensive advance survey of possible accidents and list all the things that can possibly go wrong. But the key point is that the linkage between appearance and reality is neither conceptual nor causal: it is the product of a pragmatic policy in the management of information, a ground rule of presumption that governs our epistemic practice.

The rational legitimation of a presumptively justified belief lies in the consideration that some generic mode of "suitably favorable indication" speaks on it behalf while no as-yet available counterindication speaks against it. When, after a careful look, I am under the impression that there is a cat on the mat, I can (quite appropriately) base my acceptance of the contention "There is a cat on the mat" not on certain preestablished premises, but simply on my experience—on my visual impression. The salient consideration is that there just is no good reason why (in *this* case) I should not indulge my inclination to endorse a visual indications of this kind as veridical. (If there were such evidence—if, for example, I was aware of being in a wax museum—then the situation would, of course, be altered.)

With presumption we *take* to be so what we could not otherwise *derive*. This idea of such presumptive "taking" is a crucial aspect of our language-deploying discursive practice. For presumptively justified beliefs are the raw materials of cognition. They represent contentions that—in the absence of preestablished counterindications—are acceptable to us "until further notice," thus permitting us to make a start in the venture of cognitive justification without the benefit of prejustified materials. They are defeasible alright, vulnerable to being overturned, but only by something else yet more secure some other preestablished conflicting consideration. They are entitled to remain in place until displaced by something better. Accordingly, their impetus averts the dire consequences that would ensue of any and every cogent process of rational deliberation required inputs which themselves had to be authenticated by a prior process of rational deliberation—in which case the whole process could never get under way.[9]

Plausibility and Presumption

Perhaps the single most important device for putting the idea of presumption to work is the natural inclination to credence inherent in the conception of plausibility.[10] For such plausibility can serve as the crucial determinant of where presumption resides. The basic principle here is that of the rule:

> Presumption favors the most plausible of rival alternatives—when indeed there is one. This alternative will always stand until set aside (by the entry of another, yet more plausible, presumption).

The operation of this rule creates a key role for plausibility in the theory of reasoning and argumentation.[11] In the face of discordant considerations, one "plays safe" in one's cognitive involvements by endeavoring to maximize the plausibility levels achievable in the circumstances. Such an epistemic policy is closely analogous to the *prudential* principle of action—that of opting for the available alternative from which the least possible harm can result. Plausibility-tropism is an instrument of epistemic prudence.

There are many bases for plausibility. For one thing, the standing of our informative sources in point of their authoritativeness affords one major entry point to plausibility. In this approach, a thesis is more or less plausible depending on the reliability of the sources that vouch for it—their entitlement to qualify as well-informed or otherwise in a position to make good claims to credibility. It is on this basis that "expert testimony" and "general agreement" (the consensus of men) come to count as conditions for plausibility.[12]

Again, the probative strength of confirming evidence could serve as yet another basis of plausibility. In this approach, the rival theses whose supporting case of substantiating evidence is the strongest is thereby the most plausible. Our evidential sources will clearly play a primary role.

The plausibility of contentions may, however, be based not on a thesis-warranting *source* but a thesis-warranting *principle*. Here inductive considerations may come prominently into play; in particular such warranting principles are the standard inductive desiderata: simplicity, uniformity, specificity, definiteness, determinativeness, "naturalness," and so on. In such an approach one would say that the more simple, the more uniform, the more specific a thesis—either internally, of itself, or externally, in relation to some stipulated basis—the more emphatically this thesis is to count as plausible.

For example, the concept of *simplicity* affords a crucial entry point for plausibility considerations. The injunction "Other things being anything like equal, give precedence to simpler hypotheses over more complex ones" can reasonably be espoused as a procedural, regulative principle of presumption, rather a metaphysical claim as to "the simplicity of nature." On such an approach, we espouse not the Scholastic adage "Simplicity is the sign of truth" (*simplex sigilium veri*), but its cousin, the precept "Simplicity is the sign of plausibility" (*simplex sigilium plausibili*). In adopting this policy we shift the discussion from the plane of the constitutive/descriptive/ontological to that of the regulative/methodological/prescriptive.

Again, uniformity can also serve as a plausibilistic guide to reasoning. Thus consider the *Uniformity Principle:*

> In the absence of explicit counterindications, a thesis about unscrutinized cases which conforms to a patterned uniformity obtaining among the data at our disposal with respect to scrutinized cases—a uniformity that is in fact present throughout these data—is more plausible than any of its regularity-discordant contraries. Moreover, the more extensive this pattern-conformity, the more highly plausible the thesis.

This principle is tantamount to the thesis that when the initially given evidence exhibits a marked logical pattern, then pattern-concordant claims relative to this evidence are—*ceteris paribus*—to be evaluated as more plausible than pattern-discordant ones (and the more comprehensively pattern-accordant, the more highly plausible). This rule implements the guiding idea of the familiar practice of judging the plausibility of theories and theses on the basis of a "sufficiently close analogy" with other cases.[13] (The uniformity principle thus forges a special rule for normality—reference to "the usual course of things"—in plausibility assessment.)[14]

The situation may thus be summarized as follows. The natural bases of plausibility—the criteria operative in determining its presence in greater or lesser degree—fall primarily into three groups:

1. *The standing of sources in point of authoritativeness.* On this principle the plausibility of a thesis is a function of the reliability of the sources that vouch for it—their entitlement to qualify as well-informed or otherwise in a position to make good claims to credibility. It is on this basis that expert testimony and general agreement (the consensus of men) come to count as conditions for plausibility.

2. *The probative strength of confirming evidence.* Of rival theses, that whose supporting case of substantiating evidence is the strongest is thereby the most plausible.
3. *The impetus of principles of inductive systematization.* Other things being sufficiently equal, that one of rival theses is the most plausible which scores best in point of *simplicity*, in point of *regularity* in point of *uniformity* (with other cases), in point of normalcy, and the like. The principles of presumption of this range correlate with the parameters of cognitive systematization.[15]

These three factors (source reliability, supportive evidence, and systematicity) are the prime criteria governing the assessment of plausibility in deliberations regarding the domain of empirical fact.

In general, the more plausible a thesis, the more smoothly it is consistent and consonant with the rest of our knowledge of the matters at issue. Ordinarily, the removal of a highly plausible thesis from the framework of cognitive commitments would cause a virtual earthquake; removal of a highly implausible one would cause scarcely a tremor; in between we have to do with varying degrees of readjustment and realignment when plausible contentions have to be abandoned. In general, then, the closer its fit and the smoother its consonance with our cognitive commitments, the more highly plausible the thesis.[16]

In the cognitive scheme of things the *principles* of presumption are *postulates*. They too play a special and distinctive role in the "theory of knowledge." First of all, they are not items of *knowledge* at all. If we know something we do not need to postulate it. On the other hand they differ from mere assumptions or hypotheses. For our commitment to *such* things is extremely tentative: we generally assume or suppose simply for the sake of discussion or for the sake of argument. By their very nature as such, assumptions, hypotheses, and suppositions are matters to which we have no particular commitment and to which we are not bound in any seriously committed way. Postulates, by contrast, are claims that we endorse and to which we stand committed. But knowledge they are not—they are not the sort of thing for which we have substantiating grounds or evidence.

How then, do postulates enter into the rational order of things? Their justification does not come *a tergo*—from the rear, so to speak, on the basis of a background of supporting fact. Rather it comes *a fronte*, from the future-oriented perspectives of what they can for us.

Postulates are indeed parts of the rational fabric—they have their validation and justification. But this of a *functional and prospective* nature. Their legitimization lies in what they enable us to do.

Here the practical aspect becomes crucial. For the validation of postulates is ultimately pragmatic, rooting in the fact that without them we could not achieve our ends—or at any rate not achieve them as effectively. The validation of a postulate lies in the fact that it renders the achievement of a mandatory objective possible—or at least far easier than it otherwise would be. The rational justification of postulates, in sum, lies in their pragmatic efficacy on the basis of considerations/effectiveness and efficiency in goal attainment within the domain of deliberation to which the presumption belongs.

Presumption and Probability

The plausibility of a thesis will not, however simply be a measure of its *probability*—of how likely we deem it, or how *surprised* we would be to find it falsified. Rather, it is a matter of how *chagrined* we would be into our larger cognitive scheme of things at forgoing something that fits smoothly. The core of the present conception of plausibility is the notion of the extent of our cognitive inclination toward a proposition—of *the extent of its epistemic hold on us* in the light of the credentials represented by the bases of its credibility. The key issue is that of how readily the thesis in view could make its peace within the overall framework of our cognitive commitments.

One of the cardinal lessons of modern epistemology is that considerations of probability cannot provide a validating basis for accepting factual claims. This lesson is driven home by the so-called Lottery Paradox.

This paradox is the immediate result of a decision policy for acceptance that is based on a probabilistic threshold values. Thus let us suppose the threshold level to be 0.80, and consider the following series of six statements:

This (fair and normal) die will not come up i when tossed (where i is to be taken as 1, 2, 3, 4, 5, and 6, each in turn).

According to the specified standard, each and every one of these six statements must be accepted as true. Yet their conjunction results in the patent absurdity that there will be no result whatsoever. Moreover, the fact that the threshold was set as low as 0.80 instead of 0.90 or 0.9999 is wholly immaterial. To reconstitute the same problem with respect to a higher threshold we need simply assume a lottery wholly having enough (equal) divisions to exhaust the spectrum of possibilities with individual alternatives of sufficiently small probability. Then the probability that each specific result will *not* obtain is less then 1 minus the threshold value, and so can be brought as close to 1 as we please. Accordingly, we should, by accepting each of these claims and conjoining them into a single whole, again be driven to the impossible conclusion of resultlessness.

This, then, is the Lottery Paradox.[17] It decisively rules out a propositional acceptance rule that is based on a probabilistic threshold.[18] This line of consideration inclined both Rudolf Carnap and his inductionist school and Hans Reichenbach and his probabilist school of inductive reasoning to exile categorical acceptance processes from the sphere of inductive epistemology altogether.[19] As epistemologists of the Carnap-Reichenbach orientation saw it, we are faced with a difficult and unpleasant choice. We must either give up an acceptance-oriented view of rational inquiry in science or else abandon the probabilistic theory of inductive reasoning.

But of course these difficulties arise only with respect to an *acceptance* that enrolls propositions on the register of consolidated truths. A mere *presumption* that enrolls them in the register of plausible answers to questions is another kettle of fish altogether.

There is a crucial difference between acceptance and presumption. Acceptance is nonlocalizable: to be accepted anywhere is to be accepted everywhere. Presumptions, by contrast, are contextual and issue relative: they arise and have their standing in the setting of particular problems, particular questions.

This fundamental difference between acceptance and presumption enables us after all to make some use of probabilistic considerations as a basis of presumption. An oversimple example can illustrate the principle at issue. Suppose that we classify people by gender (M or F) and age (y = young, m = mature). And suppose that in a group of five persons we have the following statistic:

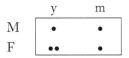

If the question before us with respect to an otherwise unidentified (presumably random) member of this group is, What is this individual's age-group? we would respond: presumably the individual is young—the majority are. Similarly with respect to the question of gender we will say, Presumably the individual is female—the majority are. But we cannot combine those two separate presumptions to justify presuming that the individual is a young female. If the question before us is: Is that individual a young female? Then we must presume the answer to be negative, seeing that most of the individuals at issue do not fall into this category. Our two earlier presumptions just do not combine. Probabilistically grounded plausibilities are not necessarily consistent—a fact which sets plausibility sharply aside from truth. (After all, plausibilities can be competitive so that not all of them can prevail.)

Two important lessons with respect to presumption emerge in the light of these considerations: (1) The presumption that can appropriately be made on probabilistic ground depends on just exactly what the question context happens to be. And (2) One cannot plausibly conjoin such presumptions across the board—detaching them from their question context and throwing them together.

The sort of difficulty that arises here will block acceptance and assertion, but will fail to impede taking the more tentative, conjectural epistemic stance that is at issue with presumption.

Presumption and Skepticism

The following objections might be offered against the idea of defeasible presumptions: "How can you speak of asserting a proposition merely as a presumption but not as a truth! If one is to assert (accept, maintain) the proposition in any way at all, does one not thereby assert (accept, maintain) *it to be true*." The answer here is simply a head-on denial, for there are different modes of acceptance. To maintain P as a presumption, as *potentially* or *presumptively* factual, is akin to maintaining P as possible or as probable. In no case are these contentions tantamount to maintaining the proposition as true. Putting a proposition forward as "possible" or "probable" commits one to claiming no more than that it is "possibly true" or "probably true." Similarly, to assert P as a presumption is to say no more than that P is potentially or presumptively true that it is a promising truth-candidate—but does not say that P is actually true, that it is a truth. Acceptance does not lie along a one-dimensional spectrum which ranges from "uncertainty" to "certainty." There are not only degrees of acceptance but also kinds of acceptance. And presumption represents such a kind: it is *sui generis*, and not just an attenuated version of "acceptance as certain."

In fact, however, the conception of a presumption does not "open the floodgates" in an indiscriminate way. Not *everything* qualifies as a presumption: the concept is to have some probative bite. A presumption is not merely something that is "possibly true" or that is "true for all I know about the matter," To class a proposition as a presumption is to take a definite and committal position with respect to it, so as to say "I propose to accept it as true insofar as no difficulties arise from doing so."

There is thus a crucial difference between an alleged truth and a presumptive truth. For allegation is a merely rhetorical category: every contention that is advanced in discussion is "allegedly true"—that is, alleged-to-be-true. But presumption—that is, warranted presumption—is an *epistemic* category: only in certain special circumstances are contentions of a *sort* that they merit to be accepted as true provisionally, "until further notice."[20]

The idea that the burden of proof always rests with the party asserting a thesis means that there is automatically a *presumption against* anything ever maintained thesis. But, of course, if the adducing of evidence in the dialectic of rational argumentation is to be possible at all, this doctrine must have its limits. Clearly, if the burden of proof inclined against *every* contention—if there were an automatic presumption of falsity against any contention whatsoever—it would become in principle impossible ever to provide a persuasive case. The rule that each contention needs evidential support through the adducing of further substantiating contentions cannot reasonably be made operative *ad indefinitum*.

In "accepting" a thesis as presumptively true one concedes it a probative status that is strictly provisional and pro tem; there is no need here to invoke the idea of *unquestionable* theses, theses that are inherently uncontestable, certain, and irrefutable.[21] To take this view would involve a misreading of the probative situation. It is to succumb to the tempting epistemological doctrine of foundationalism, that insists on the need for and ultimate primacy of absolutely certain, indefeasible, crystalline truths, totally beyond any possibility of invalidation. The search for such self-evident or protocol theses—inherently inviolate and yet informatively committal about the nature of the world—represents one of the great quixotic quests of modern philosophy.[22] It deserves stress that an epistemic quest for categories of data that are *prima facie* acceptable (innocent until proven guilty, as it were) is altogether different from this quest for absolutely certain or totally self-evidencing theses that has characterized the mainstream of epistemological tradition from Descartes via Brentano to present-day writers such as Roderick Chisholm.

The idea of presumptive truth must thus play a pivotal role in all such various contexts where the notion of a "burden of proof" applies. The mechanism of presumption accomplishes a crucial epistemological task in the structure of rational argumentation. For there must clearly be some class of claims chat are allowed at least *pro tem* to enter acceptably into the framework of argumentation, because if everything were contested then the process of inquiry could not progress at all.

In any essentially dialectical situation in which the idea of burden of proof figures, the very "rules of the game" remain inadequately defined until the issue of the nature, extent, and weight of the range of operative presumptions has been resolved in some suitable way. Burden of proof and presumption represent correlative conceptions inevitably coordinate with one another throughout the context of rational dialectic, because the recourse to presumptions affords the indispensable means by which a burden of proof can—at least provisionally—be discharged.

The reality of it is that we cannot pursue the cognitive project—the quest for information about the world—without granting certain initial presumptions. They are reminiscent of Kantian "conditions under which alone" the business of securing answers to questions about the world is even *possible*. And prominent among these conditions is the consideration that we can take our "data" about the world as *evidence*, that a presumption of experiential veridicality is in order. In matters of sense perception, for example, we presume that mere appearances ("the data") provide an indication of how things actually stand (however imperfect this indication may ultimately pose to be). That we can use the products of our experience of the world to form at least somewhat reliable views of it is an indispensable presupposition of our cognitive endeavors in the realm of fictional inquiry. If we systematically refuse, always and everywhere, to accept *seeming* evidence as *real* evidence—at least provisionally, until the time comes when it is discredited as such—then we can get nowhere in the domain of practical cognition. When the skeptic rejects any and all presumptions, he automatically blocks any prospect of reasoning with him *within* the standard framework of discussion about the empirical facts of the world. The machinery of presumptions is part and parcel of the mechanisms of cognitive rationality; abandoning it aborts the entire project at the very outset.

If it is indeed the case that rationally justified belief must *always* be based on *rationally prejustified* inputs, then skepticism becomes unavoidable. For, then the process of rationally validating our accepted beliefs can never get started. To all appearances, we here enter on a regress that is either vitiatingly infinite or viciously circular. The rational justification of belief becomes in principle impossible—as skeptics have always insisted.

But this particular skeptical foray rests on a false supposition. For, the rational justification of a belief does *not* necessarily require prejustified inputs. The idea that even as human life can come only from prior human life so rational justification can come only from prior rational justification is deeply erroneous. For, the important distinction between *discursive* and *presumptive* justification becomes crucial here in a way that skeptics conveniently overlook.

A belief is justified *discursively* when there is some other preestablished belief on whose basis this belief is evidentially grounded, that is, when the belief is substantiated by some particular item of supporting evidence. The discursive justification of a belief lies in there being an already available, prejustified belief that evidentiates it. In information-processing terms, this discursive sort of justification is not innovative but merely transformatory as a production process: there must be justified beliefs as inputs to arrive at justified beliefs as outputs. However, *discursive* justification is not the only sort there is. For there is also *presumptive* justification.

Presumptive justification—unlike discursive justification—does not proceed through the evidential meditation of previously justified grounds but directly and immediately through the force of a "presumption." A belief is justified in this way when there is a *standing presumption* in its favor and no preestablished (rationally justified) reason that stands in the way of its acceptance. Beliefs of this sort are appropriate—are rationally justified—as long as nothing speaks against them. Presumption is the epistemic analogue of "innocent until proven guilty."

The rational legitimation of a presumptively justified belief lies in the fact that some generic mode of "suitably favorable indication" speaks on it behalf while no already justified counterindication speaks against it. When, after a careful look, I am under the impression that there is a cat on the mat, I can (quite appropriately) base my acceptance of the contention "There is a cat on the mat" not on certain preestablished premises, but simply on my experience—on my visual impression. The salient consideration is that there just is no good reason why (in *this* case) I should not indulge my inclinations to endorse a visually grounded belief of this kind as veridical. (If there were such evidence—if, for example, I was aware of being in a wax museum—then the situation would, of course, be altered.)

Presumptively justified beliefs are the raw materials of cognition. They represent contentions that—in the absence of preestablished counterindications—are acceptable to us "until further notice," thus permitting us to make a start in the venture of cognitive justification without the benefit of prejustified materials. They are defeasible alright, vulnerable to being overturned, but only by something else yet more secure some other preestablished conflicting consideration. They are entitled to remain in place until displaced by something superior. Accordingly, their impetus averts the dire consequences that would ensue of any and every cogent process of rational deliberation required inputs that themselves had to be authenticated by a prior process of rational deliberation—in which case the whole process could never get under way.

But indispensability apart, what is it that justifies making presumptions, seeing that they are not established truths? The answer is that this is not so much a matter of evidentially *probative* considerations as of procedurally *practical* ones. Presumptions arise in contexts where we have questions and need answers. And when sufficient evidence for a *conclusive* answer is lacking, we must, in the circumstances, settle for a more or less *plausible* one. It is a matter of *faute de mieux*, of this or nothing (or at any rate nothing better). Presumption is a thought instrumentality that so functions as to make it possible for us to do the best we can in circumstances where something must be done. And so presumption affords yet another instance where practical principles play a leading role on the stage of our cognitive and communicative practice. For presumption is, in the end, a practical device whose rationale of validation lies on the order of pragmatic considerations.

We proceed in cognitive contexts in much the same manner in which banks proceed in financial contexts. We extend credit to others, doing so at first to a relatively modest extent. When and as they comport themselves in a way that indicates that this credit was warranted, then we extend more. By responding to trust in a "responsible" way—proceeding to amortize the credit one already has—one can increase one's credit rating in cognitive much as in financial contexts.

In trusting the senses, in relying on other people, *and even in being rational,* we always run a risk. Whenever in life we place our faith in something, we run a risk of being let down and disappointed. Nevertheless, it seems perfectly reasonable to bet on the general trustworthiness of the senses, the general reliability of our fellow men, and the general utility of reason. In such matters, no absolute guarantees can be had. But, one may as well venture, for, if venturing fails, the cause is lost anyhow—we have no more promising alternative to turn to. There is little choice about the matter: it is a case of "this or nothing." If we want answers to factual questions, we have no real alternative but to trust in the cognitively cooperative disposition of the natural order of things. We cannot preestablish the appropriateness of this trust by somehow demonstrating, in advance of events, that it is actually warranted. Rather, its rationale is that without it we remove the basis on which alone creatures such as ourselves can confidently live a life of effective thought and action. In such cases, pragmatic rationality urges us to gamble on trust in reason, not because it cannot fail us, but because in so doing little is to be lost and much to be gained. A general policy of judicious trust is eminently cost-effective in yielding good results in matters of cognition.

Of course, a problem remains: Utility is all very good but what of validity? What sorts of considerations *validate* our particular presumptions as such: how is it that they become *entitled* to this epistemic status? The crux of the answer has already been foreshadowed. A twofold process is involved. Initially it is a matter of the generic need for answers to our questions: of being so circumstanced that if we are willing to presume we are able to get . . . anything. But ultimately we go beyond such this-or-nothing consideration, and the validity of a presumption emerges *ex post facto* through the utility (both cognitive and practical) of the results it yields. We advance from "this or nothing" to "This or nothing that is determinably better." Legitimation is thus available, albeit only through experiential *retrovalidation,* retrospective validation in the light of eventual experience.[23] It is a matter of learning that a certain issue is more effective in meeting the needs of the situation than its available alternatives. Initially we look to promise and potential but in the end it is applicative efficacy that counts.

The fact is that our cognitive practices have a fundamentally economic rationale. They are all cost-effective within the setting of the project of inquiry to which we stand committed (by our place in the world's scheme of things). Presumptions are the instrument through which we achieve a favorable balance of trade in the complex trade-offs between ignorance of fact and mistake of belief—between unknowing and error.

How Presumption Works: What Justifies Presumptions

But just what sorts of claims are presumptively justified? The ordinary and standard probative practice of empirical inquiry stipulates a presumption in favor of such cognitive "sources" of information as the senses and memory—or for that matter trustworthy personal or documentary resources such as experts and encyclopedias. And the literature further contemplates such cognitively useful alternatives as:

1. *Natural inclination*: a "natural disposition" to accept (e.g., in the case of sense observation).
2. *Epistemic utility* in terms of the sorts of things that would, if accepted, explain things that need explanation.
3. *Analogy* with what has proved acceptable in other contexts.
4. *Fit*: coherence with other accepted theses.

Even a weak reed like analogy—the assimilation of a present, problematic case to similar past ones—is rendered a useful and appropriate instrumentality of presumption through providing our readiest source of answers to questions.[24] However, the salient feature of a viable basis of prescription is that it must have a good track record for providing useful information.

Our cognitive proceedings accordingly incorporate a host of fundamental presumptions of reliability, such as:

- Believe the evidence of your own senses.
- Accept at face value the declarations of other people (in the absence of any counterindications and in the absence of any specific evidence undermining the generic trustworthiness of those others).
- Trust in the reliability of the standardly employed cognitive aids and instruments (telescopes, calculating machines, reference works, logarithmic tables, etc.) in the absence of any specific indications to the contrary.
- Accept the declarations of recognized experts and authorities within the area of their expertise (again, in the absence of counterindications).

All in all, presumption favors the usual and the natural—its tendency is one of convenience and ease of operation in cognitive affairs. For presumption is a matter of cognitive economy—of following "the path of least resistance" to an acceptable conclusion. Its leading principle is: introduce complications only when you need to, always making do with the least complex resolution of an issue. There is, of course, nothing sacrosanct about the result of such a procedure. The choice of the easiest way out may fail us, that which serves adequately in the first analysis may well not longer do so in the end. But it is clearly the sensible way to begin. At this elemental level of presumption we proceed by "doing what comes naturally."

The salient point for validating a presumption is that there is reason to think—or experience to show—that in proceeding in this way we fare better than we would do otherwise. The justification of these presumptions is not the factual one of the substantive generalization. "In proceeding in this way, you will come at correct information and will not fall into error." Rather, it is methodological justification. In proceeding in this way, you will efficiently foster the interests of the cognitive enterprise; the gains and benefits will, on the whole, outweigh the losses and costs. Principles of this sort are integral parts of the operational code of agents who transact their cognitive business rationally.

Presumption is a matter of cognitive economy—of following "the path of least resistance" to an acceptable conclusion. Its leading principle is: introduce complications only when you need to, always making do with the least complex resolution of an issue. There is, of course, nothing sacrosanct about the result of such a procedure. The choice of the easiest way out may fail us, that which serves adequately in the first analysis may well not longer do so in the end. But it is clearly the sensible way to begin. At this elemental level of presumption we proceed by "doing what comes naturally." What is fundamental here is the principle of letting appearance be our guide to reality—of accepting the evidence *as evidence* of actual fact, by taking its indications as decisive until such time as suitably weighty counterindications come to countervail against them.

The justifactory rationale for a policy of presumption lies in the human need for information. It is undoubtedly true that information is a good thing: that other things equal it is better to have information—to have answers to our questions—than not to do so. But the reality of it is much more powerful than that. For us humans, information is not just a mere desideratum but an actual need.

This role of presumptions is absolutely crucial for cognitive rationality. For this mode rationality has two compartments, the discursive (or conditionalized) and the substantive (or categorical). The former is a matter of hypothetical reasoning—of adhering to the conditionalized principle that *if* you accept certain theses, *then* you should also accept their duly evidentiated consequences. But of course this conditionalized principle cannot yield anything until one has already secured some acceptable these from somewhere or other. And this is where substantive rationality comes in, by enabling us to make categorical moves. Presumptions determine our "starter-set" of initial commitments, enabling us to make a start on whose basis further "inferential" reasoning may proceed.

While the mechanism of a *presumption* affords a most useful cognitive and communicative resource, it is only recently that it has begun to interest logicians, semanticists, and epistemologists. Their discussions have, in the main, focused on the relation of presumption to the issues of benefit of doubt and burden of proof.

But what is it that justifies making presumptions since they are neither established truths nor probabilities in the most standard sense? The answer is that this is not so much a matter of evidentially *probative* considerations as of procedurally *practical* ones. Presumptions arise in contexts where we have questions and need answers. And when sufficient evidence for a *conclusive* answer is lacking, we must, in the circumstances, settle for a more or less *plausible* one. It is a matter of *faute de mieux*, of this or nothing (or at any rate nothing better). Presumption is a thought instrumentality that so functions as to make it possible for us to do the best we can in circumstances where something must be done. And so presumption affords yet another instance where practical considerations play a leading role on the stage of our cognitive and communicative practice.

The obvious and evident advantage of presumption as an epistemic device is that it enables us vastly to extend the range of questions we are able to answer. It affords an instrument that enables us to extract a maximum of information from communicative situations. Presumption, in sum, is an ultimately pragmatic resource. To be sure, its evident disadvantage is that the answers that we obtain by its means are given not in the clarion tones of knowledge and assertion but in the more hesitant and uncertain tones of presumption and probability. We thus do not get the advantages of presumption without an accompanying negativity. Here, as elsewhere, we cannot have our cake and eat it too.

What justifies a presumption initially is a need: that is, the fact that we require an information gap to be filled and that the presumption accomplishes this for us. But this is only a start. For the question remains: What sorts of considerations *validate* our particular presumptions as such: how is it that they become *entitled* to this epistemic status? The basis of the answer has already been indicated. A twofold process is involved. Initially it is a matter of the generic need for *true* presumptions continued with the mere convenience of *these* presumptions. But ultimately a this-or-nothing recognition of the validity of a presumption emerges *ex post facto* through the utility (both cognitive and practical) of the results it yields. Legitimation is thus available but only through experiential *retrovalidation,* retrospective validation in the light of eventual experience.[25] For what validates presumptions is something quite different, namely *experience*.

All the same, the concession of presumptive status to our presystematic indications of credibility is the most fundamental principle of cognitive rationality. The details of presumption management are clearly negotiable in the light of eventual experience—and improvable over the course of time. But without a programmatic policy for presumption cognitive rationality cannot get under way at all.

For present purposes, then, the salient point is that presumption provides the basis for letting appearance be our guide to reality—of accepting the evidence *as evidence* of actual fact, by taking its indications as decisive until such time as suitably weighty counterindications come to countervail against them. What is at issue here is part of the operational code of agents who transact their cognitive business rationally.

It is thus clear that all such cognition practices have a fundamentally economic rationale. They are all cost-effective within the setting of the project of inquiry to which we stand committed (by our place in the world's scheme of things).

To be sure, such a validation of our presumptions is not really theoretical but practical. It does not show that cannot sometimes or often go awry when we endorse the presumptions at issue. Instead it argues only that when we indeed need or want here-and-now to resolve issues ad fill informative gaps then those presumptions represent the most promising ways of doing so, affording us those means to accomplish our goals which—as best we can tell—offer the best prospect of success. Such a process of validation is not a matter of theoretical adequacy as of practical efficacy. It is a practical or pragmatic validation in terms of a rational recourse to the best available alternative.[26]

Accordingly, presumption, like our cognitive practices in general, has a fundamentally economic rationale. Like the rest, it too is a cost effective within the setting of the project of inquiry to which we stand committed by our place in the world's scheme of things. They are characteristics of the cheapest (most convenient) way for us to secure the data needed to resolve our cognitive problems—to secure answers to our questions about the world we live in. On this basis, we can make ready sense of many of the established rules of information development and management on economic grounds. By and large, they prevail because this is maximally cost effective in comparison with the available alternatives.

And just here the idea of presumption connects with the idea of trust.

Chapter 6

Trust and Cooperation in Pragmatic Perspective

Synopsis

- *Knowledge is power. But the hoarding of knowledge—monopolization, secretiveness, collaboration avoidance—is generally counterproductive.*
- *In anything like ordinary circumstances, mutual aid in the development and handling of information is highly cost effective.*
- *The way in which people build up epistemic credibility in cognitive contexts is structurally the same as that in which they build up financial credit in economic contexts.*
- *And this is so in particular with regard to our cognitive presumptions.*
- *Considerations of cost effectiveness—of economic rationality, in short—operate to ensure that any group of rational inquirers will in the end become a community of sorts, bound together by a shared practice of trust and cooperation.*

The Cost Effectiveness of Sharing and Cooperating in Information Acquisition and Management

The pragmatic aspect of knowledge is an unavoidable fact of life. In many ways, knowledge is power. Its possession facilitates efficacy and influence in the management of public affairs. It enables those who have "inside" information to make a killing in the marketplace. It opens doors to the corridors of power in corporations. It maintains experts in the style to which the present century has accustomed them, and assures that the U.S. government supports more think tanks than aircraft carriers.

Since information is power, there is a constant temptation to monopolize it. But information monopolies, however advantageous for some few favorably circumstanced beneficiaries, exact a substantial price from the community as a whole. In this regard, sixteenth- and seventeenth-century science affords an admonitory object lesson. The secretiveness of investigators in those times—Isaac Newton preeminent among them—in matters of mathematics and astronomy assured that the development of natural science would be slow and difficult. In protecting the priority of their claims through secretiveness and mystification, and adepts of the day greatly impeded the development and dissemination of knowledge.[1] Only with the emergence of new means of information sharing to facilitate the diffusion of knowledge, such as academies and learned societies with their meetings and published proceedings, could modern science begin it sure and steady march. In particular, the open scientific literature can be seen as an effective and productive system for the authentication and protection of the state of the creative scientist in the "intellectual property" created by his innovative efforts.[2]

While there are, of course, exceptions, in most circumstances of ordinary life, and above all in the sciences, it pays all concerned to share information. From an economic point of view, we confront the classic format of a cooperation-inviting situation, where the resultant gain in productivity creates a surplus in which all can share to their own benefit. The evident advantages for the scientific community and its members of creating a system that provides inducement for people to promulgate their findings promptly, while at the same time imposing strong sanctions against cheating, falsification, and carelessness, militate powerfully in this direction. The open exchange of information in science benefits the work of the community, except possibly in these cases where secrecy can confer an economic advantage that can serve as a stimulant to creative effort.

To be sure, secrecy survives even now in various nooks and crannies of the scientific enterprise, and good claims can sometimes be made on its behalf. For example, editors of scientific journals do not let authors of submitted papers know the identity of those who review their submissions. Or again, such journals do not publish lists of authors whose submissions have been declined for publication. These practices obviously serve the interests of the journal's effectiveness by helping to maintain pools of willing reviewers and submitters.

The fundamental principles are the same either way, however. Both the general policy of information sharing in science, and the specific ways in which particular practices standardly depart from it, have a perfectly plausible rationale in cost-benefit terms. Broadly economic considerations of cost and benefit play the determinative role throughout.

The Advantages of Cooperation

Even people who do not much care to cooperate and collaborate with others are well advised in terms of their own interests to suppress this inclination. This point is brought home by considering the matter from the angle presented in Table 6.1. By hypothesis, each of the parties involved prioritized the situation where they are trusted by the other while they themselves need not reciprocate. And each sees as the worst case a situation where they themselves trust without being trusted. Each, however, is willing to trust to avert being mistrusted themselves. Relative to these suppositions, we arrive at the overall situation of the interaction matrix exhibited in Table 6.2. (Here the entry 2/2—for example—indicates that in the particular case at issue the outcome ranks 2 for me and 2 for you, respectively.) In this condition of affairs, mutual trust is the best available option—the only plausible way to avert the communally unhappy result 3/3.[3] In this sort of situation, cooperative behavior is obviously the best policy. (We are, after all, going to end up acting alike since, owing to the symmetry of the situation, whatever constitutes a good reason for your to act in a certain way does so for me as well.)

It is easily seen that a skeptical presumption—one which rejects trust and maintains a distrustful stance toward the declarations of others—would confront us with an enormously complex (and economically infeasible) task for the project of interpersonal communication. For suppose that, instead of treating others on the basis of *innocent until proven guilty,* one were to treat them on lines of *not*

TABLE 6.1
A Preferential Overview of Trust Situations

I Trust You	You Trust Me	My Preference Ranking	Your Preference Ranking
+	+	2	2
+	−	4	1
−	+	1	4
−	−	3	3

TABLE 6.2
An Interaction Matrix for Trust Situations

	You Trust Me	You Do Not Trust Me
I trust you	2/2	4/1
I do not trust you	1/4	3/3

trustworthy until proven otherwise. It is clear that such a procedure would be vastly lass economic. For we would now have to go to all sorts of lengths in independent verifications. The problems here are so formidable that we would obtain little if any informative benefits from the communicative contributions of others. When others tend to respond in kind to one's present cooperativeness or uncooperativeness, then no matter how small on deems the chances of their cooperation in the present case, one it nevertheless well advised to act cooperatively. As long as interagents react to cooperations with some tendency to reciprocation in future situations, cooperative behavior will yield long-run benefits.

From the angle of economy there are, accordingly, substantial advantages to collaboration in inquiry, particularly in scientific contexts. For the individual inquirer, it decreases the chances of coming up completely empty-handed (though at the price of having to share the credit of discovery). For the community, it augurs a more rational division of labor through greater efficiency by reducing the duplication of effort.

BUILDING UP TRUST: AN ECONOMIC APPROACH

The process through which mutual trust in matters of information development and management is built up among people cries out for explanation by means of an economic analogy that trades on the dual meaning of the idea of credit. For we proceed in cognitive matters in much the same way that banks proceed in financial matters. We extend credit to others, doing so at first to only a relatively modest extent. When and if they comport themselves in a manner that shows that this credit was well deserved and warranted, we proceed to give them more credit and extend their credit limit, as it were. By responding to trust in a responsible way, one improves one's credit rating in cognitive contexts much as in financial contexts. The same sort of mechanism is at work in both cases: recognition of credit worthiness engenders a reputation on which further credit can be based; earned credit is like money in the bank, well worth the measures needed for its maintenance and for preserving the good name that is now at stake.

And this situation obtains not just in the management of information in natural science but in many other settings as well, preeminently including the information we use in everyday-life situations. For example, we constantly rely on experts in a plethora of situations, continually placing reliance on doctors, lawyers, architects, and other professionals. They too must so perform as to establish credit, not just as individuals but, even more crucially, for their profession as a whole.[4]

Much the same holds for other sources of information. The example of our senses is a particularly important case in point. Consider the contrast between our reaction to the data obtained in sight and dreams. Dreams, too, are impressive and seemingly significant data. Why then do we accept sight as a reliable cognitive source but not dreams—as people were initially minded to do? Surely not because of any such substantive advantages as vividness, expressiveness, or memorability. The predisposition to an interest in dreams is clearly attested by their prominence in myth and literature. Our confident reliance on sight is not a consequence of its intrinsic preferability but is preeminently a result of its success in building up credit in just the way we have been considering. We no longer base our conduct of affairs on dreams simply because it does not pay all that well.

Again, a not dissimilar story holds for our information-generating technology—for telescopes, microscopes, computing machinery, and so on. We initially extend some credit because we simply must, since they are our only means for a close look at the moon, at microbes, and so on. But subsequently we increase their credit limit (after beginning with blind trust) because we eventually learn, with the wisdom of hindsight, that it was quite appropriate for us to proceed in this way in the first place. As we proceed, the course of experience indicates, retrospectively as it were, that we were justified in deeming them creditworthy. And in this regard the trustworthiness of people and the reliability of instrumental resources are closely analogous matters.

Trust is, of course, something that we can have not only in people but in cognitive sources at large. For a not dissimilar story holds for our information-generating technology—for telescopes, microscopes, computing machinery, and so on. We initially extend some credit because we simply must, since they are our only means for a close look at the moon, at microbes, and so on. But subsequently we increase their credit limit (after beginning with blind trust) because we eventually learn, with the wisdom of hindsight, that it was quite appropriate for us to proceed in this way in the first place. As we proceed, the course of experience indicates, retrospectively as it were, that we were justified in deeming them creditworthy.

To be sure, the risk of deception and error is present throughout our inquiries: our cognitive instruments, like all other instruments, are never failproof. Still, a general policy of judicious trust is eminently cost effective. In inquiring, we cannot investigate everything; we have to start somewhere and invest credence in something. But of course our trust need not be blind. Initially bestowed on a basis of mere hunch or inclination, it can eventually be tested, and can come to be justified with the wisdom of hindsight. And this process of testing can in due course put the comforting reassurance of retrospective validation at our disposal.

In trusting the senses, in relying on other people, *and even in being rational,* we always run a risk. Whenever in life we place our faith in something, we run a risk of being let down and disappointed. Nevertheless, it seems perfectly reasonable to bet on the general trustworthiness of the senses, the general reliability of our fellow men, and the general utility of reason. In such matters, no absolute guarantees can be had. But, one may as well venture, for, if venturing fails, the cause is lost anyhow—we have no more promising alternative to turn to. There is little choice about the matter: it is a case of "this or nothing." If we want answers to factual questions, we have no real alternative but to trust in the cognitively cooperative disposition of the natural order of things. We cannot preestablish the appropriateness of this trust by somehow demonstrating, in advance of events, that it is actually warranted. Rather, its rationale is that without it we remove the basis on which alone creatures such as ourselves can confidently live a life of effective thought and action. In such cases, pragmatic rationality urges us to gamble on trust in reason, not because it cannot fail us, but because in so doing little is to be lost and much to be gained. A general policy of judicious trust is eminently cost-effective in yielding good results in matters of cognition.

It is sometimes said that an epistemology based on trust is contrastive to and distinct from one that is based on evidence.[5] But this is a very questionable standpoint. For trustworthiness is something we may initially presume but must eventually evidentiate through experience. Ongoing trust is only appropriate in the case of trustworthy sources and trust must be earned through the evidentiation of trustworthiness. When a factual claim is based on trust in a source we may not have any (independent) evidentiation of that fact but we should and do require evidence for the reliability of the source.

To be sure, the risk of deception and error is present throughout our inquiries: our cognitive instruments, like all other instruments, are never failproof. Still, a general policy of judicious trust in eminently cost effective. In inquiring, we cannot investigate everything; we have to start somewhere and invest credence in something. But of course our trust need not be blind. Initially bestowed on a basis of mere hunch or inclination, it can eventually be tested, and can come to be justified with the wisdom of hindsight. And this process of testing can in due course put the comforting reassurance of retrospective validation at our disposal.

With trust, matters can of course turn out badly. In being trustful, we take our chances (though of course initially in a cautious way). But one must always look to the other side of the coin as well. A play-safe policy of total security calls for not accepting anything, not trusting anyone. But then we are left altogether empty-handed. The quest for absolute security exacts a terrible price in terms of missed opportunities, forgone benefits, and lost chances. What recommends those inherently risky cognitive policies of credit extension and initial trust to us is not that they offer risk-free sure bets but that, relative to the alternatives, they offer a better balance of potential benefits over potential costs. It is the fundamentally *economic* rationality of such cognitive practices that is their ultimate surety and warrant.

Trust and Presumption

We know that various highly convenient principles of knowledge production are simply false:

- What seems to be, is.
- What people say is true.
- The simplest patterns that fit the data are actually correct,
- The most adequate currently available theory will work out.

We realize full well that such generalizations do not hold, however nice it would be if they did. Nevertheless we accept the theses at issue as principles of presumption. We follow the metarule: In the absence of concrete indications to the contrary, proceed as though such principles were true. Such principles of presumption characterize the way in which rational agents transact their cognitive business. Yet we adopt such practices not because we can somehow establish their validity, but because the cost-benefit advantage of adopting them is so substantial. The justification of trust in our senses, in our fellow inquirers, and in our cognitive mechanisms ultimately rests on considerations of economic rationality. And this sort of situation prevails in many other contexts. For example, the rationale of reputations for ability, as well as those for reliability, lies in the cost effectiveness of this resource in contexts of hiring, allocating one's reading time, and so on.[6]

Our standard cognitive practices incorporate a host of fundamental presumptions of initial credibility, in the absence of concrete evidence to the contrary:

- Believe in your own senses.
- Accept at face value the declarations of other people (in the absence of any counterindications and in the absence of any specific evidence undermining their generic trustworthiness).
- Trust in the reliability of established cognitive aids and instruments (telescopes, calculating machines, reference works, logarithmic tables, etc.) in the absence of any specific indications to the contrary.
- Accept the declarations of established experts and authorities within the area of their expertise (again, absent counterindications).

The justification of these presumptions is not the factual one of the substantive generalization, In proceeding in this way, you will come at correct information and will not fall into error. Rather, it is methodological justification. In proceeding in this way, you will efficiently foster the interests of the cognitive enterprise; the gains and benefits will, on the whole, outweigh the losses and costs.

It is clear that all such cognitive practices have a fundamentally economic rationale. They are all cost effective within the setting of the project of inquiry to which we stand committed by our place in the world's scheme of things. They are characteristics of the cheapest (most convenient) way for us to secure the data needed to resolve our cognitive problems—to secure answers to our questions about the world we live in. Accordingly, we can make ready sense of many of the established rules of information development and management on economic grounds. By and large, they prevail because this is maximally cost effective in comparison with the available alternatives.

A Community of Inquirers

Only through cooperation based on mutual trust can we address issues whose effective resolution makes demands that are too great for any one of us alone. In the development and management of information, people are constantly impelled toward a system of collaborative social practices—an operational code of incentives and sanctions that consolidates and supports collective solidarity and mutual support. In this division of labor, trust results from what is, to all intents and purposes, a custom consolidated compact to conduct their affairs in friendly collaboration.

If its cognitive needs and wants are strong enough, any group of mutually communicating, rational, dedicated inquirers is fated in the end to become a *community* of sorts, bound together by a shared practice of trust and cooperation, simply under the pressure of its evident advantage in the quest for knowledge.[7]

However, this cooperative upshot need not ensue from a moral dedication to the good of others and care for their interests. It can emerge for reasons of prudential self-interest alone because the relevant modes of mutually helpful behavior—sharing, candor, and trustworthiness—are all strongly in everyone' interest, enabling all members to draw benefit for their own purposes—the agent himself specifically included. Cooperation emerges in such a case not from morality but from self-interested considerations of economic advantage. In science, in particular, the advantages of epistemic values like candor, reliability, accuracy, and the like, are such that everyone's interests as well served by fostering adherence to the practices at issue.

The pursuit of knowledge in science can play a role akin to that of a pursuit of wealth in business transactions. The financial markets in stocks or commodities futures would self-destruct if the principle, *my word is my bond*, were abrogated, since no one would know whether a trade had actually been made. In just this way, too, the market in information would self-destruct if people's truthfulness could not be relied on. Thus in both cases, unreliable people have to be frozen out and exiled from the community. In cognitive and economic contexts alike, the relevant community uses incentives and sanctions (artificially imposed costs and benefits) to put into place a system where people generally act in a trusting and trustworthy way. Such a system is based on processes of reciprocity that advantage virtually everyone.

It is no wonder that common practice of the scientific community involves severe sanctions for background of trust. Strong incentives in matters of developing and exchanging information induce powerfully to the general advantage. And on the other side, data forging, credit grabbing, plagiarism, and the like are all significantly injurious to an economy of information development and exchange among rational agent. Strong disincentives against such practices are clearly sensible.

Several recent studies illuminate the extent to which we actually depend on others in our beliefs.[8] The experiments of Solomon Asch have dramatized people's tendency to conform to erroneous public judgments on matters where they would never make mistakes by themselves.[9] His subjects had only to specify which of three lines was closest in length to a given line. People made this judgment unerringly, except when they knew that all the others who were asked the same question concurred in giving a different answer.[10] Commenting on Asch's experiments, Sabini and Silver report: "All (or nearly all) subject reacted with signs of tension and confusion. Roughly one-third of the judgments subjects made were in error. Nearly 80 percent of the subject gave the obviously wrong answer on at least one trial. The perception that a few other people made an absurd judgment of a clear, unambiguous physical matter was a very troubling experience, sufficient to cause doubt, and in some cases conformity."[11] Such experiments actually reveal (in their own dishonest way) the extent to which people incline to trust others. A recent study of American juries arrived at very similar findings.[12] On examining more than 250 jury deliberations, the investigators found that in no case was a hung jury caused by a single dissenter. Unless someone who disagreed with the majority found support by at least two others, the dissenters relaxed their reservations and came around to the majority view. And the rationale for this sort of thing is validated by sound economic considerations, a trusting relationship reduces current interaction costs in return for past investments in its buildup. Knowing whom one can trust is worth a great deal. Outsiders who come as strangers into an established social framework generally have to pay for the benefit of learning that agents are trustworthy—and generally find this information well worth paying for.

Such considerations militate for a universally advantageous modus operandi, under whose aegis people can trust their fellows in a setting of communal cooperation. And the harsh measures used to uphold the integrity of science—the destruction of careers through ostracism from the community—are thus not devoid of rational justification. Cheating is worth eliminating at great cost, because its toleration endangers and undermines the fabric of mutual trust, in whose absence the whole enterprise of collaborative inquiry becomes infeasible. Establishing and maintaining a community of inquirers united in common collaboration by suitable rewards and sanctions is a mode of operation that is highly cost effective. Individual probity and mutual helpfulness are virtues whose cultivation pays ample dividends for the community of inquirers. In these matters, the cold iron hand of individual and communal interest lies behind the velvet glove of etiquette and ethics. The commodity of information illustrates rather than contravenes the division of labor that results from Adam Smith's putative innate human "propensity to truck, barter, and exchange." The market in knowledge has pretty much the same nature and the same motivation as any other sort of market—it is a general-interest arrangement. The establishment of conditions that foster cooperation and trust are critical to the cognitive enterprise of productive inquiry. Cooperation evolves because what is in the interests of most is, in most cases, in the interests of each.

As these deliberations indicate, our cognitive practices of trust and presumption are undergirded by a justificatory rationale whose nature is fundamentally economic. For what is at issue throughout is a system of procedure that assures for each participant the prospect of realizing the greatest benefit for the least cost. The commitment to impersonal standards and interpersonal generality that characterizes the cognitive enterprise is thus an inherently practical matter. Our standard cognitive policies and procedures are validated by consideration of practical reason, seeing that they are substantiated and sustained on what is, in the end, a matter of economic rationality.[13]

Part II

Rational Inquiry and the Quest for Truth

Chapter 7

Foundationalism versus Coherentism

Synopsis

- *The most prominent and historically most influential model of cognitive systematization is the Euclidean model of a linear, deductive exfoliation from basic axioms.*
- *Such an approach leads to foundationalism.*
- *But the network model of cyclic systematization affords a prime alternative to this traditional axiomatic approach.*
- *These two different models of cognitive systematization give rise to two rival and substantially divergent epistemic programs for the authentication of knowledge, namely foundationalism and coherentism.*
- *The inherent difficulties and limitations of the foundationalist program indicate the advisability of a closer look at the coherentist approach.*

Hierarchical Systemization: The Euclidean Model of Knowledge

The model of knowledge canonized by Aristotle in the *Posterior Analytics* saw Euclidean geometry as the most fitting pattern for the organization of anything deserving the name of a science (to put it anachronistically, since Euclid himself postdates Aristotle). Such a conception of knowledge in terms of the geometric paradigm views the organization of knowledge in the following terms. Certain theses are to be basic or foundational: like the axioms of geometry, they function as used for the justification of other theses without themselves needing or receiving any intrasystematic justification. Apart from these fundamental postulates, however, every other thesis of the system is to receive justification of a rather definite sort. For every nonbasic thesis is to receive its justification along an essentially linear route of derivation or inference from the basic (axiomatic, unjustified) thesis. There is a step-by-step recursive process, first of establishing certain theses by immediate derivation from the basic theses, and then of establishing further theses by sequential derivation from already established theses. In the setting of the Euclidean model every (nonbasic) established thesis is extracted from certain basic theses by a linear chain of sequential inferences.

On this approach to cognitive systematization, one would, with J. H. Lambert, construe such a system on analogy with a building whose stones are laid, tier by successive tier, on the ultimate support of a secure foundation.[1] Accordingly, the whole body of knowledge obtains the layered makeup reminiscent of geological stratification: a bedrock of basic theses surmounted by layer after layer of derived theses, some closer and some further removed from the bedrock, depending on the length of the (smallest) chain of derivation that links this thesis to the basic ones. In this way, the system receives what may be characterized as its *foundationalist* aspect: it is analogous to a brick wall, with a solid foundation supporting layer after successive layer. A prominent role must inevitably be allocated here to the idea of "relative fundamentality" in the *systematic* order—and hence also in the *explanatory* order of things the systematization reflects.[2]

In the setting of this Euclidean model of cognitive systematization, as we shall call it, every (nonbasic) established thesis is ultimately connected to certain basic theses by a linear chain of sequential inferences. These axiomatic theses are the foundation on which rests the apex of the vast inverted pyramid that represents the total body of knowledge.

With virtual unanimity, the earlier writers on cognitive systems construed the idea in terms of such a linear development from ultimate premises (or "first principles") that are basic both in fundamentality and in intelligibility, so that the order of exposition (or understanding) and the order of proof (or presupposition) run parallel.[3] The axiomatic development of our knowledge is seen in terms of both a deepening and a confirming of our knowledge, subject to the principle that clarification parallels rational grounding so that explanation replicates derivation.[4]

It does not matter for the fundamental structure of this Euclidean mode of systematization whether the inferential processes of derivation are deductive, and necessitarian, or somehow "inductive" and less stringently compelling. In this regard the label "Euclidean Model" is somewhat misleading. Nothing fundamental is altered by permitting the steps of derivative justification to proceed by means of probabilistic or plausibilistic nondeductive inferences. We are still left with the same fundamental pattern of systematization: a "starter set" of basic theses that provide the ultimate foundation for erecting the whole cognitive structure on them by the successive accretion of inferential steps. And so while modern epistemologists generally depart from a traditional Euclideanism in admitting nondeductive (e.g., probabilistic) arguments—abandoning the idea that the only available means of linking conclusions inferentially to premises is by means of steps that are specifically deductive in character—they still continue for the most part to accept at face value Aristotle's argumentation to the following effect:

> Some hold that, owing to the necessity of knowing the primary premisses, there is no scientific knowledge, Others think that there is, but that all truths are demonstrable. Neither doctrine is either true or . . . necessary. . . . The first school, assuming that there is no way of knowing other than by demonstration, maintain that an infinite regress is involved. . . . The other party agrees with them as regards knowing, holding that it is only possible by demonstration, but they see no difficulty in holding that all truths are demonstrated, on the ground that demonstration may be circular and reciprocal. Our own doctrine is that not all knowledge is demonstrative; on the contrary, knowledge of the immediate premisses is independent of demonstration. (*Posterior Analytics*, I, 3; 72b5–24 (tr. W. D. Ross)

The road thus indicated by Aristotle is followed by all those later epistemologists—by now legion—who feel constrained to have recourse to cognitive ultimates to serve as the basic, axiomatic premises of all knowledge. Accordingly, they commit themselves to a category of basic beliefs which, though themselves unjustified—or perhaps rather *self-justifying* in nature—can serve as a justifying basis for all the other, nonaxiomatic beliefs: the unmoved (or self-moved) movers of the epistemic realm, as Roderick Chisholm has characterized them.[5] With these epistemologists, axiomlike foundations still play a central role in the criteriology of truth, even when they are no longer used to provide a rigidly deductive basis.

It is almost impossible to exaggerate the influence and historical prominence this Euclidean model of cognitive systematization has exerted throughout the intellectual history of the West. From Greek antiquity through the eighteenth century it provided an ideal for the organization of information whose influence was operative in every field of learning. From the time of Pappus and Archimedes and Ptolemy in antiquity to that of Newton's Principia and well beyond into modern times the axiomatic process was regarded as the appropriate way of organizing scientific information. And this pattern was followed in philosophy, in science, and even in ethics—as the more geometrico approach of Spinoza vividly illustrates. For over two millennia, the Euclidean model has provided virtually the standard ideal for the organization of knowledge. Most early theorists of cognitive systematization viewed the geometric or Euclidean model as being so obviously appropriate that it can virtually be taken for granted. And a rigid insistence on this linear and hierarchical aspect of cognitive systems has continued to characterize the thinking of most recent writers on cognitive systematization. A particularly vivid example is the work of the German philosopher Hugo Dingler,[6] who characterized this linearity as the principle par-excellence of systematically ordered thought ("des geordneten Systemdenkens") and calls it the System Principle *tout court*. Such a view fails decisively to recognize that this approach characterizes only one particular mode of systematic thought—to be sure a very important one.

Now on the basis of a Euclidean model of cognitive systematization, a piece of knowledge must be fitted within the derivational scaffolding: it must either be "immediate knowledge" (forms a part of the axiomatic basis), or it must be "derived knowledge" (be justified by derivation from the axioms). Knowledge becomes a complex structure erected on a suitable foundation of basic facts. And even when they depart from Euclideanism in giving up the idea that the only available means of linking conclusions inferentially to premises is by means of steps that are specifically *deductive* in character, modern epistemologists still continue for the most part to accept at face value the argument of Aristotle's *Posterior Analytics* to the following effect:

> Some hold that, owing to the necessity of knowing the primary premisses, there is no scientific knowledge. Others think that there is. But that all truths are demonstrable. Neither doctrine is either true or . . . necessary. . . . The first school, assuming that there is no way of knowing other than by demonstration, maintain that an infinite regress is involved. . . . The other party agrees with them as regards knowing, holding that it is only possible by demonstration, but they see no difficulty in holding that all truths are demonstrated, on the ground that demonstration may be circular and reciprocal. Our own doctrine is that not all knowledge is demonstrative: on the contrary: knowledge of the immediate premisses is independent of demonstration.[7]

The road thus indicated by Aristotle is followed by epistemologists whose name is by now legion, who feel constrained to have recourse to cognitive ultimates to serve as the basic, axiomatic premises of all knowledge. Accordingly, they commit themselves to a category of basic belief's which, though themselves unjustified—or perhaps rather *self-justifying* in nature—can serve as justifying basis for all the other, nonaxiomatic beliefs. Thus axiomlike foundations still play a central role for these epistemologists, even when they are no longer used to provide a rigidly *deductive* basis. Committed to a quest for ultimate bedrock "givens" to provide a foundational basis on which the rest of the cognitive structure can be erected, these theorists are thus generally characterized as *foundationalists*. This foundationalist approach to cognitive justification views certain theses as self-evident—or immediately self-evidencing—and then takes these as available to provide a basis for the derivative justification of other beliefs (which can then, of course, serve to justify still others in their turn). It is committed to a quest for ultimate bedrock "givens" capable of providing a foundational basis on which the rest of the cognitive structure can be erected.

The foundationalist approach to knowledge has it that every discursive (i.e., reasoned) claim to truth requires truths as inputs. If a presumptively true result is to be obtained, the premises on which it rests must themselves be true (or assumed to be so). This foundationalist approach to epistemology is deep-rooted throughout the Western tradition from Aristotle through Descartes to the present day. It implements an ancient and enduring idea—based ultimately on the Greek concept of science as a Euclidean system—that truth is a structure that must have foundations. There must be a starter set of primitive (ungrounded, immediate, "intuitive") truths and, outside this special category, truths can only be established from or grounded upon other truths. We are given an essentially recursive—stepwise reductive—picture of the epistemic process of thesis justification. There is a special set that is the axiomatic starter-set of truths, and a grounding process for validating certain truth-claims in terms of others. The overall domain of truths is then to be built up by recursion.

Accordingly this foundationalist approach is subject to certain characteristic commitments, as follows:

1. There are two fundamentally distinct sorts of truths, the immediate and the derivative.
2. There is some privileged epistemic process which, like the cataleptic perception of the ancient Stoics or Descartes' clear and distinct intuitions of the mind, is capable of providing immediately evident truths. These initial "givens" are wholly nondiscursive and fixed invariants, which are sacred and nowise subject to reappraisal and revision.
3. All discursive epistemic processes—be they inductive or deductive—require an input of truths if truths are to be an output (which is exactly why an immediate, non-discursive route to justification must also be available.) Accordingly:
4. Nothing whatever that happens at the epistemically later stages of the analysis can possibly affect the starting-point of basic truths. They are exempt from any retrospective reevaluation in the light of new information or insights.

To a really surprising extent, modern epistemologists have, notwithstanding massive departures from classical modes of thought, continued remarkably faithful in their attachment to the central themes of the Euclidean model of cognitive systematization. They often begin with an adherence (at however schematic a level) to the ancient conception of knowledge as true, justified belief,[8] and join to this view the doctrine that the deductive approach to systematization affords the appropriate means of justification needed to implement this principle. They thus arrive at an essentially foundationalist view of the structure of knowledge.

The idea of immediate or basic or "protocol" truths of fact has a long and distinguished philosophical history that goes back to Aristotle and beyond. Such truths—it is held—are to be apprehended in some direct and fundamental way, typified by the immediate sensory apprehension of phenomenal colors or odors. Within the epistemic structure of our knowledge of truth, such basic truths are to serve as a foundation; other truths are made to rest on them, but they rest on no others: like the axioms of a deductive system they provide the ultimate support for the entire structure. Many epistemologists have held that truths—nay even mere probabilities—can be maintained only on a basis of certainty. One influential twentieth-century philosopher puts it as follows:

> If anything is to be probable [let alone definitely true), then something must be certain. The data which eventually support a genuine probability [for a warranted truth-claim], must themselves be certainties. We do have such absolute certainties, in the sense data initiating belief.[9]

This essentially axiomatic concept of truth finds its formal articulation in the Euclidean model with its pivotal reliance on a starter-set of basic truths. It underlies Aristotle's thesis that perception provides an ultimate stopping point where the generalizations of inductive reasoning can find a secure axiomatic foothold. It is the mainstay of the Stoic doctrine of "cataleptic" perception which provides failproof certainty. It provides the motive for Descartes' search—basic to the whole Cartesian program of methodological skepticism—for a secure, Archimedean point to serve as fulcrum for the lever of knowledge acquisition (the "clear and distinct" apprehensions of the mind). And modern epistemologists have continued remarkably faithful in their attachment to the central themes of the Euclidean model of cognitive systematization. It recurs again and again in recent epistemology—in Franz Brentano, in C. I. Lewis, in A. J. Ayer, in Roderick Chisholm and in many others.[10] Foundationalism, in sum, represents the predominant and most strikingly prominent approach to Western epistemology.

Cyclic Systematization: The Network— An Alternative to the Euclidean Model

But while the mainstream of the Western tradition in the theory of knowledge has unquestionably cast mathematics—and, in particular, geometry—in this paradigmatic role nevertheless there has almost from the very first there has been a succession of rebels sniping from the sidelines and advocating discordant views as to the proper systematic structure for the organization of scientific knowledge regarding how things work in the world. A small but constantly renewed succession of thinkers have steadfastly maintained that the traditional geometric model is not of sufficiently general applicability, and insisted that we must seek elsewhere for our governing paradigm of scientific systematization.

The major alternatives to the Euclidean model that have been supported most prominently have certain general features in common. The present discussion will focus on these shared features to portray what might count as a common-denominator version of these models. This common-denominator theory will be referred to as the network model. As we shall see, its approach to cognitive systematization also has an ancient and respectable lineage.

This network model sees a cognitive system as a family of interrelated theses, not necessarily arranged in a hierarchical arrangement (as with an axiomatic system), but rather linked among one another by an interlacing network of connections. These interconnections are inferential in nature, but not necessarily deductive (since the providing of "good explanatory accounts" rather than "logically conclusive grounds" is ultimately involved). What matters is that the network links theses in a complex pattern of relatedness by means of some (in principle variegated) modes of probative interconnections.

Some particularly vivid illustrations of the network approach to organizing information come from the social sciences, for example, the problem of textual interpretation and exegesis. Here there is no limit of principle to the width of the context of examination and the depth of analysis of detail. The whole process is iterative and cyclical; one is constantly looking back to old points from new perspectives, using a process of feedback to bring new elucidations to bear on preceding analyses. What determines correctness here is the matter of overall fit, through which every element of the whole interlocks with some others. Nothing need be more fundamental than anything else: there are no absolutely fixed pivot-points about which all else revolves. One has achieved adequacy when everything stands in mutual coordination with everything else.

An important advantage of a network system over an axiomatic one inheres in the former's accommodation of relatively self-contained subcycles, This absence of a rigidly linear hierarchical structure is a source of strength and security. In an axiomatic system a change anywhere ramifies into a change everywhere—the entire structure is affected when one of its supporting layers is removed. But with a network system that consists of an integrated organization of relatively self-sufficient components, certain of these components can generally be altered without dire repercussions for the whole.[11]

Another vivid illustration of the network approach to organizing information comes from textual interpretation and exegesis. Here there is no rigid, linear pattern to the sequence of consideration. The whole process is iterative and cyclical; one is constantly looking back to old points from new perspectives, using a process of feedback to bring new elucidations to bear retrospectively on preceding analyses. What determines correctness here is the matter of overall fit, through which every element of the whole interlocks with some others. Nothing need be more fundamental or basic than anything else: there are no absolutely fixed pivot-points about which all else revolves. One has achieved adequacy when—through a process that is continually both forward- and backward-looking—one has reached a juncture where everything stands in due mutual coordination with everything else.

Again, think of explanation-patterns incorporated in a (single-language) dictionary. Not every word of a language can be defined explicitly in that language. Ultimately this would lead to a circle. For if one starts with any finite list of words, by the time one comes to the last word, all the earlier words in the list will have been "used up," so that a circularity will have to appear in the final definition (if not before). Since standard dictionaries do actually define all the words they use in their definitions, they are guilty of circularity. (Dictionary makers attempt to counteract this problem by confining circularity as much as possible to definitions of words whose generally accepted meaning will be clear and unequivocal to the average reader.) Such circularity ultimately does no harm—most readers have a partial understanding of at least some units of the cycle, and a consideration of the interrelationships serves to clarify the whole series.

Two very different conceptions of explanatory procedure are thus at issue. The Euclidean approach is geared to an underlying conception of fundamentality or logical dependency in the Aristotelian sense of *priority,* in terms of what is supposed to be "better understood." Its procedure is one *reduction by derivation:* reducing derivative, "client" truths to their more fundamental "master" truths. By contrast, the network appeal is unreductive. Its mission is not explanation by derivation but explanation by interrelation. It merely seeks to coordinate the facts that are at issue. To speak figuratively, it views the structure of facts not as akin to an organizational manual for a military or bureaucratic organization, but to a novel that traces out a complex web of diversified mutual interrelationships among its cast of characters. In terms of practical advice about scientific procedure, the network model shifts the perspective from unidirectional dependency to reciprocal interconnection: Do not worry about discerning an ordering of fundamentality or dependency; worry about establishing interlinkages and mutual connections. Just find relationships among your parameters, and forget about which are the dependent and which the independent variables. Bear in mind the dictum from Goethe's "To Natural Science": "*Natur hat weder Kern noch Schale/Alles ist sie mit einem Male.*" The network theory does not deny that a cognitive system must have a structure (how else could it be a system!). But it recognizes that this structure need not be of the form of a rank ordering—that it can provide for the more complex interrelationships that embody a reciprocity of involvement. Its motto is Forget about establishing any all-embracing order of fundamentality in your explanatory efforts. It is no longer geared to the old hierarchical world picture that envisages a unidirectional flow of causality from fundamental to derivative orders of nature.

But, of course, our present concern is with network-patterned justifications rather than with network-patterned elucidations. A paradigmatic illustration is afforded by the explanations at issue with the workings of a closed physical or biological system, where every aspect of the modus operandi traces back into some of the others. Again, the sort of explanatory justifications at issue with solving a crossword-puzzle or breaking a code or interpreting an ancient document afford other appropriate examples. The key operative idea is that of explanation through systematization—that is, solving the puzzle by "getting all the pieces to fit properly" so that a "comprehensive picture emerges which 'makes sense' by putting everything into place."

The most critical points of difference that separate this network model of cognitive systematization from its Euclidean counterpart are as follows:

1. The network model dispenses altogether with the need for a category of basic (self-evident or self-validating) protocol theses capable of playing the role of axiomatic supports for the entire structure.
2. In the network model the process of justification need not proceed along a linear path. Its mode of justification is in general nonlinear, and can even proceed by way of (sufficiently large) cycles and circles.
3. The structure of the arrangement of theses within the framework of the network model need not be geological: no stratification of theses into levels of greater or lesser fundamentality is called for. (Of course, nothing blocks the prospect of differentiation; the point is that this is simply not called for by the *modus operandi* of the model.)
4. The network model accordingly abandons the conception of priority or fundamentality in its arrangement of theses. It replaces such *fundamentality* by a conception of *enmeshment*—in terms of the multiplicity of linkages and the patterns of interconnectedness with other parts of the net.

On the network-model approach to the organization of information, there is no attempt to erect the whole structure on a foundation of basic elements, and no necessity to move along some unidirectional path—from the basic to the derivative, the simple to the complex, or the like. One may think here of the contrast between the essentially linear order of an expository book, especially a textbook, and the inherently network-style ordering of an entire library or an encyclopedia. Again, the contrast between a taxonomic science like zoology or geology) and a deductive science (like classical celestial mechanics) can also help to bring out the difference between the two styles of cognitive organization.

A network system does, however, dispense with one advantageous feature that characterizes Euclidean systems par excellence. Since everything in a deductive system hinges on the axioms, these will be the only elements that require any independent support or verification. Once they are secured, all else is supported by them. The upshot is a substantial economy of operation: since everything pivots about the axioms, the bulk of our epistemological attention can be confined to them. A network system, of course, lacks an axiomatic basis, and so lacks this convenient feature of having one delimited set of theses to carry the burden of the whole system upon its shoulders. On the network model, the process of justification need not proceed along a linear path. Its mode of justification is in general nonlinear, and can even proceed by way of (sufficiently large) cycles. To be sure, while a network system gives up Euclideanism at the global level of its overall structure, it may still exhibit a locally Euclidean aspect, having local neighborhoods whose systematic structure is of such a format. Some of its theses may rest on others, and even do so in a rigorously deductive sense. For a network system may well contain various deductive compartments based on locally operative premises rather than globally operative axioms.

Notwithstanding its sharp differences from the axiomatic mode of development, a network approach to cognitive systematization does share some important features in common with the Euclidean organization of a "body of knowledge." The most important of these is that a network system can also exhibit the crucial facet of an axiom system in possessing more content than is overtly explicit. Parts of the network need not be explicitly presented, but may be demanded by systematic considerations to round out the systematic structure of otherwise available interrelationships. Like axiom systems, network systems can have both explicit and implicit components.

Moreover, neither approach is smoothly attuned to the linear narrative order of written exposition (whose main implicit connective is "and"). Axiomatic exfoliation requires constant flashbacks to a plurality of earlier stages; network exposition requires not only this, but also attention to spiderweblike interlinkages with heretofore unexamined elements. Both procedures invite a recourse to diagrammatic techniques that break through the confines of verbal sequentialism and suggest the use of mathematical structures rather than merely literary resources.

A heavy charge can be laid against the Euclidean model on grounds of the enormous hold it has established on philosophical and scientific thought in the West. Its exclusion of circles and cycles on grounds of their violating the prohibition of Aristotelian logic against "circular" inferences and reasonings impeded the conceptualization of reciprocal causal models in science for over two thousand years. For we know from Aristotle's own strictures that the core idea of coherentism was astir in his day. He criticizes those who hold of knowledge that

> it is only possible by demonstration, but they see no difficulty in holding that all truths are demonstrated, on the ground that demonstration may be circular and reciprocal. (*Posterior Analytics*, bk. I, ch. 3; 72b5–24)

Not until the present century have reciprocity cycles and feedback mechanisms become prominent not only in the domain of causal explanation but in information processing contexts as well. The growing prominence of the Network Model is attributable in no small part to the growing prominence of such examples of its operation in practice.

The two models indicated by the analogy of walls and networks, respectively, represent the two principal alternative lines of strategy in the systematization of knowledge that have been prominent in the epistemological tradition of the West. To be sure, these strategies both lend themselves to a virtually infinite variety in their detailed implementation. At the level of generality presently at issue, we are dealing with alternative program frameworks rather than concrete procedures. But even in their general aspect, the two programs envisage very different lines of approach. For foundationalism finds its rival—and its opponent(in the doctrine of coherentism for which the network model of cognitive systematization is determinative.

THE CONTRAST BETWEEN FOUNDATIONALISM AND COHERENTISM

The coherentist program of epistemology takes as its index of acceptability the overall fit of a presumptively acceptable thesis with the rest of what is presumptively acceptable. On its approach, the standard of the acceptability of theses is not their deductive derivability from some sacrosanct basis, but their systematic connectability with one another. For the coherentist, the network systematizability of best-fit considerations comes to provide the key testing-standard of the acceptability of truth-claims. (After all, scientists have always in fact tended to give weight in considering the acceptability of theories not merely to the status of their evidential support as distinct items considered in their own right, but also to the pattern of their connections with the rest of our knowledge.)

The two approaches to cognitive systematization are thus correlative with two quite distinct programs of confirmatory argumentation or "inductive reasoning." On the Euclidean approach, a thesis derives its evidential support from premises to which it is linked by deductive or probabilistic inference. On the network approach its security of probative standing is largely a matter of the systematic enmeshment of its overall interlinkage with other elements of the system. Very different theories of supportive reasoning are thus at issue: the one gives exclusive recognition to the weight of supportive evidence, the other goes beyond this in also recognizing—and indeed emphasizing—the probative efficacy of systematic interconnection.

The coherentist approach is thus quite prepared to dispense with any requirement for self-evident protocols to serve as the foundations of the cognitive system. The justification of a thesis of the systems will not proceed by derivations from the *axioms,* but comes to obtain through the pattern of its interrelationships with the rest. On a coherence approach, the truth is not seen as a treelike structure supported by a firm-rooted trunk, as it is on the foundationalist theory. Rather, it appears like a multitude of tied objects thrown into water; some of them rise to the surface themselves or are dragged there by others, some of them sink to the bottom under their own weight or through the pull of others.

And so, unlike foundationalism, coherentism dispenses with any appeal to basic, foundational truths of fact, diametrically opposing the view that knowledge of the actual, and even of the probable, requires a foundation of certainty. The coherence approach maintains that truth is accessible in the extralogical realm on the basis of best-fit considerations, without any foundation of certainty. (The qualifier "factual" occurs here because the instrumental need for the resources of logic is, of course, conceded, seeing that they are needed as a mechanism for best-fit judgments, since logic must be used in determining what does and does not "fit.") This entire procedure goes wholly counter to the classical epistemologists' axiomatic quest for basic or foundational truths.

Foundationalism might be caricatured as an essentially "aristocratic" view of truth: truths as such are not equal; there are certain "master" truths on which the other "client" truths are totally dependent. Negating the need for any axiomatic truths, the coherence theory sets out to implement a rather more "democratic" concept of treating all the truth-candidates not necessarily as equal but at any rate as all more or less plausible. The possibilities rendered available by the data at our disposal are treated with a complete "equality of opportunity"; truthfulness is determined from them only through a process of interaction—that is, by considerations of a best fit in terms of *mutual* accord and attunement (rather than their falling into the implicative captivity of certain basic prior truths).

The coherentist's approach effectively *inverts* that of the foundationalist. The foundationalist begins his epistemological labors with a very small initial collection of absolutely certain truths from which he proceeds to work *outward* by suitably *additive* procedures of supplementation to arrive at a wider domain of truth. By contrast, the coherentist begins with a very large initial collection of insecure pretenders to truth from which he proceeds to work *inward* by suitably *reductive* procedures of elimination to arrive at a narrower domain of truth. The expansive approach of the foundationalist is the very opposite of the contractive approach of the coherentist. The foundationalist is forced to a starting point of few but highly secure items, and immediately faces the dilemma of security *vs.* content. The coherentist bypasses this difficulty altogether. He begins with too many items—far too many since "the data" generally stand in a conflict of logical incompatibility—but he proceeds to undo the damage of this embarrassment of riches by suitably reductive maneuvers. This approach avoids altogether the characteristic perplexity of foundationalist epistemology in finding appropriate candidates to supply the requisitely secure foundation.

For the coherentist, knowledge is not a Baconian brick wall, with block supporting block upon a solid foundation; rather, an item of knowledge is like a node of a spider's web which is linked to others by thin strands of connection, each alone weak, but all together adequate for its support.

The essential difference between the coherence theory and any foundationalist approach to acceptance-as-true lies in the fact that on the latter line of approach every discursive (i.e., reasoned) claim to truth requires *truths* as inputs. If a (presumptively) true result is to be obtained, the premises on which it rests must themselves be true (or assumed to be so). The only strictly originative provider of *de novo* truths is the process that yields the "immediate" truths of the starter set. The decisive difference of the coherence theory is its capacity to extract (presumptive) truths discursively from a basis that includes no conceded truths whatsoever—that is, from data that are merely truth-candidates and not truths. The foundationalist requirement for basic truths is something that the coherence theory—proceeding as it does from a basis of data that need be neither compatible nor true—has been designed to overcome. The analysis seeks to provide a procedure for arriving at output truths without requiring any input truths as an indispensable staring basis. The motto Truth without true foundations may properly be inscribed on the banner of the coherence theory of truth.

The contrast between a foundationalist *Aufbau* of the domain of truth and the approach of the coherence analysis is set out graphically in Figure 7.1. This diagram makes plain the basic similarities between the two approaches, but also brings out their substantial differences, which are as follows:

1. On the foundationalist approach there are two distinct sorts of knowledge, the immediate and the derivative, while for the coherentist all knowledge is essentially of a piece.
2. On the foundationalist approach experience is called on to provide basic knowledge (in the form of certain immediately evident truths), while for the coherentist it only provides the "raw" data for knowledge.
3. On the foundationalist approach all discursive—inductive or deductive—processes require an input of known truths if truths are to be an output (which is exactly why an immediate, nondiscursive route to truth must be postulated). The coherence analysis differs fundamentally in this regard.
4. On the foundationalist approach the initial "givens" are wholly nondiscursive and fixed invariants, while on the coherence approach the "data" represent a *mixture* of experiential and discursive elements. (The "raw" data are, to be sure, nondiscursive, but for the coherentist they are only one part of the total data and are by no means fixed and

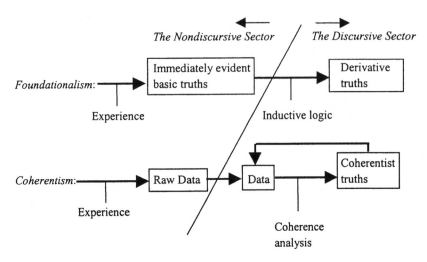

FIGURE 7.1 Foundationalism versus Coherentism in Factual Inquiry

sacred but subject to a cyclic process of reappraisal and revision.) In consequence:

5. On the foundationalist approach nothing whatever that happens at the epistemically later stages of the analysis can possibly affect the starting point of basic truths, while on the coherentist approach there is a feed-back loop through which the data themselves can be conditioned by the outcome of a coherence analysis (in other contexts) and their status is subject to reevaluation in the light of new insights regarding their plausibility.
6. Unlike the foundationalist approach, the coherence analysis does not require a sharp disparity in the treatment of particular and general propositions (between "observation statements" and "laws"). Its "data" for factual inquiry are subject to no particularity stipulations, unlike the usual "directly evident" givens of the foundationalists.[12]
7. On the foundationalist approach the body of "evidence" from which the reasoning proceeds must be self-consistent. The coherence analysis has no need for this unrealistic supposition.

The diagram also brings to the fore one further facet of coherentist method that merits special emphasis. Foundationalist inductivism adopts the basically *linear* systematization of reasoning typical of mathematics; here once a result is obtained one simply passes on to other matters—there is no need ever to return to the reappraisal or resubstantiation of something that has already been "established." But with coherentist inductivism the case is quite otherwise, as the feedback loop of the diagram illustrates graphically. Here there is a definite place for a dialectical process of cyclical structure, where one returns repeatedly to an item already "established." For the process of confirmation is now more complex, and a thesis might first appear on the status of a mere datum of low plausibility, later as one of higher plausibility, and ultimately even as a validated truth.

A coherence epistemology of truth-claims recognizes that the grounding of such claims cannot be evidential "all the way down," because evidentiation has to come to a stop. And it cannot be probabilistic "all the way down" because probabilities must ultimately come to rest on some proportionately factual basis or other. Rather, so coherentism insists, the validating process is not linear but cyclic. And coherentism's network conception of grounding offers a perfectly viable alternative here. Our concern with the validation of cognitive claims is not—or need not be—an issue of extracting "real knowledge" from "mere beliefs" by use of a philosopher's stone of epistemic justification. This entire reductive or extractive impetus is absent from coherentism's preoccupation with the *rational structure* of knowledge rather than with its *heuristic origin*. And here the network approach is not only a possible, but an emphatically attractive alternative to the foundationalist's axiomatic methodology.

The prospect of alternative modes of cognitive systematization has far-reaching implications. For only a short and easy step separates a program of cognitive systematization (be it of the Euclidean or of the network type) from a full-fledged criteriology of knowledge, namely that of accepting a thesis as an item of proper knowledge if it can be accommodated in smooth systematic fit with the remainder of our purported (or putative) knowledge. This step amounts to what might be called "the Hegelian Inversion"—the step from the implication claim KNOWABLE → SYSTEMATIZABLE, or its cognate [PRESUMABLY] KNOWN → [DULY] SYSTEMATIZED, to the reverse implication claim: [DULY] SYSTEMATIZED → [PRESUMABLY] KNOWN. Given this inversion, the criterion for knowledge becomes a matter of being "duly fitted into a systematization of the candidates for cognition." Systematicity becomes the controlling standard of truth and its work shifts from justification to validation.

The inherent difficulty of applying approach to the systematization of factual knowledge relates to the question of the sorts of theses that are to serve as basic (in the role of ultimate axioms). On the one hand, they must be very secure (certain, "self-evident" or self-evidencing) to be qualified to dispense with all need for further verification. But on the other hand, they will have to be enormously content-rich, since they must carry on their back the whole structure of knowledge. These two qualifications for the axiomatic role (content and security) clearly stand in mutual conflict with each other. This tension makes for a weak point, an Achilles heel that critics of the Euclidean model of knowledge have always exploited.

Thus consider—for the sake of illustration the sorts of theses needed for common-life claims about the realm of everyday experience. On the one hand, if I claim "I see a book" (say) or "I see an apple," my claims go far beyond their evidential warrant in sense perception. A real apple must, for example, have a certain sort of appearance from the (yet uninspected) other side, and a certain sort of (yet uninspected) subcutaneous makeup. And if anything goes wrong in these respects my claim that it was *an apple* I saw (rather than, say, a clever sort of apple substitute, or something done with mirrors and flashes of colored light) must be retracted. The claim to see *an apple*, in short, is not sufficiently secure. Its content extends beyond the evidence in hand in such a way that the claim becomes vulnerable and defeasible in the face of further evidence. If, on the other hand, one "goes for safety"—and alters the claim to "*It seems to me* that I see an apple" or "I *take myself* to be seeing an apple"—this resultant claim in the language of appearance is effectively immune from defect and so secure enough. But such assertions purchase this security at the price of content. For no amount of claims in the language of appearance—however extensively they may reach in terms of how things "appear" to me and what I "take myself" to be seeing, smelling, and so on—can ever issue in any theoretically guaranteeable result regarding what is *actually* the case in the world. While they themselves are safe enough, appearance-theses will fall short on the side of objective *content*. This dilemma of security versus content represents the Achilles heel of foundationalist theory of factual knowledge.

In the face of such difficulties in the foundationalist program, it deserves emphatic stress that the foundationalist approach to justification *does not represent* the only way of implementing the justificatory process inherent in the approach to knowledge as true, justified belief. It is important to recognize that the coherentist network model of cognitive systematization affords a perfectly workable alternative approach.

Problems of Foundationalism

The difficulty with any sort of foundationalism lies in the matter of foundations. If those foundations are phenomenal ("I seem to see a steel dagger before me," "I deem it highly likely that he will come") then they are in the final analysis autobiographical rather than objectively factual. If on the other hand they are factual ("There is an apple on the table"—which, of course, cannot have a golden interior—after all, apples just don't—even though I have not checked this) then it is hard to see how they can be foundational, since their own status is problematic. The foundationalist approach to factual knowledge faces a dilemma: the sort of claims at issue in objective knowledge lack cognitively unproblematic foundations, while cognitively unproblematic contentions do not yield objective facts.

On the one hand, they must be very secure (certain, "self-evident" or self-evidencing) to be qualified themselves to dispense with all need for further substantiation. But on the other hand, they will have to be enormously content-rich, since they must carry on their back the whole structure of knowledge. These two qualifications for the axiomatic role—fullness of content and probative security—obviously stand in mutual conflict with each other. This tension makes for a weak point which critics of the Euclidean model of knowledge have always exploited.

Insofar as any statement provides information about the world (e.g., "I now see a cat there" which entails, *inter alia*, that a cat is there), it is not invulnerable to a discovery of error; insofar as it is safeguarded (e.g., by some guarding-locution such as "I take it that . . .", so that we have "It now seems to me that I see a cat there" or "I am under the impression that I see a cat there"), the statement relates to appearance instead of reality and becomes denuded of objective content. Egocentric statements of phenomenological appearances may have the requisite security, but are void of objective information; claims regarding impersonal reality are objectively informative, but are in principle vulnerable.[13] The quest for protocol statements as a foundation for empirical knowledge has always foundered on this inherent tension between the two incompatible objectives of indubitable certainty on the one hand and objective factual content on the other.

Coherentism sets out to resolve this problem. Rather than proceeding linearly, by fresh deductions from several premises, it proposes to cycle round and round the same given family of prospects and possibilities, sorting out, refitting, refining until a more sophisticatedly developed and more deeply elaborated resolution is ultimately arrived at. The information-extracting process developed along these lines is one not of advance into new informative territory, but one of a cyclic reappraisal and revision of the old, tightening the net around our ultimate conclusion as we move round and round again, gaining a surer confidence in the wake of more refined reappraisals. This cyclic process of reappraisal is such that one can even—in suitable circumstances—dispense with the need for "new" data-inputs in an endeavor to squeeze more information out of the old. It is readily observed that this repeated reappraisal of claims is in fact closer to the processes of thought one generally employs in scientific reasoning.

The coherentist unhesitatingly endorses the traditional thesis that knowledge must be *justified* belief, construing this as tantamount to claiming that the known is that whose acceptance-as-true is adequately warranted *through an appropriate sort of systematization*. However, since the systematization at issue is viewed as being of the network type, the impact of the thesis is drastically altered. For we now envisage a variant view of justification, one which radically reorients the thesis from the direction of the foundationalists' quest for an ultimate basis for knowledge as a quasi-axiomatic structure. Now "justified" comes to mean not "derived from basic (or axiomatic) knowledge," but rather "appropriately interconnected with the rest of what is known."

The second prime difficulty of foundationalism relates to the grounding relationship which governs the derivational process. If this is to have the strength of deductive validity, affording a failproof guarantee of the conclusion relative to the premisses, then it will be unable to lead us beyond the logical information-content of the premisses. If, on the other hand, these inferential processes are in any way inductive or probabilistic—if the inferential derivations at issue are to be able (as they should be) to lead significantly beyond the information-content of the premisses themselves—then we are hard put to validate the claims to certainty-preservation that must be established on their behalf.

In the face of such difficulties in the foundationalist program, it deserves emphatic stress that the Euclidean approach to cognitive systematization does not represent the only way of implementing the justificatory demands of an approach to knowledge as true, *justified* belief. It is important to recognize that the network model of cognitive systematization affords the prospect of a workable alternative approach—one which leads straightaway to the coherentist program of epistemology. Above all, this variant approach would relieve us of the unhappy burden of a commitment to primary, absolutely sacrosanct propositions on the order of the "first principles" of traditional epistemology. A closer examination of the coherentist program is accordingly very much in order.[14]

Chapter 8

The Pursuit of Truth: Coherentist Criteriology

Synopsis

- *Coherentism bases its criteriology of truth in "coherence," that is, on cognitive systematicity. But what is it that gets systematized?*
- *The answer here lies in the idea of data—claims that are substantially plausible in the epistemic circumstances at issue, not certified truths but promising truth-candidates.*
- *The ultimate validation of a coherentist approach to truth assessment lies in the pragmatic utility of its results. For in the end it is the efficacy of its applications—alike in theoretical and practical matters—that serves as the arbiter of our cognition—the controlling monitor of adequacy for truth-estimation.*
- *The idea of truth with regard to general matters of fact as we deal with them in science is in fact something of an idealization.*
- *It is not the accomplished product of current inquiry but the hypothesized product of idealized inquiry. The idea of truth as idealized coherence accordingly provides a cogent specification for the nature of truth.*

The Coherentist Approach to Inquiry

Coherentism, as we have seen, views the systemic-interrelatedness of factual theses as the criterial standard of their acceptability. But just how does such a coherence criteriology work?

The overall stance of the theory is to be articulated in terms somewhat as follows:

> Acceptance-as-true is in general not the starting point of inquiry but its terminus. To begin with, all that we generally have is a body of *prima facie* truths, that is, presumption-geared and thereby merely plausible propositions that qualify as potential—perhaps even as promising—*candidates* for acceptance. The epistemic realities being as they are, these candidate-truths will, in general, form a mutually inconsistent set, and so exclude one another so as to destroy the prospects of their being accorded in toto recognition as truths pure and simple. The best that can be done in such circumstances is to endorse those as truths that best "cohere" with the others so as to "make the most" of the data as a whole in the epistemic circumstances at issue. Systemic coherence thus affords the criterial validation of the qualifications of truth-candidates for being classed as genuine truths. Systematicity thus becomes not just the organizer but the test of truth.

A coherentist epistemology thus views the extraction of knowledge from the plausible data by means of an analysis of best-fit considerations. Its approach is fundamentally holistic in judging the acceptability of every purported item of information by its capacity to contribute toward a well-ordered, systemic whole.

In general terms, the coherence criterion of truth operates as follows. One begins with a datum-set $S = \{P_1, P_2, P_3, ...\}$ of suitably "given" propositions. These data are not necessarily true nor even consistent. They are not given as secure truths, in a foundationalist's manner of theses established once and for all, but merely as *presumptive* or *potential* truths, that is, as plausible truth-*candidates*—and in general as *competing* ones that are mutually inconsistent. The task to which a coherentist epistemology addresses itself is that of bringing order into S by separating the sheep from the goats, distinguishing what merits acceptance as true from what does not.

The process at issue such a coherence analysis accordingly calls for the following epistemic resources:

1. *"Data"*: theses that can serve as *acceptance-candidates* in the context of the inquiry, plausible contentions which, at best, are merely *presumptively* true (like the "data of sense"). These are not certified truths (or even probable truths) but theses that are in a position to make some claims on us for acceptance: They are *prima facie* truths in the sense that we would inline to grant them acceptance-as-true *if* (and this is a very big IF) there were no countervailing considerations on the scene. (The classical example of "data" in this sense are those of perception and memory.)

2. *Plausibility ratings:* comparative evaluations of our initial assessment (in the context of issue) of the relative acceptability of the "data." This is a matter of their relative acceptability "at first glance" (so to speak) and *in the first analysis*, prior to their systematic evaluation. The plausibility-standing of truth-candidates is thus to be accorded without any prejudgments as to how these theses will fare *in the final analysis*.[1]

The process of coherence analysis consists in first classifying those relevantly plausible data into plausibility categories that reflect the nature of their source or substance and then using this as a rating index of plausibility. (On the side of substance, the various parameters of systematicity—generality, uniformity, and so forth—can themselves serve to provide indices of plausibility.) In cases of a conflict of inconsistency the next step is then a process of "primacy" that abandons sufficiently many of the data so that that consistency is restored, being grounded in this process by plausibility considerations in such a way as to preserve as much plausibility as possible by sacrificing what is less plausible to what is more so—exactly along the lines envisioned in the earlier chapters.

The governing injunction of a plausibilistic coherentism is: Maintain as best you can the overall fit of mutual attunement by proceeding—when necessary—to make the least plausible competitors give way to the more plausible. On this approach, a truth candidate comes to make good its claims to recognition as a truth through its consistency with as much as possible from among the rest of the data. The situation arising here resembles the solving of a jigsaw puzzle with superfluous pieces that cannot possibly be fitted into any maximally orderly, systemically integrated picture representing the "correct solution."

Accordingly, the general strategy of the coherence theory lies in three-step procedure:

- To gather in all of the relevant "data" (in the present technical sense of this term).
- To inventory the available conflict-resolving options that represents the alternative possibilities for achieving overall consistency.
- To choose among these alternatives by using the guidance of plausibility considerations, subject to the principle of minimizing implausibility.

In this way, the coherence theory implements F. H. Bradley's dictum that *system* (i.e., systematicity) provides a test-criterion most appropriately fitted to serve as arbiter of truth.

With plausible data as a starting point, the coherence analysis sets out to sift through these truth-candidates with a view to minimizing the conflicts that may arise. Its basic mechanism is that of best-fit considerations, which brings us to the second main component of coherentism, the machinery of "best fit" analysis:

> That family embracing the truth candidates which are maximally attuned to one another is to count—on this criterion of over-all mutual accommodation—as best qualified for acceptance as presumably true, implementing the idea of compatibility screening on the basis of "best-fit" considerations. Mutual coherence becomes the arbiter of acceptability which make the less plausible alternatives give way to those of greater plausibility. The acceptability-determining mechanism at issue proceeds on the principle of optimizing our admission of the claims implicit in the data, striving to maximize our retention of the data subject to the plausibilities of the situation.[2]

The process of deriving useful information from imperfect data is a key feature of the coherence theory of truth, which faces (rather than, like standard logic, evades) the question of the inferences appropriately to be drawn from an inconsistent set of premises. On this approach, the coherence theory of truth views the problem of truth-determination as a matter of bringing order into a chaos comprised of initial "data" that mingle the secure and the infirm. It sees the problem in transformational terms: incoherence into coherence, disorder into system, candidate-truths into qualified truths.

The interaction of observation and theory provides an illustration. Take grammar. Here one moves inferentially from the phenomena of actual usage to the framework of laws by the way of a best-fit principle (an "inference to the best systematization" as it were), and one checks that the cycle closes by moving back again to the phenomena by way of their subsumption under the inferred rules. Something may well get lost en route in this process of mutual attunement—for example, some of the observed phenomena of actual language use may simply be dismissed (say as "slips of the tongue"). Again, the fitting of curves to observation points in science also illustrates this sort of feedback process of discriminating the true and the false on best-fit considerations. The crucial point for present purposes is simply that a systematization can effectively control and correct data—even (to a substantial extent) the very data on which it itself is based.

The coherentist criterion accordingly assumes an entirely *inward* orientation; it does not seek to compare the truth candidates directly with "the facts" obtaining outside the epistemic context; rather, having gathered in as much information (and this will include also *misinformation*) about the facts as possible, it seeks to sift the true from the false *within* this body. On this approach, the validation of an item of knowledge—the rationalization if its inclusion alongside others within "the body of our knowledge"—proceeds by way of exhibiting its interrelationships with the rest: they must all be linked together in a connected, mutually supportive way (rather than having the form of an inferential structure built up on a footing of rock-bottom axioms). On the coherentist's theory, justification is not a matter of derivation but one of systematization. We operate, in effect, with the question: "justified" = "systematized." The coherence approach can be thought of as representing, in effect, the systems-analysis approach to the criteriology of truth.

THE CENTRAL ROLE OF DATA FOR A COHERENTIST TRUTH-CRITERIOLOGY

As coherentist methodologists see the process of inquiry, all that we generally have to begin with is a body of data, that qualify as potential—perhaps even as promising—*candidates* for truth. The epistemic realities being as they are, these candidate-truths, will in general, form a mutually inconsistent set so as to destroy their prospects of their being accorded in toto recognition as truths pure and simple. We are accordingly well advised to make the most of the data as a whole. Coherence thus becomes the critical *test* of the qualifications of truth candidates for gaining endorsement as genuine truths.

The concept of a *datum*, whose role is pivotal in coherentist methodology, is something of a technical resource. To be sure, the idea is one not *entirely* unrelated to the ordinary use of that term, nor to its (somewhat different) use among philosophers; yet it is significantly different from both. A datum is a plausible *truth-candidate*, a proposition to be taken not as true, but as potentially or *presumptively* true. It is a *prima facie* truth in exactly the sense in which one speaks of *prima facie* duties in ethics—a thesis that we would in the circumstances, be prepared to class as true provided that no countervailing considerations are operative. A datum is thus a proposition that one is to class as true *if one can*, that is, if doing so does not generate any difficulties or inconsistencies.

A datum is not *established* as true; it is backed only by a *presumption* that it may turn out true "if all goes well." It lays a claim to truth, but it may not be able to make good this claim in the final analysis. (A datum is thus something altogether different from the basic, protocol truths that serve as foundation for an *Aufbau* of truth in the manner of the logical positivists.)

In taking a proposition to be a datum we propose to class it as true *whenever possible*, recognizing that this may not be possible because some data may contradict one another, so that we must consider some of them as nontruths.[3] A datum is to be carried across the line from datahood to truth automatically whenever such a transfer is *unproblematic* (i.e., in no way involves a contradiction), and member of a group of data that meets with no rivalry from its fellows can immediately be accepted as true. Our stance toward data is unashamedly that of fair-weather friends: we adhere to them when this involves no problems whatsoever, but are prepared to abandon them in the face of difficulties—that is, whenever consistency maintenance so requires. But it is quite clear that such *loose* attachment to a datum is by no means tantamount to no attachment at all.

The following objection might be made: How can one speak of asserting a proposition merely as a datum but not as a truth? If one is to assert (accept, maintain) the proposition in any way at all, does one not thereby assert (accept, maintain) it *to be true?* The answer to be made here is simply a head-on denial. To maintain *P* as a datum, as *potentially* or *presumptively* factual, is akin to maintaining *P* as *possible* or as *probable:* in no case are these tantamount to maintaining the proposition as true. Putting a proposition forward as "possible" or "probable" commits one to claiming no more than that it is *"possibly* true" or *"probably* true." Similarly, to assert *P as a datum* is to say no more than that *P* is *potentially* or *presumptively* true—that it is a truth-candidate—but does not say that *P* is *actually* true, that it is a truth. As with assertions of possibility or probability, a claim of datahood definitely stops short—far short—of a claim to truth.

We in general *know* that data cannot be identified with truths—that some of them must indeed be falsehoods—because they are generally incompatible with one another. Truth-candidates—like rival candidates for public office—can work to exclude one another: they are mutually exclusive and victory for one spells defeat for the others. Candidate-truths are not truths pure and simple because it is of the very nature of the case that matters must so eventuate that some of them are falsehoods.

The conception of a datum certainly does not "open the floodgates" in an indiscriminate way. Not *everything* is a datum: the concept is to have *some* logicoepistemic bite. To be a datum is not just to be a proposition that *could conceivably* be claimed to be true but to be a proposition that (under the circumstances) can be claimed to be true with at least *some* plausibility: its claim must be well-founded. A proposition will not qualify as a datum without *some* appropriate grounding. Data are propositions that have a proper claim on truth, and we must distinguish between truth-claims that can reasonably be made from those that are merely theoretically possible. (Not every human being is a possible winner in a race but only those who are genuinely "in the running.") A datum is a proposition which, given the circumstances of the case, is a *real prospect for truth* in terms of the availability of reasons to warrant its truth-candidacy. A datum is not merely something that is "possibly true" or that is "true for all I know about the matter." To class a proposition as a datum is to take a definite and committal position with respect to it, so as to say "I propose to accept it as true in so far as this is permitted by analogous and possibly conflicting commitments elsewhere."

Virtually all writers on the subject take the position that, as one recent authority puts it, "acceptance as true is a necessary condition for acceptance as evidence."[4] On this view, if something can only count as genuine and actual evidence when it is established as true, the weaker conception of *potential* evidence—as contradistinguished from evidence as such (the *actual* evidence)—comes closer to our conception of the data. But although *some* basis is required for counting as a datum, this basis need not be very strong by the usual epistemic standards. To be a datum is to be a truth-*candidate*. And to count as such, a proposition need be neither an *actual* truth, nor a *probable* truth: it need only be "in the running" as a genuine candidate or a live possibility for truth. Such propositions are not truths, but make good a claim to truth that is at best tentative and provisional; by themselves they do not formulate truth but at most *indicate* it. Historically the tentatively of the experientially and mnemonically "given" has always been recognized, and the deliverances of our senses and our memory are the traditional examples of this circumstance of merely *purported* truth.

A more ambitious claim to truthfulness characterizes such a family of data than any one member of it. No imputation of truth (as opposed to presumptive or candidate truth) attaches to any *individual* datum, but there is a definite implicit claim that the "logical space" spanned by the data as a whole somewhere embraces the truth of the matter. The winner of a race must be sought among those "in the running." Taken individually, the data are merely truth-*presumptive*, but taken collectively as a family they are to be viewed as truth-embracing: admitting that they do not *pinpoint* where the truth lies, we are committed to the view that once we have all the relevant data in hand they *surround* the area where it is to be found.

One may well ask: "How can coherence with the data yield truth if the data themselves are not individually true? How can something so tentative prove sufficiently determinative?" The answer can only be found in the detailed development of the machinery; at this preliminary stage we can do no better than give a rough analogy. Consider again the earlier example of a jigsaw puzzle with superfluous pieces. It is clear here that the factor of "suitably fitting in" will be determinative of a piece's place (or lack of place) in the correct solution. Not the (admittedly tentative) status of the individual pieces but their mutual relationships of systematic accord is the determinative consideration. Thus the issue is not that of how mere truth-candidacy itself can serve to confer truth but of how it can help to determine truth on the basis of certain systematic considerations. Of course the exact nature of these considerations remains to be considered, but the key fact at this stage is that there is an important epistemic category of claims to presumptive or provisional verisimilitude that carry truth-indicative weight, while yet stopping well short of claims to truth as such.

Data—construed along the plausible lines as that of the preceding discussion—play a pivotal role in the articulation of a coherence-based criterion of truth. The entire drama of a coherence analysis is played out within the sphere of propositions, in terms of the sorts of relationships they have to one another. Now there is—in any event—enough merit in a correspondence account of truth that an appropriate consonance must obtain between "the actual facts of the matter" and a proposition regarding them that can qualify as true to the criticism "Why should mere coherence imply truth?" One can and should reply: What is at issue here is not *mere* coherence, but coherence *with the data*. It is not with bare coherence as such (whatever that would be) but with data-directed coherence that a truth-making capacity enters on the scene. But, of course, datahood only provides the building blocks for truth-determination and not the structure itself. Coherence plays the essential role because it is to be through the mediation of coherence considerations that we move from truth-candidacy and presumptions of factuality to truth as such. And the procedure is fundamentally noncircular: we need make no imputations of truth at the level of data to arrive at truths through application of the criterial machinery in view.

The concept of the datahood is the crux of the coherence theory of truth. It serves to provide an answer to the question "Coherent with what?" without postulating a prior category of fundamental truth. It provides the coherence theory with grist to its mill that need not itself be the product of some preliminary determinations of truth. A reliance on data makes it possible to contemplate a coherence theory that produces truth not *ex nihilo* (which would be impossible) but from a basis that does not itself demand any prior determinations of truthfulness as such. A coherence criterion can, on this basis, furnish a mechanism that is *originative* of truth—that is, it yields truths as outputs without requiring that truths must also be present among the supplied inputs.

A novel or science fiction tale or indeed any other sort of made-up story can be perfectly coherent. To say simply that a proposition coheres with *certain* others is to say to little. *All* propositions will satisfy this condition, and so it is quite unable to tell us anything that bears on the question of truth. It would be quite senseless to suggest that a proposition's truth resides in "its coherence" alone. Its coherence is of conceptual necessity a relative rather than an absolute characteristic. Coherence must always be coherence *with something*: the verb "to cohere with" requires an object just as much as "to be larger than" does. We do not really have a coherence theory in hand at all, until the *target domain* of coherence is specified. Once this has been done, we *may* very well find that the inherent truth-indeterminacy of abstract coherence—its potential failure to yield a unique result—has been removed. Now it must be said in their defense that the traditional coherence per se, but have insisted that it is specifically "coherence with our experience" that is to be the standard of truth.[5] The coherence theory of the British idealists has never abandoned altogether the empiricist tendency of the native tradition of philosophy.

What is clearly needed is a halfway house between coherence with *some*—that is, *any*—propositions (which would be trivial) and coherence with *true* propositions (which would be circular with a criterion for truth). Essentially, what is needed is coherence with somehow "the right" propositions. The coherence at issue in a coherence theory of truth must be construed as involving all in some way *appropriately qualified* propositions. This line of thought poses a task central to the construction of a workable coherence theory: that of specifying just what propositions are at stake when one speaks of determining the truth of a given proposition in terms of its "coherence *with others.*" Which others are at issue? It was, of course, for the sake of a satisfactory answer to this question that our approach made its crucial resort to the key concept of an experientially grounded *datum* that concerned us so elaborately in the preceding section.

It is sometimes objected that coherence cannot be the standard of truth because there we may well arrive at a multiplicity of diverse but equally coherent structures, whereas truth is of its very nature conceived of as unique and monolithic. Bertrand Russell, for example, argues in this way:

> There is no reason to suppose that only *one* coherent body of beliefs is possible. It may be that, with sufficient imagination, a novelist might invent a past for the world that would perfectly fit on to what we know, and yet be quite different from the real past. In more scientific matters, it is certain that there are often two or more hypotheses which account for all the known facts on some subject, and although, in such cases, men of science endeavour to find facts which will rule out all the hypotheses except one, there is no reason why they should always succeed.[6]

One must certainly grant Russell's central point: however the idea of coherence is articulated in the abstract, there is something fundamentally undiscriminating about coherence taken by itself. Coherence may well be—nay certainly is—a descriptive feature of the domain of truths: they cohere. But there is nothing in this to prevent propositions other than truths from cohering with one another: Fiction can be made as coherent as fact: truths surely have no monopoly of coherence. Indeed "it is logically possible to have two different but equally comprehensive sets of coherent statements between which there would be, in the coherence theory, no way to decide which was the set of true statements."[7] In consequence, coherence cannot of and by itself discriminate between truths and falsehoods. Coherence is thus seemingly disqualified as a means for *identifying* truths. Any viable coherence theory of truth must make good the claim that despite these patent facts considerations of coherence can—somehow—be deployed to serve as an indicator of truth.

But the presently envisioned theory averts these difficulties. It looks not to coherence in and of itself a criterion of truth, but to coherence with the date of experience. It thus renders Russell's objection effectively irrelevant.[8]

On Validating the Coherence Approach

But how can it be shown that the specifically coherentist approach to cognitive systematization meets the demands for an effective standard of quality control? The argument here has two stages: (1) recalling our earlier thesis that the coherentist approach to cognitive systematization can assimilate the standard mechanisms of scientific method, and then (2) noting the dramatic efficacy of science vis-à-vis any even remotely available alternative candidate as a mechanism of prediction and control over nature. One can thus invoke on coherentism's behalf the pragmatic efficacy of science, holding that coherentist accommodation of scientific method gains to the credit of a coherentist approach the dramatic success of science in realizing its conjoint purposes of explanation, prediction, and control over nature. On this perspective, the pragmatic warrant of coherentism is seen to reside in its capacity to serve as organon of scientific reasoning. But how can this capacity be made manifest?

It is sometimes suggested that coherentism's picture of interlocked circles of the theoretical and applicative validation of cognitive systems portrays the process of system-validation in the essentially timeless terms customary in epistemological discussions. However, this *static* view of system-validation needs to be supplemented—indeed corrected—by considering the issue in its *temporal and developmental* aspect. The atemporal relationships of probative justification must be augmented by examining the justificatory bearing of the historical dynamics of the matter—the evolutionary process of system development. After all, the articulation of cognitive systems is a matter of historical dynamics of the matter—the evolutionary process of system development. After all, the articulation of cognitive systems is a matter of historical development, of repeated efforts at improvements in systematizing in the light of trial and error. We are faced with a fundamentally *repetitive* process of the successive revision and sophistication of our ventures at cognitive systematization, a process that produces by way of iterative elaboration an increasingly satisfactory system, one that is more and more adequate in its internal articulation or effective in its external applicability. There are iterative cycles of tentative systematizations followed by resystematizations in the light of the feedback provided by its utilization for theoretical application and practical implementation.

This dynamic process is depicted in the diagram given in Figure 8.1, which presents the cycle at issue in an *historical* perspective, regarding it as a developmentally iterative feedback process. What is at stake is not just *retrospective reappraisal* in the theoretical order of justification, but an actual *revision or improvement* in the temporal order of development.

This sequential and developmental process of historical mutation and optimal selection assures a growing conformation between our systematizing endeavors and "the real world." In the final analysis our systematized cognition fits the world for the same reasons that our eating habits do: both are the product of an evolutionary course of selective development. It is this evolutionary process that assures the *adaequatio ad rem* of our system-based claims to knowledge. The legitimating process at issue here is not only a matter of a static pattern of relationships in the probative order of rational legitimation, it also reflects a temporal and developmental process of successive cyclic iterations

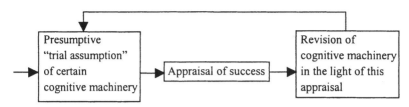

FIGURE 8.1 The Feedback Cycle of Legitimation

where all the component elements become more and more attuned to one another and pressed into smoother mutual conformation. And the pivotal feature of the process is its concern not with the substantiation of theses *or* propositions themselves but rather with the validation of the *methods* that we use for substantiating them.

This evolutionary development of intellectual methodologies proceeds by *rational* selection. As changes come to be entertained (within the society) it transpires that one "works out for the better" relative to another in terms of its fitness to survive because it answers better to the socially determined purposes of the group. Just what does "better" mean here? This carries us back to the Darwinian perspective. Such a legitimation needs a standard of survivalistic "fitness." And this normative standard is provided by considerations of *theoretical adequacy and applicative practice*, and is inherent in the use to which conceptual instrumentalities are put in the rational conduct of our cognitive and practical affairs. Our legitimation of the standard probative mechanisms of inquiry regarding factual matters began with the factor of pragmatic success and subsequently transmuted this into an issue of Darwinian survival. As the discussion has already foreshadowed at many points, it is clearly the *method of scientific inquiry* that has carried the day here. The mechanisms of scientific reasoning clearly represent the most developed and sophisticated of our probative methods. No elaborate argumentation is necessary to establish the all-too-evident fact that science has come out on top in the competition of rational selection with respect to alternative processes for substantiating and explaining our factual claims. The prominent role of the standard parameters of systematization in the framework of scientific thought thus reflects a crucial aspect of their legitimation.

The methodological directives that revolve about the ideal of systematicity in its regulative role ("Of otherwise co-eligible alternatives, choose the simplest!," "Whenever possible invoke a uniform principle of explanation or prediction!," etc.) form an essential part of the methodological framework (the procedural organon) of a science. Experience has shown these methodological principles to be rooted in the functional objectives of the enterprise, being such as to conduce efficiently to its realization of its purposes. We have every reason to think that an abandonment of these regulative principles, while not necessarily spelling an abandonment of science as such, would make hopelessly more difficult and problematic the realization of its traditional goals of affording intellectual and physical control over nature. Considerations of functional efficiency—of economy of thought and praxis—militate decisively on behalf of the traditional principles of scientific systematization.

The key considerations are *effectiveness* and *efficiency,* purposive adequacy and functional economy, acceptability of product and workability of procedure. (And systematicity is, of course, an ideal vehicle here in its stress on simplicity, regularity, uniformity, etc., all of which have to do with the minimizing of unnecessary complications and the pursuit of intellectual parsimony.) A quasi-economic dialectic of *costs* and *benefits* is operative here. And the question of system-choice can ultimately be seen as a matter of "survival of the fittest," with *fitness* ultimately assessed in terms of the theoretical and practical objectives of the rational enterprise. Legitimation is thus evidenced by the fact of survival through historical vicissitudes.

To be sure, there are a *variety* of approaches to the problem of systematizing "how things work in the world." The examples of such occult cognitive frameworks as those of numerology (with its benign ratios), astrology (with its astral influences), and black magic (with its mystic forces) indicate that alternative explanatory frameworks exist, and that these can have very diverse degrees of merit. Thus the orthodox scientific approach to cognitive systematization is simply one alternative among others, and it does not have an irrevocably absolute foothold on the very constitution of the human intellect, nor indeed any sort of abstract justification by purely "general principles." Its legitimation is not a priori and absolute, but a posteriori and experientially determined.

It is not difficult to give examples of the operation of Darwinian processes in the domain of the instrumentalities of cognitive systematization. The intellectual landscape of human history is littered with the skeletal remains of the extinct dinosaurs of this sphere. Examples of such defunct methods for the acquisition and explanatory utilization of information include: astrology, numerology, oracles, dream-interpretation, the reading of tea leaves or the entrails of birds, animism, the teleological physics of the Presocratics, and so on. There is nothing intrinsically absurd or contemptible about such unorthodox cognitive programs; even the most occult of them have a long and not wholly unsuccessful history. (Think, for example, of the long history of numerological explanation from Pythagoreanism, through Platonism, to the medieval Arabs, down to Kepler in the Renaissance.) But there can be no question at this historical juncture that science has won the evolutionary struggle among various modes of methods of cognitive procedure, and this more than anything else makes it manifest that the inherent coherentism of the orthodox scientific approach to cognitive systematization satisfies the requirement of pragmatic efficacy.

It makes perfectly good sense to ask: Why should our scientific deliberations proceed in the usual way—with reference to the pursuit of systematicity, and so on? And it is possible to answer this question along two seemingly divergent routes:

1. the pragmatic route: it is efficient, effective, successful, "it works," and so on.
2. the theoretical route: it is rationally cogent, cognitively satisfying, aesthetically pleasing,[9] conceptually "economical," and so forth.

But the divergence here is only seeming, for Darwinian considerations assure that in the course of time the two maintain a condition of convergent conformity.

The merit of entrenched cognitive tools lies in their (presumably) having established themselves in open competition with their rivals. It has come to be shown before the tribunal of bitter experience—through the historical vagaries of a Darwinian process of selection—that the accepted methods work out most effectively in actual (albeit *rational* rather than *natural* selection) practice vis-à-vis other tried alternatives.

The overall line of validation for a probative methodology of coherence involves a double circle, combining a theory-internal cycle of theoretical self-substantiation with a theory-external cycle of pragmatic validation—as is illustrated in Figure 8.2. Here Cycle I represents the theoretical/cognitive cycle of *theoretical consistency* between regulative first principles and their substantive counterparts, and Cycle II, the practical/applicative cycle of *pragmatic efficacy* in implementing the substantive results of the first principles.

Accordingly, the overall legitimation of a methodology for the substantiation of our factual beliefs must unite two distinctive elements: (1) an apparatus

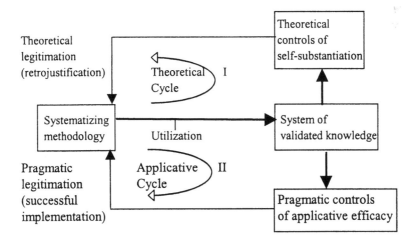

FIGURE 8.2 The Twofold Cycle of the Legitimation of Systematizing Methodology

of systematic coherence at the theoretical level (a coherence in which factual presumptions and metaphysical presuppositions both will play a crucial part), and (2) a controlling monitor of considerations of pragmatic efficacy at the practical level. Neither can appropriately be dispensed with for the sake of an exclusive reliance on the other. The proof of the theoretical pudding must, in the final analysis, lie in the applicative eating, by monitoring the adequacy of our procedures of cognitive systematization through an assessment of their applicative success in prediction and control over nature.[10]

The legitimative process at issue thus relies on an appropriate fusion of considerations of theory and praxis. It is a complex of two distinct but interlocked cycles—the *theoretical* cycle of cognitive coherence and the *pragmatic* cycle—the *theoretical* cycle of cognitive coherence and the *pragmatic* cycle of applicative effectiveness. Only if both of these cycles dovetail properly—in both the theoretical and the applicative sectors—can the overall process be construed as providing a suitable rational legitimation for the cognitive principles at issue. The symbiotic and mutually supportive nature of the enterprise is fundamental: its structure must afford a systematic union in which *both* methods and theses are appropriately interlinked. Legitimation once more inheres in an appropriate sort of systematization in which both cognitive and ontological factors play a role.

On this approach, the strictly *theoretical* aspect of explanation and understanding is coordinate in important in the teleology of science with its *pragmatic* aspect of "control over nature." Indeed control—throughout the range from prediction as minimal control (the adequate alignment of our own expectations) to the more elaborately modificatory change in the course of nature through effective interaction—comes to be seen in the pivotal role of the final arbiter of adequacy. This aspect of the cognitive centrality of control over nature leads us to an *interventionist* theory of knowledge, one which sees the issue of *monitoring the adequacy* of our theorizing to reside ultimately on the side of efficacy in application.[11]

The conformity between the regulative presumptions and other methodological instrumentalities of inquiry and its *results* is not guaranteed by a preestablished harmony. Nor is it just a matter of contingent good luck. It is the product of an evolutionary pressure that assures the conformation of our systematizing efforts and the real world under trial and error subject to controlling constraint of applicative success (pragmatic efficacy). The evolutionary process assures the due coordination of our cognitive systematizing with the "objective" workings of a nature that is inherently indifferent to our purposes and beliefs.

It is clear that if the legitimation of the regulative principles of cognitive systematization is construed as proceeding along such pragmatic/evolutionary lines, then these principles come to stand on an ultimately factual footing—one that is a posteriori and contingent. The "first principles" by whose means we constitute our factual knowledge of nature (uniformity, simplicity, and the other parameters of systematization that make up our guidelines to plausibility) are themselves ultimately of an a posteriori and factual standing in point of their controlling force. Seemingly serving merely in the role of *inputs* to inquiry, they emerge in the final analysis as its *products* as well, and accordingly have a contingent rather than necessary status. For then the legitimation of our methodological guidelines to cognitive systematization in the factual domain is ultimately not a matter of abstract theoretical principle, but one of *experience*.

Our "first principles" of cognitive systematization have no claims to inherent "necessity." Conceivably things *might* have eventuated differently—even as concerns the seemingly a priori "first principles of our knowledge." (Why didn't they so eventuate? The question "Why these principles rather than something else?" is *not* illegitimate—it is indeed answerable in terms of the overall double-circle of methodological legitimation.)

Are such first principles a priori and analytic (a part of the "conceptual schema" of our science) or are they a posteriori and synthetic (a product of scientific inquiry)? This question now looks naive—for they are both. The idea of a feedback cycle of evolutionary legitimation indicates that we are ill advised to put the question in terms of a logically tidy yes-or-no. Such first principles are thus "first" only in the first analysis. Their theory-internal absoluteness is deceptive—it represents but a single phase within the historical dialectic of evolutionary legitimation. They do not mark the dead-end of a *ne plus ultra*.

To be sure, all this is to say no more than that circumstances *could* arise in which even those very fundamental first principles that define for us the very idea of the intelligibility of nature might have to be given up. In a world sufficiently different from ours reliance on simplicity, uniformity, etc. could conceivably prove misleading and cognitively counterproductive. But to concede the *possibility* is not, of course, to grant the likelihood—let alone the reality—of the matter as regards our actual world. The first principles at issue are so integral a component of *our* rationality that we cannot even conceive of *any* rationality that dispenses with them: we can conceive *that* they might have to be abandoned, but not *how*.[12]

The very circumstance that these principles are *in theory* vulnerable is a source of their strength *in fact*. They have been tried in the history of science—tried long, hard, and often—and yet not found wanting. They are founded on a solid basis of trial in the harsh court of historical reality. Forming, as they do, an integral component of the cognitive methods that have evolved over the course of time it can be said that for them—as for all our other strictly methodological resources—"*die Weltgeschichte ist das Weltgericht*." A quasi-Lamarckian process of *rational* selection is the key to quality control in the cognitive domain.[13]

Ideal Coherence

In general terms, a coherence epistemology of truth has it that a proposition p is true iff it forms part of a suitably formed set of optimally coherent propositions. On such an approach, truth-recognition is accorded to generality-encompassing propositions regarding matters of fact propositions to reflect simply local characteristics of their own, but an ultimately global feature of their place in a larger context so that truth-qualification is not an isolated but a contextual characteristic.

A common objection to the coherence theory of factual truth is that the linkage of coherence to truth is simply too loose for coherence to provide the definitive standard of truth. As one writer put it some years ago:

> It is quite conceivable that the coherence theory is a description of how the truth or falsehood of statements comes to be known rather than an analysis of the meaning of "true." ... One might agree that a given statement is accepted as true in virtue of standing in certain logical relations to other statements; still it would not follow that in calling it true one means to ascribe to it those relations.[14]

Here we have the oft-repeated standard reservation regarding a coherence theory of truth: "Coherence may perhaps be suitable as a criterion for the true, but certainly not as a definitional standard of truth."[15]

It can, however, be shown that if one is prepared to consider coherence in an idealized perspective—as optimal coherence with a perfected database, rather than as a matter of manifest coherence with the actual data at our disposal—then an essential link between truth and coherence emerges.

Supporters of a coherentist standard of truth must be able to establish that this criterion is duly consonant with the definitional nature of truth. For there ought rightfully to be a continuity of operation between our evidential criterion of acceptability-as-true and the "truth" as definitionally specified. Any really satisfactory criterion must be such as to yield the real thing—at any rate in sufficiently favorable circumstances. Fortunately for coherentism, it is possible to demonstrate rigorously that truth is tantamount to ideal coherence—that a proposition's being true is in fact equivalent to its being optimally coherent with an ideal data base.

This circumstance has far-reaching implications. Given that "the real truth" is guaranteed only by ideal coherence, we have no categorical assurance of the actual correctness of our coherence-guided inquiries, which are, after all, always incomplete and imperfect. This is amply substantiated by the history of science, which already shows that the "discoveries" about how things work in the world secured through scientific coherentism constantly require adjustment, correction, replacement. It is only too obvious that we cannot say that our coherence-grounded scientific theorizing furnishes us with the real (definitive) truth, but just that it furnishes us with the best estimate of the truth that we can achieve in the circumstances at hand.

In characterizing a claim as true, we indicate that what it states corresponds to the facts, so that its assertion is in order. But while this factually ("stating which is the case," "corresponding to the facts") is what truth is all about, we cannot apply or implement it as such: it does not provide a basis on which the truth of claims can be determined. Thus while we have the equation

> to be true = to be (assertible as) factually correct

we also will have

> to be determinable as true = capacity to meet certain conditions that serve as adequate indicators of factual correctness.

But these conditions of truth-determination are in their very nature conditions whose full realization is a matter of idealization—that is, the conditions are such as to obtain only in ideal circumstances.

As noted above, it is of the traditional objections to coherentism in that the coherence theory is unable to deal with the "problem of error"—to be able to explain how it is that what is thought to be true might yet actually prove false. But, of course, the consideration that truth is a matter of ideal —rather than actually realized) coherence at once sweeps this difficulty aside.

Definitive knowledge—as opposed to "merely putative" knowledge—is the fruit of perfected inquiry. Only here, at the idealized level of perfected science, could we count on securing the real truth about the world that "corresponds to reality" as the traditional phrase has it. Factual knowledge at the level of generality and precision at issue in scientific theorizing is akin to a perfect circle. Try as we will, we cannot quite succeed in producing it. We do our best and call the result knowledge—even as we call that carefully drawn "circle" on the blackboard a circle. But we realize full well that what we currently call scientific knowledge is no more authentic (perfected) knowledge than what we call a circle in a geometry diagram is an authentic (perfected) circle. Our "knowledge" is in such cases no more than our best estimate of the truth of things. Lacking the advantage of a God's-eye view, we have no access to the world's facts save through the mediation of (inevitably incomplete and thus always potentially flawed) inquiry. All we can do—and what must suffice us because indeed it is all that we can do—is to do the best we can with the cognitive state of the art to estimate "the correct" answers to our scientific questions.

Truth as an Idealization

Definitive truth is realizable only by way of idealization: actual inquiry presents us with estimates of truth, in matters of scientific theorizing, the real truth as such is realizable only under ideal conditions.

We have no workable alternative to presuming that our science as it stands here and now does not present the real truth, but only estimates it. "Our truth" in matters of scientific theorizing is not—and presumably never actually will be—the final truth. However confidently science may affirm its conclusions, the realization must be maintained that its declarations are provisional, tentative—subject to revision and even to outright abandonment and replacement. But all this is not, of course, any reason to abandon the link to truth at the teleological level of aims, goals, and aspirations. The pursuit of scientific truth, like the pursuit of happiness, or for that matter any other ideal in life, is not vitiated by the consideration that its full realization is not a matter of the practicalities of this imperfect world.

The ideal of a state-of-the-art in science that attains definitive finality in empirical inquiry is pie in the sky. It represents an idealization and not a matter of the practical politics of the epistemic domain. But it affords the focus imaginarius whose pursuit canalizes and structures our actions. It represents the ultimate objective (goal) of inquiry—the destination of an incompletable journey. The conception of capital-T Truth thus serves a negative and fundamentally regulative role to mark the fact that the place we have attained falls short of our capacity actually to realize our cognitive aspirations. It marks a fundamental contrast that regulates how we do and must view our claims to have got at the truth of things. It plays a role somewhat reminiscent of the functionary who reminded the Roman emperor of his mortality in reminding us that our pretentions to truth are always vulnerable. Contemplation of this ideal enables us to maintain the ever-renewed recognition of the essential ambiguity of the human condition as suspended between the reality of imperfect achievement and the ideal of an unattainable perfection.

We must suppose that science does not and cannot ever actually attain an omega-condition of final perfection. The prospect of fundamental changes lying just around the corner can never be eliminated finally and decisively.

A basic analogy obtains as per the following proportion:

putative knowledge : actual inquiry :: genuine knowledge : ideal inquiry

Rational inquiry is the pursuit of an unattainable ideal—the ideal of "the real truth" about laws of nature as yielded by perfected science. Actual inquiry is no more than our best effort in this direction, and the information (the putative knowledge) it yields is no more than our best available estimate of the real truth of things.

To abandon this conception of the truth as such—rejecting the idea of an "ideal science" which alone can properly be claimed to afford a grasp of reality—would be to abandon an idea that crucially regulates our view as to the nature and status of the knowledge we lay claim to. We would then no longer be constrained to characterize our truth as merely ostensible and purported. And then, did our truth not exhibit any blatant inherent imperfections, we would be tempted to view it as real, authentic and final in a manner which as we at bottom realize it does not deserve.

The lesson of these deliberations is clear: "the real truth" in scientific matters is not the actual product of current inquiry, but the hypothesized product of idealized inquiry. The conditions of objectivity and definitiveness we have in view in relation to "the real truth" are not satisfiable in the circumstances in which we do and unavoidably must labor. In this regard we have no realistic alternative but to regard the truth in these matters as having an idealization.

Conceptions such as definitive knowledge and truth in matters of scientific theorizing are idealizations geared to the idea of completed and perfected science. And this—all too obviously—is not something we have in hand. It is no more than a regulative ideal that guides and directs our efforts in question-resolving inquiry.

Such an ideal is not (or should not be) something that is unhealthily unrealistic. It should produce not defeatism and negativity toward our efforts and their fruits but rather a positive determination to do yet better and fill a half-full barrel yet fuller. It should given not our expectations of realized achievement but our aspirations, and should give not our demands but our hopes. Endowing us with a healthy skepticism toward what we actually have in hand, it should encourage our determination to further improvements and should act not as an obstacle but a goad.

And here, as elsewhere, we must reckon appropriately with the standard gap between aspiration and attainment. In the practical sphere—in craftmanship, for example, or the cultivation of our health—we may strive for perfection, but cannot ever claim to attain it. And the situation in inquiry is exactly parallel with what we encounter in such other domains—ethics specifically included. The value of an ideal, even of one that is not realizable, lies not in the benefit of its attainment (obviously and ex hypothesi!) but in the benefits that accrue from its pursuit. The view that it is rational to pursue an aim only if we are in a position to achieve its attainment or approximation is mistaken; it can be perfectly valid (and entirely rational) if the indirect benefits of its pursuit and adoption are sufficient—if in striving after it, we realize relevant advantages to a substantial degree. An unattainable ideal can be enormously productive. And so, the legitimation of the ideas of "perfected science" lies in its facilitation of the ongoing evolution of inquiry. In this domain, we arrive at the perhaps odd-seeming posture of an invocation of practical utility for the validation of an ideal.

In any case, this view of truth as idealization enables us to establish a smoothly consonant continuity between a truth-criterion based on the standards of coherentism and the classic definitional view of the nature of truth as based on that correspondence with fact (*adaequatio ad rem*) which is then assured under altogether idealized circumstances.

And so the defense of coherentism against traditional objections comes clearly to view.

The objection that coherence does not determine truth—that "truth" is not to be defined in terms of coherence—is met by accepting coherence as a criterion for (rather than definition of) truth.

The objection that all sorts of fictions can be just as coherent as the truth is to be met with the acknowledgement that what is at issue is not merely coherence as such but coherence *with* something, namely. with the plausible data—with information that has initial credibility on the basis of its inherent plausibility or the reliability of its sources.

The objection that alternative ways of forming what complexes is not with an instance or construing coherence in terms of a systematization that prioritizes factors like economy, simplicity, uniformity, elegance, and so forth.

The objection that the parameters of systematization that determine coherence must themselves be validated is met by taking a pragmatic turn toward those methodologies and procedures that produce results that exhibit applicative adequacy—ones that can be utilized effectively in matters not just of explanation but above all in matters of prediction and operational (i.e., technological) application.

The objection that there is no binding linkage between coherence and the truth as such is met by insisting that ideal coherence—coherence perfected under ideal conditions—is a definitive hallmark of the truth. For the connection at issue is not one of hermeneutics (or meaning-explication) but of criteriology.

Chapter 9

Cognitive Relativism and Contextualism

Synopsis

- *Relativism is the position that any group's standard of knowledge is on a par with any other's, seeing that there is no "higher," asitutational standard from which those groups standards themselves can be assessed.*
- *Actually, however, this sort of position gives a misleading interpretation of the situation because the only standard it makes sense to apply is our own standard—the one that we ourselves accept as the reasonable standard of epistemic appraisal.*
- *Objectivity is not a matter of a "God's-eye view" but one of what we ourselves can see as reasonable insofar as we try to be reasonable about it.*
- *The rational primacy of our own position is—or should be—the pivot point.*
- *And here as elsewhere, experience will provide us with crucial guidance if we are willing to listen to its teachings.*
- *Accordingly relativism is a profoundly unreasonable practice.*
- *To be sure, we have to operate within our own context, but contextualism is something very different from an indifferentist relativism, seeing that it pivots on making good rational sense of what one does in context.*
- *Relativism's Achilles' heel is that it falls afoul of considerations of functional efficacy in relation to the aims and purposes that define the enterprise of rational inquiry.*

Cognitive Realism

The fact that our knowledge, and especially our scientific knowledge, rests on the continually growing database of ever-expanding observational horizons means that our present claims to truth in scientific matters—where precision and generality are paramount—involve an element of speculative hope. No doubt "the real truth" is the aim of the cognitive enterprise, but in this real and imperfect world of ours we have to accept the limitations of imperfection and face the fact that different inquirers living at different times and in different circumstances do have—and are bound to have—different ideas about the nature of things.

Relativism is the doctrine that people make their judgments by standards and criteria that have no inherent validity or cogency because their standing and status lies solely and wholly in the sociological fact of their acceptance by the group. The norms of different schools of thought differ, but those different norms are entirely on a par with one another on a basis of "to each their own." Beliefs are a matter of local custom, pretty much as is the case with ways of eating or greeting.

In its general form, this relativist position has two prime components:

1. *Basis Diversity.* Judgments of the sort at issue—be it the true, or for that matter the good, the right, and so on—are always made relative to a potentially variable normative basis: a standard, criterion, or evaluative perspective regarding acceptability that change from one group to another.
2. *Basis Egalitarianism.* Any and every basis of normative appraisal—any and every such standard of assessment—is intrinsically as good (valid, appropriate) as any other.

Egalitarian relativism in general holds that it is not rational to opt for one normatively ratified alternative than for any of its rivals because all norms are subject to change and variation. Relativism accordingly insists that the difference between fact and opinion may be all very nice in theory but is simply unimplementable in practice. Whatever we may see as constituting "our knowledge" is simply a matter of opinion through and through. And one set of group opinions is every bit as justified as any other. It is all just a matter of what people think. To believe anything else is no more than a recourse to the myth of the God's-eye point of view—a point of view which, in the very nature of things cannot possibly be ours.

Relativism thus understood does not deny that those who have a particular commitment (who belong to a particular school or tendency of thought) do indeed have a standard of judgment of some sort. But it insists that only custom speaks for that standard—that it is nothing more than just another contingent characteristic of the cognitive position of that particular group. It is all a matter of the parochial allegiances of the community—there is no larger, group-transcending "position of impersonal rationality" on whose basis one particular standard could reasonably and appropriately be defended as inherently superior to any other. The cognitive relativist, in particular, insists that no actual or possible group of inquirers whatsoever is in a privileged epistemic position. Every group's resolutions in this regard are final; each is its own final arbiter. There is no higher court of appeal—no inherently cogent basis of cognitive acceptability that has any real claim to validity.[1]

Over the last century, indifferentist relativism of this sort has gathered strength from various modern intellectual projects. As the sciences of man developed in the nineteenth century—especially in historical and sociological studies—the idea increasingly gained acceptance that every culture and every era has its own characteristic fabric of thought and belief, each appropriate to and cogent for its own particular context. Historicist thinkers from Wilhelm Dilthey onward have lent the aid and comfort of their authority to cultural relativism. And the aftermath of Darwinian biology reinforced this doctrinal tendency in giving currency to the idea that our human view of reality is formatively dependent on our characteristically human cognitive endowments—as opposed to those of other possible sorts of intelligent creatures. Not only are the data about the world that we can acquire something that comes to us courtesy of the biological endowment of our senses, but the inferences we can draw from those data will be analogously dependent on the biological endowment of our minds. Along these lines, William James wrote:

> Were we lobsters, or bees, it might be that our organization would have led to our using quite different modes from these [actual ones] of apprehending our experiences. It *might* be too (we cannot dogmatically deny this) that such categories, unimaginable by us to-day, would have proved on the whole as serviceable for handling our experiences mentally as those we actually use.[2]

Different cultures and different intellectual traditions, to say nothing of organically different sorts of creatures, will, so it is contended, describe and explain their experience—their world as they conceive it—in terms of concepts and categories of understanding substantially different from ours but in principle every bit as good. And anthropological and sociological investigations militated toward relativism on similar grounds. Moreover, psychology-inspired thinkers like F. S. C. Schiller and John Dewey were also impelled in very much the same direction. In this way, investigations in various modern fields of inquiry have conspired to provide aid and comfort for cognitive relativism.

All the same, it is a deeply flawed position.

What's Wrong with Relativism

I think *X*, you think *Y*. I hold *A* to be correct, you opt for *B*. And this, as the cognitive relativist sees it, is simply the end of the matter. To refute such a position it must be shown that there is a sound basis for claiming that it is *not* the end of the matter. It has to be established that it is, or can be, the case that you *should* think as I do, and conversely, because certain beliefs represent the objectively right and appropriate thing to think.

It must be emphasized from the very outset that to ask for it to be shown that my view is the correct one is not to ask whether—and, if so, how—I can succeed in convincing you that my view is correct. It is not necessarily *you* that I need to convince here. Rather, it is the uninvolved, unprejudiced, reasonable bystander. It is in relation to such suitable third parties that the issue of cognitive appropriateness must be pursued. The real question is not what you or I *do* think but rather what, in the circumstances, any sensible individual *ought* to think.

Ought you to think as I do? At this point we will have to take a closer look at me (and ultimately you as well). The pivotal question becomes one of how I proceed in matters of thinking. Clearly, if I am a frivolous, careless, sloppy thinker there is no reason why you should think as I do. And so I cannot, without further ado, plausibly take the line, "What I accept, everyone ought to accept." But of course I can, if I so choose, reverse matters here. That is, I can choose to heed the demands of objective reason by proceeding on the principle and policy of myself accepting just exactly that which anyone and everyone ought to accept in the circumstances. I can, in sum, be reasonable about this matter of acceptance. And so the idea "I think *X*, therefore you should think so too" is (or can be) sensible and acceptable—but only if I myself am rational by proceeding on the basis of seeking to accept only that which people-in-general ought to accept.

What is now at issue is not the cognitive egomania of the idea "Everyone ought to think as I do" but the cognitive modesty of the idea "I ought to think as any reasonable person would in the circumstance." The line is not: "I think *X* and so you should do so too because you ought to be guided by me." Rather, it is: "I think *X* because it's the right think to do in the circumstances and *that's* the reason why you should think *X* too." It is precisely this policy that best accommodates the demands of impersonal reason. And of course if I proceed on this basis, then the approval of those who are—by hypothesis—*reasonable* bystanders is (ipso facto) assured.

The most promising prospect for refuting relativism accordingly lies in the recourse to reason. For rational acceptance carries with it an inherent claim to universality. The rational person accepts exactly what he is convinced that everyone *ought* to accept because accepting it is, in the circumstances, the rationally appropriate thing to do.

To be sure, when I think *X* and am appropriately convinced that you should think so too, I do not thereby revoke your right to think non-*X*. You are a free agent and are at liberty to do as you please. But this only means that I concede you the right to fall into error. And of course in granting you this right (in the sense of entitlement) I do not and need not think that it is right (in the sense of appropriate) that you should fall into error.

At this stage, an objector may well complain as follows: "The argumentation pivots on the principle 'if it is universally reasonable to think X then everyone should do so.' But you have given us no way of discerning or identifying that which it is reasonable to think. It is surely one thing to *claim* objective validity, and another actually to *achieve* it."

Here it must, of course, be acknowledged that we have no automatic process—no *mechanical algorithm*—for achieving cognitive adequacy. But the difficulty of attaining generality here does not mean that this desideratum is beyond our grasp in the various particular cases that confront us. It is certainly possible to produce examples of things every reasonable person could reasonably be expected to accept. Examples abound, witness Descartes' "I exist"; G. E. Moore's "This is a human hand"; Dr. Samuel Johnson's "This is a stone"; C. S. Peirce's "This stone will fall to the ground when I let go of it." Reasonable people are prepared to give credit to the "normal" sources of cognition. They can be expected to concede the ordinary presumptions of truth, and we will certainly expect them to take the line "innocent until proven guilty" toward sources like the testimony of the senses and the reflections of commonsense thinking.

There are certainly no guarantees that matters will always turn out for the good—or, rather, for the true—when we do the reasonable thing in matters of belief. But experience indicates that we will then be right more often than not in doing so. And this is not only a matter of experience but of the inherent sense of things. For, after all, that right-indicative tendency is exactly the determinative function that makes for reasonableness in matters of belief. It simply makes no sense to take the line that any group's standard is as good as any others.' For the only position that we can reasonably take is that the only standard it makes sense to apply is that which we ourselves accept as reasonable.

THE CIRCUMSTANTIAL CONTEXTUALISM OF REASON

Objectivity is something that pivots on what one *ought* to think—that which is right and proper to think in the circumstances. And that little expression "in the circumstances" is critical to a relativism-constraining objectivity. For that which one ought to want *in the circumstances*—is exactly that which any sensible person think if they were "in one's shoes," that is, if the circumstances were the same. Reason is (circumstantially) universal, and it is objectivity's coordination with rationality that links it to universality. That which (as best one can tell) is the sensible thing for us to do in the circumstances is thereby the reasonable thing for *anybody*—any rational individual—to do *in those circumstances*. The objectivity at issue accordingly comes down to rationality. If it is reasonable for you to *A* in circumstance *X*, then it is so for anybody else—and conversely. Reason is agent-indifferent.

Of course, different people are differently situated, and this is something that rationality requires us to take into account. In cognitive matters, for example, different people have different information at their disposal: it is only the deeply underlying "fundamental principles" that are uniform. And so reason plays no favorites, for while people's circumstances differ—vastly and endlessly—nevertheless that which is reasonable for *X* in his circumstances is ipso facto also reasonable for anyone who is (or would be) situated in circumstances that are identical in the relevant respects.

To be sure, it must be granted to relativists that, as William James insisted, "There is no point of view absolutely public and universal."[3] The "God's-eye view" on things is unavailable—at any rate to us. Whatever we can judge we must judge from the vantage point of a position in space, time, and cultural context. But, of course, it is not the absoluteness of an unrealizable point of view from nowhere or from everywhere-at-once or from God's vantage point that is at issue with objectivity. Objectivity is a matter of how we *should* proceed—and how otherwise reasonable people *would* proceed if they were in our shoes in the relevant regards. It is a matter of doing not what is impossibly non-circumstantial but what is circumstantially appropriate.

The principles operative in the rational economy of things are all objective and universal—though of course their application will bear differently on differently situated individuals. What it was rational for Galen to believe—given the cognitive "state of the art" of his day regarding medical matters—is in general no longer rational for us to believe today. Obviously, what it is rational for someone to do or to think *hinges on the particular details of how this individual is circumstanced*—and the prevailing circumstances of course differ from person to person and group to group. To be objective in one's proceedings is to do what any sensible person would do in one's place and we do not all stand on the same spot. The resolution of an issue is objective if it is arrived at without the introduction of any resources (be they substantive or methodological) that would not be deemed as acceptable *in the circumstances* by any rational and reasonable individual. Accordingly, the rulings of rationality are universal, alright, but conditionally universal, indeed subject to a person-relativity geared to the prevailing conditions. This sort of contextualism does not engender corrosive do as you please "relativism," but represents a deep central fact about our procedural situation. To reemphasize: rationality is universal alright, but it is *circumstantially* universal—universal in its principled generality but particularized in its specific prescriptions.

The universalized aspect of rationality turns on its being advisable by person-indifferent and objectively cogent standards for *anyone in those circumstances* to do the "rationally appropriate" things at issue. The standards of rational cogency are general in the sense that what is rational for one person is also rational for others—but here we have to add: for those others who are *in his shoes*. Both aspects, the situational and the universal, are inseparable facets of rationality as we standardly conceive it. But what about this "in my place" business? What can they bring along in getting there! What those others can bring along and what of mine are they allowed to displace? Clearly what they are bringing along is their rationality, and reasonableness, their common sense and good judgment. But my circumstances and conditions, my commitments and interests are things they have to leave in place. Clearly they are not to substitute their predilections and preferences, their values and affinities for mine, their beliefs and desires for mine. *Everything* must remain as was except for those characteristics that go against the dictates of reason: phobias, groundless anxieties, delusions, senseless antipathies, and irrationalities of all sorts. These must be erased, so to speak—and left blank. In making that suppositional transfer, one has to factor out all those psychic aberrations that stand in the way of a person's being sensible or reasonable.

The circumstances of human life are such that, like it or not, we need knowledge to guide our actions and to satisfy our curiosity. Without knowledge-productive inquiry we cannot resolve the cognitive and practical problems that confront a rational creature in making its way in this world. But in matters of knowledge production life is too short for us to proceed on our own. We simply cannot start at square one and do everything needful by ourselves. We must—and do—proceed in the setting of a larger community that extends across the reaches of time (via its cultural traditions) and space (via its social organization). This requires communication, coordination, collaboration. And so even as the pursuit of objectivity is aided by an agent's recourse to the resources of the environing community, so conversely, is objectivity an indispensably useful instrumentality for the creation and maintenance of intercommunicative community. For there can be no community where people do not understand one another, and it is the fact that I endeavor to proceed as any rational person would in my place that renders my proceedings efficiently intelligible to others. The commonality of rational procedure provides the crucial coordination mechanism that renders people understandable to one another. It is, accordingly, a key instrumentality that positions each of us to benefit by a mutually advantageous commerce that is indispensable for the cooperation and collaboration without which our cognitive enterprise—and other social enterprises in general—would be infeasible.

But can objectivity manage to achieve a universalized impersonality? Can we expect other cultures to conform to our standards? Do not those other cultures have their own way of doing things—and thus also their own rationality?

Anthropologists, and even, alas, philosophers, often say things like "The Wazonga tribe has a concept of rationality different from ours, seeing that they deem it rationally appropriate (or even mandatory) to attribute human illness to the intervention of the rock-spirits."[4] But there are big problems here; this way of talking betokens lamentably loose thinking. For, compare:

1. The Wazonga habitually (customarily) attribute . . .
2. The Wazonga think it acceptable (or perhaps even necessary) to attribute . . .
3. The Wazonga think it rationally mandatory to attribute . . .

Now, however true and incontestable the first two contentions may be, the third is untenable. For compare (3) with

4. The Wazonga think is *mathematically* true that dogs have tails.

No matter how firmly convinced the Wazonga may be that dogs have tails, thesis 4, taken as a whole, remains a thesis *of ours*, and *not* of theirs! Accordingly, it is in deep difficulty unless the (highly implausible) condition is realized that the Wazonga have an essentially correct conception of what is, and, moreover, are convinced that the claim that dogs have tails belongs among the appropriate contentions of this particular realm. Analogously, one cannot appropriately maintain (3) unless one is prepared to claim both that the Wazonga have an essentially correct conception of what rationality is (correct, that is, by our lights), and furthermore that they are convinced that the practice in question is acceptable within the framework of this rationality project. And this concatenation is highly implausible in the circumstances.

The fact is that different cultures do indeed implement a rational principle like "Be in a position to substantiate your claims" very differently. (For one thing, there are different standards as to what constitutes proper "substantiation.") But, they cannot simply abandon such a characterizing principle of rationality. For if they were to convert to "It's all right to maintain anything that suits your fancy" they would not have a *different* mode of cognitive rationality but rather, in this respect at any rate, are simply *deficient* in cognitive rationality.

But this anthropological route to a relativism of rationality, is, to say the least, highly problematic. There is no difficulty whatever about the idea of different belief systems, but the idea of different *rationalities* faces insuperable difficulties. The case is much like that of saying that the tribe whose counting practices are based on the sequence: "one, two, many" has a different arithmetic from ourselves. To do anything like justice to the facts one would have to say that they do not have *arithmetic* at all—but just a peculiar and very rudimentary way of counting. And similarly with the Wazonga. On the given evidence, they do not have a *different* concept of rationality, but rather, their culture has not developed to the point where they have any *conception* of rationality whatsoever. Rationality is, after all, a definite sort of enterprise with a characteristic goal structure of its own—the pursuit of appropriate ends by appropriate means. Its defining principles make for an inevitable uniformity.

To be sure, the question "What is the rational thing to believe or to do?" must receive the indecisive answer: "That depends." It depends on context and situation—on conditions and circumstances. At the level of the question "What is rational; what is it that should be believed or done?" a many-sided and pluralistic response is called for. The way in which people proceed to give a rational justification of something—be it a belief, action, or evaluation—is unquestionably variable and culture relative. We mortal men cannot speak with the tongues of extramundane angels. The means by which we pursue our ends in the setting of any major project—be it rationality, morality, communication, or nourishment—are "culture dependent" and "context variable." But the fact remains that certain crucial uniformities are inherent in the very nature of the projects in which we are engaged. With rationality as with swimming, different cultures and different people may go at it differently but the object of the enterprise is uniformly the same, fixed by the definitive conception of what is at issue. And the sensible way to view the issue of objectivity is that this is not a matter of a "God's eye view" but rather one of the view which we ourselves can see as reasonable insofar as we are being reasonable about it.

A Foothold of One's Own: The Primacy of Our Own Position

Despite its fashionable pervasiveness, however, cognitive relativism accordingly has serious and indeed gravely debilitating defects. And in the main, these inhere in its commitment to basis egalitarianism.

The crucial question, however, is not Are there indeed different norms and standards of rationality?, where the answer is an immediate and emphatic *yes*. (Just as *autres temps, autres moeurs*, so also, other cultures, other standards.) Rather, the salient question is: Are we well advised—perhaps even rationally obligated—to see all those various alternative norms and standards as equally appropriate, equally correct? Must we adopt that Principle of Basis Egalitarianism: All of the various standards of judgment have equivalent justifications. *Ours* is on an equal footing with *theirs* in point of acceptability. Is it a matter of indifference which basis we adopt—each one is every bit as good (or poor) as the next? And here the answer is an immediate and emphatic *no*. There is no good and sufficient reason to see this principle as plausible. Basis indifferentism is daring and exciting—but also absurd. For, at this point we must turn relativism against itself by asking: *Indifferent to whom?* Certainly not to us! For, we have in place our own basis of rational judgment, and it speaks loud and clear on its own behalf. Nor yet by parity of reasoning is *ours* equally acceptable to *them*. From just what "angle of consideration" is it that claim to merit equivalency going to be made? Not from *ours* surely—for this, after all, is ours precisely because we deem it superior. And by parity of reasoning not from *theirs* either. (From God's? Well ... perhaps. But he of course is not a party to the present discussion—and if he were, then what price relativism?)

Perhaps from the point of view of the universe all experiential perspectives are of equivalent merit; and perhaps they are equal before the World Spirit—or even God. But we ourselves cannot assume the prerogative of these mighty potencies. We humans can no more contemplate information with our minds without having a perspectival stance than we can contemplate material objects with our bodily eyes without having a perspectival stance. We ourselves do and must occupy a particular position, with particular kinds of concerns and particular practical and intellectual tools for dealing with them. And for us, there indeed is one particular set of standards for making such appraisals and adjudications, namely, our own—those relevantly operative in the conditions and circumstances of our existence. (And were we to trade this set of standards in for another, then of course that other one would automatically become ours.)

In appraising positions we have, of course, no alternative to doing so from the perspective of our cognitive posture—our own cognitive position and point of view. (It wouldn't *be* our point of view if we didn't use it as such.) We cannot cogently maintain a posture of indifference. Each thinker, each school, is bound to take a strongly negative stand toward its competitors: belittling their concerns, deploring their standards, downgrading their ideals, disliking their presuppositions, scorning their contentions, and so on.

And so, in discussions about "alternative modes of rationality" we do indeed have a decisively "higher standpoint" available to us—namely our own. And this is rationally justified by the consideration that no real alternative is open to us—we have to go on from where we are. Accordingly, while one must recognize the reality of alternative cognitive methodologies, one certainly need not see them as equally valid with one's own. "You have your standards and I have mine. There are alternatives." True enough. But this fact leaves me unaffected. For, I myself have no real choice about it: I *must* judge matters by my own lights. (Even if I turn to you as a consultant, I must ultimately appraise the acceptability of your recommendations on my own terms.)

Cognitive (or evaluative or aesthetic, etc.) standard—perspectives do not come to us *ex nihilo*. From the rational point of view such standards themselves require validation. And this process—is itself something that is standard presupposing. For, of course, we cannot assess the adequacy of a standard-perspective in a vacuum, it must itself be supported with reference to standards of some sort. But in this world we are never totally bereft of such a basis: in the order of thought as in the world's physical order we always have a position of some sort. By the time one gets to the point of being able to think at all, there is always a background of available experience against which to form one's ideas. And just there is where one has to start. It is precisely because a standard-deploying certain position is appropriate *from where we ourselves actually stand* that makes this particular position of ours appropriate *for us*.

The salient point is that we are entitled—indeed, rationally constrained—to see our own criteriological basis of rational judgment as rationally superior to the available alternatives. If we did not take this stance—if we did not deem our cognitive posture effectively optimal—then we could not sensibly see ourselves as rationally justified in adopting it. We cannot responsibly deem some variant scheme as co-meritorious with our own, because (if rational) it is precisely because we deem that scheme superior that we have made it ours. It would, of our view here different, *ipso facto* fail to be our real position—contrary to hypothesis.

In the final analysis, one can and should turn the weapons of relativism against itself. If indeed each group has its own standards against which there is no further, external, higher appeal, then we ourselves have no viable alternative but to see our conception of rationality as decisive for any judgments that we conscientiously make on these matters.

If we are going to be rational we must take—and have no responsible choice but to take—the stance that our own standards (of truth, value, and choice) are the appropriate ones. Be it in employing or in evaluating them, we ourselves must see our own standards as definitive because just exactly this is what it is for them to *be* our own standards—their being our standards *consists in our seeing them in this light*. To say that we are not entitled to view our standards as definitive could be to our having standards at all. And of course, someone who denies us this right—who says that we are not entitled to adopt those standards of ours—does no more than insist that it is by *his* standards (who else's) inappropriate for us to have these standards, and thus is simply pitting his standards against ours. We have to see our standards in an absolutistic light—as the uniquely right appropriately valid ones—because exactly this is what is at issue in their being our standards of authentic truth, value or whatever. To insist that we should view them with indifference is to deny us the prospects of having any standards at all. Commitment at this level is simply unavoidable.

Recognizing that others see such matters differently from ourselves should not and need not daunt us in attachment to our own views if we take due care in their formation. It may give us "second thoughts"—may invite us to rethink—but this is not to move us to admit their standards but rather to make more careful and conscientious use of our own. Our cognitive or evaluative perspective would not actually be our perspective did we not deem it rationally superior to others.

THE ARBITRAMENT OF EXPERIENCE

Even if one grants that different people (groups or schools of thought) have different standards and that they can do so appropriately given their differential emplacement in the cognitive scheme of things, it certainly does not follow from this that one should fail to prioritize his own standards (any more than it follows from my acknowledging that your spouse to be appropriate *for you* constitutes on my part an act of disloyalty toward my own). To concede that certain standards make good rational sense for someone in your shoes is not to say that they make sense for one, who is differently situated.

The stance espoused throughout the present discussion has quite emphatically been that of a *perspectival rationalism* (or *contextualism*). Such a preferentialist position combines a pluralistic acknowledgment of distinct alternatives with a recognition that a group's choice among them is *not* rationally indifferent, but rather constrained by the probative indications of the *experience* that provides them with both the evidential basis and the evaluative criteria for effecting a rational choice.

Of course, people's experience differs. Even if we are pluralists and accept a wide variety of normative positions as being (abstractly speaking) available, still, if we have a doctrinal position at all—that is, if we are actually committed to solving our cognitive and normative problems—we have no serious alternative to seeing our own position as rationally superior. Faced with various possible answers to our questions, the sensible and appropriate course is clearly to figure out, as best we can, which one it is that deserves our endorsement. But, of course, there is—or should be—a good rational basis for effecting such a choice, namely the normative perspective, whatever it be, that is based on the probative indications of our own experience. For there indeed is an objectifying impetus to avert our problem-resolutions from becoming a mere matter of indifferent choice based on arbitrary preferences. And throughout the sphere of rational inquiry this objectifying impetus lies in the appropriate utilization of the lessons of experience. Empiricism is our appropriate and optimal policy. We have to go on from where *we* are and proceed on the basis of our experiential endowment. For us, a perspectival egalitarianism makes no sense in making our decisions regarding theoretical, practical, or evaluative matters.

Different individuals, different eras, different societies all have different bodies of experience. This being so, then whether discordant thinkers are at issue is not *their* perspectival accommodation of their experience just as valid for them as ours is for us? No doubt, the answer here has to be affirmative. But it must be followed by an immediate What of it? The fact that others with different bodies of experience might resolve the matter differently is simply *irrelevant* to our own resolution of the issues. We have to go on from where *we* are and proceed on the basis of our perspective. For us, a perspectival egalitarianism makes no sense. Indifferentism is ruled out by the fact that it is experience that is the determinative factor and for us, the experience at issue is *our* experience and cannot be someone else's.

In matters of cognition as elsewhere, our normative orientations do not come to us *ex nihilo* but emerge from experience. And in this world we are never totally bereft of an experiential basis: in the order of thought as in the world's physical order we always have a position of some sort. By the time one gets to the point of being able to think at all, there is always a background of available experience against which to form one's ideas. And just there is where one has to start. It is precisely because a certain position is appropriate *from where we stand* that makes this particular position of ours appropriate for us. The posture that emerges from this way of approaching the issue is thus that of a contextualistic rationalism:

> Confronted with a pluralistic proliferation of alternative positions you have your acceptance-determination methodology, and I have mine. Yours leads you to endorse P; mine leads me to endorse not-P. Yours is just as valid for you (via your methodology of validating principles) as mine is for me. The situational differences of our contexts simply lead to different rational resolutions. And that's the end of the matter.

The fact that the cognitive venture viewed as a whole incorporates other positions does nothing to render a firm and fervent commitment to one's own position somehow infeasible, let alone improper.

It is, in the eyes of some, a fatal flaw of pluralism that it supposedly undermines one's commitment to one's own position. But this is simply fallacious. There is no good reason why a recognition that others, circumstanced as they are, are inclined—and indeed rationally entitled *in their circumstances*—to hold a position of variance with ours should be construed to mean that we, circumstanced as we are, need feel any rational obligation to abandon our position. Once we have done our rational best to substantiate our position, the mere existence of alternatives need give us no pause. A pluralism of the availability of incompatibly rival perspectives is one thing but a pluralism that maintains the *for-us appropriateness* of these incompatibly rival perspectives is absurd.

Jean Paul Sartre deplored the attempt to secure rationally validated knowledge, which he saw as a way of avoiding responsibility for *making* something of oneself, for "choosing one's own project," seeing that the real truth is not something one can make up as one goes along but is something one regards as entitled to one's recognition (to a subordination of sorts on one's part). But this view turns the matter topsy-turvy. Not the pursuit of truth but its *abandonment* represents a failure of nerve and a crisis of confidence. The avoidance of responsibility lies in an indifferentism that sees merit everywhere and validity nowhere (or vice versa), thereby relieving us of any and all duty to investigate the issues in a serious, workmanlike way.

Admittedly, there are cognitive postures different from ours—different sorts of standards altogether. But what does that mean *for us?* What are *we* to do about it? Several stances toward those various bases are in theory open to us:

1. accept none: reject ours
2. accept one: retain ours
3. accept several: conjoin others with ours

4. "rise above the conflict": say "a plague on all your houses" and take the path of idealization invoking the "ideal observer," the "wise man" of the Stoics, the "ideally rational agent" of the economists, or the like

The first option is mere petulance—a matter of stalking off in "fox-and-grapes" fashion because we cannot have it all our own preferred way. The third option is infeasible: different bases do not combine, they make mutually incompatible demands, and in *conjoining* them we will not get something more comprehensive and complete—we will get a mess. The fourth option is utopian and unrealistic. We have no way to get there from here. Only the second alternative makes sense: to have the courage of our convictions and stand by our own guns. In *evaluating* contentions and positions (of any sort) we just have no plausible alternative to doing so from the perspective of our cognitive values—our own conscientiously adopted cognitive point of view.

To obtain informative guidance from inquiry, it is not enough to *contemplate* cognitive standards—be it as historical actualities or as theoretical possibilities. We must actually *commit* ourselves to one. We can only get viable answers to substantive questions if we do our inquiring in the doctrinalist manner—only if we are willing to "stick our necks out" and take a position that endorses some answers and rejects others, to be sure, in a principled way that is in line with standards for whose due validation we have made appropriate provisions. A perspectively grounded position is good enough for sensible individuals precisely because that position-determinative perspective of ours is by hypothesis OUR perspective.

For us, our own experience (vicarious experience included) is something unique—and uniquely compelling. You, to be sure, are in the same position—your experience is compelling, *for you*—but that's immaterial *to me*. But isn't such an experiential absolutism just relativism by another name—is it not itself just a relativism of a particular sort—an experiential relativism? Whatever relativity there may be is a relativization to evidence, so that relativism's characteristic element of indifference is lacking. It is just this, after all, that distinguishes indifferentist *relativism* from a rationalistic *contextualism*. And there is nothing *corrosive* about such a contribution: it does not dissolve any of our commitments. Its absolutism lies in the fact that, for us, our own experience is bound to be altogether compelling.

But would not such a contextualistic pluralism put everyone's position on a par? Does it not underwrite the view that all the alternatives ultimately lie on the same level of acceptability? The question again is: Acceptable to whom? Rational inquiry as such maintains a certain olympian indifference—a noncommittal neutrality. However, this certainly does *not* mean that my position need be just as acceptable to you as to me. A sensible version of contextualistic pluralism will flatly refuse to put everyone's position on a par—save from the unachievable olympian point of view of the community at large which is, of course, by its very nature unavailable to any single individual. For each individual stands fully and decidedly committed to his own orientation on the basis of his or her own experience, so that there is no question of a relativistic *indifferentism* in acknowledging the pivotal role of a cognitive perspective. A pluralism of contextually underwritten cognitive positions does not lend to indifferentism precisely because a normative position is in the very nature of things something that a sensible person cannot view with indifference.

It makes no sense to hold that all normative perspectives are equally acceptable, because where experiential bases of judgment are at issue, the pattern of our own experience is—for us at any rate—altogether decisive. After all, rationality requires that we attune our beliefs and evaluations to the overall pattern of our experience. For us, our own experience is rationally compelling. We could not (rationally) deviate from its dictates—and it would really make no sense for us to want to do so. We can no more separate from the indications of our own experience than we could separate ourselves from our own shadows.

Such a position accordingly leaves no room for indifferentism. It pivots on the idea of contextual appropriateness—appropriateness in the context that is delimited and defined by the specific experiential circumstances of one's situation. Recognizing that pluralism prevails—that other standards are used by others—we nevertheless do and (appropriately) can deem our own standards of rational cognition as *appropriate* for ourselves. Even when conceding the prospect of someone's having another position, we cannot see it as appropriate for ourselves.

AGAINST RELATIVISM

And so, while rationally appropriate knowledge claims and rationally appropriate actions and even the accepted criteria of appropriateness themselves vary across times and cultures, the determinative principles of rationality do not. However, this interesting circumstance does not so much reflect a fact about different times and cultures as the fact that what *counts* as a "standard of rationality" at all is something that rests with us, because we are the arbiters of the conceptual makeup of an issue within the framework of our discussions of the matter. What sort of thing *we ourselves propose to understand by "rationality,"* becomes determinative for our own discussions of the matter. And this uniform conception of "what rationality is" suffices to establish and render uniform those top-level, metacriteriological standards by whose means each of us can judge the rationality of another's resolutions relative to that other's own basis of appraisal. For, those "deeper principles" of reason are inherent in the very conception of what is at issue. If you "violate" certain sorts of rules then—for merely *conceptual* reasons—you simply are not engaged in the evidential enterprise at all. The most basic principles of knifehood or evidence or rationality are "culture dependent" only in the sense that some cultures may not pursue a particular project (the cutting project, the evidence project, the rationality project) at all. It is not that they can pursue it in a different way—that they have learned how to make knives without blades, to evidentiate without grounds, or make rational deliberations without subscription to the fundamental principles we take as definitive of what rationality is all about. In such matters we do—and must—take our position to be pivotal.

However, the crux here is not the factual issue of what we ourselves do accept with respect to something we see as settled but the normative issue of what we think we *should* accept with respect to something we see as a matter of inquiry. Nor are we here endeavoring to legislate for others. For we are not involved here in subordinating others to ourselves, but rather in subordinating ourselves to what we can conscientiously regard as resolution processes appropriate for use by anybody.

These deliberations accordingly lead to a result that might be characterized as criteriological egocentrism. We can and indeed must see our own standards as optimal with respect to the available alternatives. If we did not so see them, then—insofar as we are rational—they would thereby cease to be our standards.

Indifferentist relativism insists that a choice between different bases of judgment must, if made at all, be based on extrarational considerations—taste, custom, fashion, or the like. But rationality simply blocks the path to this destination in its demand that we attune our judgments to the structure of available experiences and its insistence that doing anything else would be *irrational*. Relativism thus has its limits. The implications of our own conception of rationality, truth, and inquiry are absolutely decisive for our deliberations. We ourselves must be the arbiters of tenability when the discussion at issue is one that we are conducting. And so, we cannot at once maintain our own rational commitments as such while yet ceasing to regard them as results at which all rational inquirers who proceed appropriately also ought to arrive given the same circumstances. In this sort of way, the claims of rationality are inherently universal and, if you will, absolute.[5]

But is it not possible for someone to go out and get another normative standard? It certainly is. But on what basis would one do this? You might *force* me to change standards. Or you can, perhaps, brainwash me. But you cannot *rationally persuade* me. For, rational persuasion at the normative level has to proceed in terms of norms that I accept and, by the norms I actually have, my present standards are bound to prevail, if I am rational in the first place.[6] Only when the course of experience shows that our standards conflict with *our own values*—that they do not take us where we wish to go—can there be any rational pressure toward a change of standards.

And so, even while acknowledging that other judgments regarding matters of rationality may exist, we have no choice but to see our standards as appropriate for us. (In using someone else's with "no questions asked" we would, *ipso facto*, be making them ours!) To use another standard (categorically, not hypothetically) is to make it ours—to make it no longer *another* standard. Like "now" and "here," our standards follow us about no matter where we go. Of course, to see our own standards as the appropriate ones for us to use here and now is not to deny the prospect of a change of standards. But when this actually occurs, our stance toward the new standards is identical to our present stance towards the presently adopted ones. A commitment to the appropriateness of his present standards follows the rational man about like his own shadow.

Relativity ends where charity begins—at home. For, our discourse is governed by *our* conceptions which are absolute at any rate *for us*. One care for the *concepts* involved is bound to chasten any impetus to relativism by enjoining absolutistic constraints.

To be sure, such a criteriological egocentrism can and should be tempered by a posture of criteriological humility. The wisdom of hindsight and the school of bitter experience teaches us the chastening lesson that our cognitive standards—and the judgments we base on them—are by no means necessarily perfect. All the same, we have no real alternative to using our standards—to doing the very best we can with the means at our disposal. While we have to bear in mind the sobering thought that our best just may not be good enough, we are nevertheless bound to see the standards we have adopted in the pursuit of rationality as superior to the *available* alternatives and to regard ourselves as rationally entitled to do so here and now. (Future improvement "from within" can of course be envisioned along the above indicated lines.) To refrain from making this commitment is simply to opt out of the project of rational inquiry altogether. In the pursuit of rational cognition we must, as with any other pursuit, begin from where we are.

To acknowledge that other people hold views different form ours, and to concede the prospect that we may, even in the end, simply be unable to win them over by rational suasion, is emphatically not to accept an indifferentism to the effect that their views are just as valid or correct as ours are. To acquiesce in a situation where others think differently from oneself is neither to endorse their views not to abandon one's own. In many departments of life—in matters of politics, philosophy, art, morality, and so on—we certainly do not take the position that the correctness of our own views is somehow undermined and their tenability compromised by the circumstance that others do not share them. And there is no good reason why we should see matters all that differently in matters of inquiry—or even evaluation.

But how, short of megalomania, can one take the stance that one's own view of what is rational is right—that it ought to be binding on everyone? How can I ask for this agreement between my position and that of "all sensible people"? Not, surely, because I seek to impose *my* standard on *them*, but because I do—or should!—endeavor to take account of their standards in the course of shaping my own. Coordination is achieved not because I insist on their conforming to me, nor yet because I supremely adjust my views to theirs, but rather because I have made every *reasonable* effort to make mine only that which (as best I can tell) ought to be everyone's. In the end, I can thus insist that they should use the same standard that I do because it is on this very basis of a commitment to impersonal cogency that I have made that standard my own in the first place. One's commitment to one's own rational standards is—or ought to be—produced not by magalomania but by objectivity. It is a matter of seeking—to realize a situation where I can reasonably expect that others will see matters in the same light as myself because I have found my light in the illumination of theirs.

Thus, nothing stands in the way of a realization that ours is not inherently and inevitably the best conceptual scheme—a kind of *ne plus ultra*—and that we can *nevertheless* stand rationally committed to it. Consider how cognitive progress happens. We can admit *that* the scientists of the future will have a better science, an ampler and more adequate understanding of the natural universe, and thus a better conceptual scheme—though, admittedly, we cannot anticipate just *how* this is to be so. We need not take the stance that our own conceptual scheme is somehow the last word. Our recognition *that* our scheme is imperfect, though correct and appropriate in the interests of realism, is, to be sure, of rather limited utility. A realization of the *en gros* deficiency of our conceptual machinery unhappily affords no help toward its emendation in matters of detail. A rational commitment to our position is just exactly a commitment to accepting its claims to be the proper basis for accepting its rulings as proper and appropriate. It is a matter of doing the best we can in the cause of responsible truth estimation—of carrying on the business of rational inquiry as best we can in the circumstances in which the world's realities have emplaced us.

Contextualistic Pluralism is Compatible with Commitment on Pursuing "The Truth"

Regrettably, many people have—under the influence of relativism—simply given up on the truth. The very idea of "the truth" is of small interest to various theorists nowadays. Heidegger, for one, regarded those so-called absolute truths as no more than "remnants of Christian theology in the problem-field of philosophy."[7] Of themselves, truth and falsity, correctness and incorrectness, adequacy and inadequacy, reason and unreason, sense and nonsense—approached as issues of logic, semantic theory, or epistemological explication—simply do not interest the hermeneuticist. He wants to know what role the *ideas* about these issues have in the sphere of authentic human experience; he does not ask about what these ideas *mean* but about what people *do* with them. Truth as such is something he is eager to abandon.

And his is not alone. In stressing the pluralism of philosophizing, William James wrote:

> *The* Truth: what a perfect idol of the rationalistic mind! I read in an old letter—from a gifted friend who died too young—these words: "In everything, in science, art, morals, and religion, there *must* be one system that is right and *every* other wrong." How characteristic of the enthusiasm of a certain stage of young! At twenty-one we rise to such a challenge and expect to find the system. It never occurs to most of us even later that the question "What is *the* Truth?" is no real question (being irrelative to all conditions) and that the whole notion of *the* truth is an abstraction from the fact of truths in the plural.[8]

Inspired by James, various contemporary pragmatists are quite prepared to abandon concern for truth.

But this reaction is gravely misguided. Epistemological pluralism has no substantive consequences for the nature of truth as such. The fact that "our truth," the truth as we see it, is not necessarily that of others—that it is no more than the best *estimate* of the real truth that we ourselves are able to make—should not disillusion us in our inquiries and should not discourage us in "the pursuit of truth." In inquiry as in other departments of human endeavor we are well advised simply to do the best we can. Realizing that there are no guarantees we have little sensible choice—*pro tem* at least—but to deem the best we can do as good enough.

Recognizing that others see matters differently in other settings and contexts of inquiry need not daunt us in attachment to our own views of the matter where we stand here and now. There is clearly no conflict between our commitment to the truth as we see it and a recognition that the adoption of a variant probative perspective leads others to see the truth differently. Given that we ourselves occupy our perspective, we are bound to see *our* truth as *the* truth. But we nevertheless can and do recognize that others who operate in different times and conditions see matters in a different light. The circumstance that different people see something differently does not destroy or degrade the thing as such.

An experiential pluralism of cognitive positions is no impediment to a commitment to the pursuit of truth. There is no reason that the mere existence of different views and positions should leave us immobilized like the ass of Buridan between the alternatives. Nor are we left with the gray emptiness of equilitarianism that looks to all sides with neutrality and uncommitted indifference. In important matters like inquiry (or evaluation, etc.) we cannot—and in good conscience should not—bring ourselves to view disagreement in the light of a "mere divergence of opinion."

We have no choice but to shape our estimate of the truth ourselves, to pursue *the* truth by way of cultivating *our* truth; we have no direct access to truth unmediated by the epistemological resources of rational inquiry. And, given the ground rules of rational inquiry, this means that one's view of the truth is bound to be linked to one's cognitive situation. To say that this is not good enough and to give up on our truth—to declare petulantly that if we cannot have *the* absolute, capital-*T* Truth that of its very nature constrains everyone's allegiance, then we will not accept anything at all—is automatically to get nothing and to abandon the pursuit of truth as such. It is foolishness to reject an orientation-bound position as not worth having in a domain where a position is only to be had on this basis. The only truth-claims worth staking are those that we deem to qualify for acceptance as universally cogent and rational on the standards that we ourselves endorse. And the only usable and useful contrast is not that between absolute and relative truth, but that between what is true and what is not.

THE ACHILLES' HEEL OF RELATIVISM

But are we really entitled—rationally entitled—to place reliance on our own standards? Assuming that we indeed have done what reason—as best we can understand it—demands for their substantiation (and this is a long story best reserved for another occasion), the answer is and has to be an emphatic *yes*. Inquirers should not and need not be intimidated by the fact of disagreement—it makes perfectly good sense for a person to do the best possible toward securing evidentiated beliefs and justifiable choices without undue worry about whether or not others disagree.

There is nothing admirable in relativism's inclination to noncommittal detachment with its concomitant reluctance to trust one's personal judgment in matters of human significance. Relativism reflects a regrettable unpreparedness to take intellectual responsibility—to say: "I've investigated the matter as best I can, and this is the result at which I have arrived. Here I stand, I can do no other. If you wish to stand with me, then welcome to you; if not, then please show me how my position is untenable." They represent recourse to an uncritical open-mindedness that comes down to empty-mindedness. In matters of rational inquiry, as in politics and religion, one does well to prefer someone who has views and sticks by them to those who reject the whole project or (equally wrongly) try to ride off in every direction at once.

The decisive weakness of a philosophical relativism with its commitment to basis egalitarianism in matters of cognition lies deep in the nature of the human condition. For the characteristic stance of relativism is to insist on the *rational indifference* of alternatives. Be it contentions, beliefs, doctrines, practices, customs or whatever that is at issue, the relativist insist that "it just doesn't matter" in point of rationality. People are led to adopt one alternative over another by *extra-rational* considerations (custom, habituation, fashion, or whatever); from the rational point of view there is nothing to choose—all the alternatives stand on the same footing. The fatal flaw of this position roots in the fact that our claims, beliefs, doctrines, practices, customs, all belong to identifiable departments of purposive human endeavor—identifiable domains, disciplines, and the like. For all (or virtually all) human enterprises are at bottom teleological—conducted with some sort of end or objective in view. In particular, the cognitive enterprise of rational inquiry has the mission of providing implementable information about our natural and artificial environments—information that we can use to orient ourselves cognitively and practically in the world. After all, explanation, prediction, and effective intervention constitute definitive enterprise-characterizing goals of science. (And the moral enterprise is also purposive, its mission being to define, teach and encourage modes of action that bring the behavior of individuals into alignment with the best overall interests of the group.) Human endeavors in general have an inherent teleology seeing that Homo sapiens is a goal-pursuing creature. Now in this context, the crucial fact is that some claims, beliefs, doctrines, practices, customs, and so forth are bound to serve the purposes of their domain better than others. For it is pretty much inevitable that in any goal-oriented enterprise, some alternative ways of proceeding serve better than others with respect to the relevant range of purpose, proving themselves more efficient and effective in point of goal-realization. And in the teleological contexts *thereby* establish themselves as rationally appropriate with respect to the issues. It lies in the nature of the thing that the quintessentially rational thing to do is to give precedence and priority to those alternatives that are more effective with respect to the range of purposes at issue.

There is no doubt that such a position qualifies as a version of "pragmatism." But it is crucially important to note that it is *not* a version of practicalism. It stands committed to the primacy of purpose, but it certainly does not endorse the idea that the only possible (or only valid) sort of human purpose is that of the type traditionally characterized as "practical"—that is, one that is geared to the physical, and "material" well-being of people. Purposive enterprises are as diverse and varied as the whole spectrum of legitimate human endeavor at large, and such purposes can relate not only to the "material" but also to the "spiritual" side of people (their knowledge, artistic sensibility, social dispositions, etc.). To decipher the position from the classical range of specifically *practical* purposes such a pragmatism might perhaps better be characterized as *functionalism*.

Such a functionalist perspective is decisive in its impetus against relativism. For relativism with its commitment to basic egalitarianism is flatly indifferentistic. Presented with alternatives of the sort at issue—be it cognitive, moral, or whatever—the relativist insists that at bottom it just does not matter—at any rate as far as the rationality of the issue is concerned. But once we see the issues in a purposive perspective, this line just doesn't work. In a purposive context, alternatives are not in general portrayable as rationally indifferent. Rationality not only permits but requires giving preference to purposively effectual alternatives over the rest (at any rate as long as other things are anything like equal). It is quintessentially rational to prefer that which—as best we can tell—actually works.

To the relativist one must grant a pluralism of contextualized variation—the lack of uniformity and the availability of alternatives. But this concession does not "give away the store" because it does not authorize irrationalistic indifferentism, seeing that we need not—and rationally cannot—regard that plurality of alternatives as equivalently valid. On the contrary, two higher-level metanorms—to wit, conformity to one's standards and conduciveness to one's goals—operate so as to assure a purchase hold for rational appropriateness at the procedural level (even though they do not otherwise engender substantive uniformity).

As this line of reflection indicates, relativism stubs its toe against the pervasively purposive nature of the human situation, the fact that our proceedings—be it in inquiry, interpersonal interaction, or whatever—fall within the scope of purposive ventures that have an end in view. And the issue of what effectively serves a given purpose is never a matter of arbitrary indifference because in this context any way of proceeding is as good as any other. Nothing could be more mistaken. The teleological aspect provides a basis for rationality in a way that puts relativism into an altogether problematic and dubious light. It is, accordingly, functionalism—the impetus to purposive efficacy—that bars the way to indifferentistic relativism.[9]

Chapter 10

The Pragmatic Rationale of Cognitive Objectivity

Synopsis

- *The circumstance that our ideas are human artifacts does not prevent them from applying to the real world.*
- *For in fact the social and communicative practice of an interactive group of intelligent agents is bound to manifest an impetus to objectivity.*
- *The problem-solving objectives that are unavoidable for a community of intelligent agents constrain them to the objectivity that characterizes the modus operandi of rational beings.*
- *For without a commitment to objectivity rational agents cannot meet their needs in an efficient and effective way.*
- *Objectivity is an essential factor in realizing the collective self-interest of an intelligent agent.*

Objectivity and the Circumstantial Universality of Reason

Consider the following contention: "There are no objective facts—or at least none that we can formulate by the use of language. For the man-made character of all our human contrivances means that everything that we can manage to produce is a cultural artifact within the course of human history in the setting of a particular place and time. And this is also emphatically the case with respect to our language and whatever we produce by its means—all of our statements, claims, and assertions included. How, then, can anything that people say possibly be objectively independent of all they think and believe? If objectivity is something impersonally factual and thought-independent, then how can we ever get there from here?"[1]

It is readily seen that this sort of worry is simply a mass of confusions. Thus take a shovel. It, too, is a cultural artifact made in the course of human history. But it can touch and move materials that are nowise man-made (earth and sand, say). Even so, statements are artifacts that can present and discuss facts that are nowise man-made (mountains on the moon, for example).

Consider the following contention:

> No truths or purported truths are ever actually objective in the sense of being wholly independent of what people think or believe. Consider the (presumptive) fact that "copper conducts electricity." Clearly if people thought differently about the meaning of words—and specifically about the meaning of "copper," say by letting it stand for what we call *wood*—then this would no longer be true.

This argument trades on a vitiating ambiguity. Two very different items are in play here, namely:

- copper-1 = "copper" as we ourselves actually understand it
- copper-2 = "copper" in its modified understanding on the hypothesis at issue

However true it might be to say that copper-2 fails to conduct electricity, it would be patently false to say that copper-1 fails to do so. Indeed the status of copper-1 is entirely untouched by bringing copper-2 on the stage of consideration. (And it is, after all, the meaning of the prevailing terminology as we actually have it and use it that is at issue in the communication we presently conduct by its means.)

"But if facts are something independent our human thoughts and beliefs—if they obtain irrespective of anything that we do in this mind-connected realm—then how can human thought and belief ever succeed in managing to consider, present, or convey them? How can thought and discourse come into (cognitive) contact with something entirely outside itself?" To ask this is to be obtuse. It is like asking "Since numbers are nonphysical abstractions how can they ever be used to count sheep?" Numbers by their very nature are thought-devised counting instruments that can be used to enumerate collections of objects. Similarly statements are thought-derived claim instruments that can be used to characterize objective arrangements in the world. That's simply how things work. If numbers could not be used to count real things they would not be what they are. And similarly if words, ideas, and concepts could not be used to characterize objectively real things they would not be what they are and communication would be blocked at the starting gate.

"But surely when I say or think 'The cat is on the mat' I am managing to achieve no more than to indicate and avow that 'I think/believe that the cat is on the mat.'" Wrong! These two remarks are very different. The one is something about the world (viz., that the cat is on the mat). The other is something about you and your thoughts (viz., that you think or believe that the cat is on the mat). And these are very different issues. And no sort of statement specifically about you and your beliefs can ever be equivalent with a claim regarding the you-independent arrangement of the world.

The long and short of it is that the circumstance that our ideas and the words in which we formulate them are human artifacts does not prevent them from applying the real world. Neither our physical nor our mental artifacts are severed from reality in virtue of their man-made condition.

THE BASIS OF OBJECTIVITY

Our commitment to objectivity is a fundamental aspect of the cognitive project. But it is certainly not an arbitrary commitment. For it has a rationale that reaches deep into the regions both of our cognitive needs and our cognitive performances.

A key problem of epistemology is that of effecting an inferential transition from subjectivity (the "inner" realm of personal experience) to objectivity (the "outer" realm of impersonal fact), from how things seem to someone to how things actually are. Deeming oneself to be seeing a cat on the mat may be a good reason for someone *to think* that this is so, but it really does not suffice to establish that claim itself. The inferential transit from what someone thinks remains problematic in this context. What we still require is a reliable way to effect the transit from what one subjectively thinks and what is objectively the case.

In the endeavor to achieve this desideratum let us consider adopting Robert Brandom's social-practice approach by exposing the argument:[2]

Argument A

The entire social group R of people take r to be a good reason for thinking that p.

Therefore: r indeed is a good reason for thinking that p, and thereby a good reason for p itself.

Such argumentation could, at the level of entire groups, clearly enable us to effect the needed transit from subjectivity to objectivity. But unfortunately this is still problematic since neither facts nor reasons are validated simply by what people happen to think—even though this be at the level of groups rather than individuals.

However, that case would be very different if our group R were to consist specifically of *rational* people. For such a normatively constituted group we would arrive at:

Argument B

The entire social group R of (fully) rational people think that r is a good reason for thinking that p.

What (fully) rational people think with respect to rationality is correct.

Therefore: r indeed is a good reason for thinking that p.

Does this argument do the job? Almost ... but not quite. For the idea of the group of "rational people" still operates at the level not of subjectivity but of evaluative objectivity. After all, the cogency of this argumentation hinges crucially on the fact that that group R is itself constituted not descriptively but normatively.

And herein lies a tale. For the fact of it is that the idea of a community of rational people can be construed in two ways, depending on the sort of "rationality" that is at issue:

1. *Descriptively factual rationality*: The group R_{DF} consist of individuals who "care for reason" and always or generally "endeavor to provide *what they see* as cogent reasons for their beliefs."
2. *Normatively evaluative rationality*: The group R_{NE} consists of individuals who "are rational" and always or generally "provide cogent— that is, genuinely good reasons for their beliefs."

The trouble with Argument B is that it rests on an equivocation in that its cogency hinges on construing R as R_{NE} whereas its capacity to effect a subjectivity-to-objectivity transit hinges on R's being R_{DF}.

But now we can raise the question: What if in the context of Argument B's crucial premiss we could effect the requisite transition from

1. The entire group R_{DF} of *DF*-rational people that think that r is a good reason for thinking that p.

to

2. The entire group R_{NE} of *NE*-rational people think that r is a good reason for thinking that p?

If this step could be validated, then we would be home free. And so what we clearly need here is a Bridging Thesis to enable us to effect an inferential transit from (1) to (2). Can such a bridge be secured?

There is reason to think that it indeed can. However, the linking connection at issue is not so much logicoconceptual as sociological. For I submit we can get what we need from the following contention:

> It is determinable as a matter of empirical fact that communities of descriptively rational people—people who care about giving and having reasons for their beliefs—will in the course of time tend (under the pressure of "rational selection") increasingly to offer and require good reasons for beliefs (i.e., reasons of a kind that will, at least by and large, be available when and only when beliefs that are based upon them are true).

Given that the purpose for having beliefs at all is to secure information—to obtain the *true* beliefs required for the effective guidance of action—rational people will, for this very reason, prefer processes that achieve this goal to those that do not, and will, over time, tend to adopt, teach, and promulgate processes that prove effective in this regard. In this perspective it is not only theoretically plausible but (no less important) empirically confirmable that *DF*-rational communities will gravitate toward *NE*-rational belief-substantiating processes.

To be sure, this argumentation will effect its requisite fact-to-nature transit only if there are (descriptively factual) tests for good reasons—that is, reasons that are cogent in the normative/evaluative mode. This is something that will pivot on the idea (the "concept" if you will) of what a "good reason for a belief" is. This is going to be something that can be spelled out along the following lines, namely, that: "a reason for holding a belief is a good (cogent) one iff the beliefs substantiated by it are generally true." It is clear that goodness of *this* sort will be effectively determinable in the descriptive/factual order of deliberations.

The upshot of this line of reasoning is that we do indeed have at our disposal the Bridging Thesis requisite for validating a transit in the manner of Argument *B* from subjective to objective rationality. And this transit is provided by what is ultimately an empirical (factually determinable) process of social evolution proceeding under the aegis of a rational selection that impels de facto *reason-concerned* communities in the direction of a commitment to *normatively cogent* reasons. The social and communicative practices of an interactive group of intelligent agents are bound to manifest an impetus to objectivity.

The Problem of Validating Objectivity

But what is it that validates acting under the aegis of rationality and the objectivity that goes with it? Proceeding rationally is certainly not a guarantor of adequacy: for example, there is no way to establish on general principles that rationally conducted inquiries will (let alone *must*) achieve the actual truth. Admittedly there are no guarantees here. But nevertheless, we *can* say with confidence is that proceeding rationally and objectively enables us to avert a variety of recognizable sources of error, so that in conducting our inquiries on objective principles it becomes less likely that we will fall into error.

However, paradoxical though it may seem, to eliminate error is *not* necessarily to get closer to the truth (save in the case of omniscient knowers). Where the answer to a numerical problem has to be an odd number one can eliminate as erroneous 2, and 4, and 6, and so on ad infinitum through the entire series of even numbers, without drawing any nearer to getting the correct result. And that, in fact, is all that cognitive objectivity ever does for us—it eliminates various common avenues to error. Being objective—avoiding bias, prejudice, and the like—is thus no guarantor of appropriate results in the absence of cognitive competence. It helps us only to the extent that our methodology of inquiry is otherwise effective. It is an aid to rational inquiry but no magic wand.

But we live in a world without guarantees. After all, why do we endorse the world-descriptions of the science of the day seeing how often that of the past has gone awry? Why do we follow the medical recommendations of the physicians of the day, or the policy recommendations of the economists of the day? Because we know them to be correct—or at any rate highly likely to be true? Not at all! We know or believe no such thing—historical experience is too strongly counterindicative. Rather, we accept them as guides only because we see them as more promising than any of the *identifiable* alternatives that we are in a position to envision. We accept them because they afford us the greatest available subjective probability of success—discernibly the best bet. We do not proceed with unalloyed confidence, but rather with the resigned recognition that we can do no better at the moment. Similarly, the recommendations of reason afford not *assurance* of success, but merely what is, to all achievable appearances, the *best overall chances* of averting failure with respect to our goals. We act, in short, on the basis of *faute de mieux* considerations: in real-world situations reason trades in courses of action whose efficacy is a matter of hope and whose rationalization is a matter of this-or-nothing-better argumentation.

Rationality foregoes guarantees because upon recognizing their unavailability it ceases to ask for the impossible. It is content to play the odds. And specifically as regards objectivity it does so with caution. The considerations at issue here can be posed at a very general level.

The rational person is, by definition, someone who uses intelligence to maximize the probability—that is, *the responsibly formed subjective probability*—that matters will eventuate favorably for the promotion of his real interests. Rationality calls for adopting the overall best (visible) alternative—the best that is, in practice, available to us in the circumstances. It is thus *actually* rational to do the *apparently* rational thing, provided that those appearances reflect the exercise of due care. And it is here that the impetus to objectivity lies.

In pursuing our goals, we do well to emulate the drowning man in clutching at the best available prospect, recognizing that even the most rationally laid scheme can misfire. In this world of imperfect inquirers reality is not always and inevitably on the side of the strongest arguments. Heeding the call of reason, constructive though it is, affords no categorical guarantees of success, but only the reassurance of having made the best rational bet—of having done as well as one could in the circumstances of the case. In this world we have no guarantees—no means are at our disposal for preestablishing that following rationality's counsel actually pays. We have to be content to settle for the best alternative that is available as a pathway toward realizing our goals. The situation is such that even cognitive rationality must ultimately be validated in the prudential order of reason.

Why be objective? Because it is the sensible, the rationally appropriate thing to do. And why is this so? Because it is quintessentially rational to pursue one's best or real interests in the way that looks to be the most promising, and the use of reason—which carries objectivity in its wake—is part and parcel of this endeavor.

We do not (or need not) pursue objectivity for its own sake,[3] its value emerges in the context of ulterior purpose within the methodology of inquiry. The merit of a commitment to objectivity is utilitarian and lies in the positive things that it can do for us in the pursuit of normative adequacy in matters of belief, evaluation, and practical action.

In the specifically cognitive domain, the assets of objectivity preeminently include:

- *Robustness*. Objectivity provides us with checks and balances. It facilitates the aims of inquiry by providing a safeguard against biases and misimpressions. It is an instrument of verification.
- *Communication*. Objectivity open up avenues of exchange that serve productive ends. In proceeding objectively (as others would) we make ourselves intelligible to them and lay the basis for mutual understanding. And not just understanding but also—
- *Community and collaboration*. Objectivity facilitates an eminently useful sharing of concerns. Life is too short and the self too limited; to get pretty much anything done in this world we must proceed communally.

This last item is particularly important. No man is an island, an only by standing on ground where others too are at work are the possibilities of communication, cooperation, and (above all) collaboration available. Only by joining with others—by shifting from I to we (and indeed even a cross generational we!) can I manage to accepted the tasks that life in this world sets before us.

The issue of caring about the views of others is thus rather different with respect to an epistemic virtue such as objectivity from what it is with respect to a moral virtue such as courtesy. For the prudential rationale of objectivity means that there is a matter of due case for *our* interests while the moral aspect of courtesy means that it is a matter of due case for *their* interests. With objectivity, I seek to align my proceedings with theirs in order to realize *my* interests; with courtesy I do so to cater to *their* sensibilities—at any rate in the first analysis. All the same, both modalities alike avail self-aggrandizement. Neither is a matter of insisting that others shall do as I do, but rather one of conforming my own proceedings to theirs.

To be sure, being objective in matters relating to the public domain affords no failproof guarantee of success. Throughout our cognitive and practical affairs we have to conduct our operations under conditions of risk. We have to do the plausible thing—to "play the odds." And so, when we do the rational thing but it just does not pan out, we simply have to "grin and bear it." The matter is one of calculated risks and plausibly expectable benefits. Rationality—to reemphasize—affords no guarantees. By the very nature of what is involved in rational procedure, the determinable odds are in its favor. But that may still be cold comfort when things go wrong. Then, all we have is the satisfaction of having done our best. The long and short of the matter is that nothing "obliges" us to be rational except our rationality itself. It lies in the nature of things that reason is on the side of rationality. Admittedly, reason offers us no categorical guarantees; yet, if we abandon reason there is no better place where we can (rationally) go.

The imperative "Act as a rational agent ought" does not come from without (from parents or from society or even from God). It roots in our own self-purported nature as rational beings—in the fact of its being an integral part of what our reason demands of us in the context of its own cultivation. (In Kant's words, it is a *dictamen rationis*, a part of the "internal legislation of reason.") And this obligation to rationality carries the obligation to objectivity in its wake, so that the validation of rationality and objectivity are of a piece. The problem-solving objectives that are unavoidable for a community of intelligent agents constrain them to the concern for objectivity that characterizes the modus operandi of rational beings.

What Is Right with Objectivism

The preceding deliberations provide a clear indication of what is wrong with a relativism that takes the line that "There's no good reason to expect that you should think as I do." For when my thinking proceeds as it ought—when it is grounded in what are standardly taken to be evidentially sound reasons—then this very fact constitutes a good reason why you should think as I do. It is the factor of *objective cogency* that constitutes the crucial coordinating principle that links together the conclusions and acceptances of reasonable people.

But this line of thought still leaves some important open questions on the agenda—in particular: What's right with objectivism? Why be objective? To ask this is to ask for cogent reasons for cognitive objectivity. But of course there is no point in plunging into a discussion in the setting of Why-do-something? questions if one is not going to cast rationality to the winds here. So we have to construe Why be objective? as: Give me a good reason—an impersonally and objectively good reason—for being objective. This circumstance alone suffices to show that objectivity has to be respected even by those who are, for reasons of their own, concerned to call it into question. It itself attests the appropriateness of objectivity.

But just what does objectivity require? How is one to understand what is involved here? To achieve a reasonable approximation we shall suppose that, more or less by definition: To be objective is to proceed as *any* reasonable person would do in the circumstances at issue.

But what does their reasonableness demand of reasonable people? The answer here is that reasonable persons are optimizers—individuals who do the best they can with the means available in the prevailing circumstances, endeavoring to do that which is the best thing possible to do (in the circumstances).

And at this point, the answer to our initial question Why be objective? lies before us clear and plain. For the very fact that something is the best that can be done in this circumstance is ipso facto a good reason for doing it—for anyone. The challenge to defend objectivity is now met in a manifestly cogent way by the simple and straightforward answer: Because this is the rational and reasonable way to proceed.

Rational belief, action, and evaluation are possibly only in situations where there are *cogent grounds* (and not just compelling personal motives) for what one does. And the cogency of grounds is a matter of objective standards: cogency is not something variable and idiosyncratic. The idea of rationality is in principle inapplicable where one is at liberty to make up one's rules as one goes along—effectively to have no predetermined rules at all. For a belief, action, or evaluation to qualify as rational, the agent must (in theory at least) be in a position to "give an account" of it on whose basis others can see that "it is only right and proper" for him to resolve the issue in that way.

Objectivity thus pivots on rationality. But the rationality at issue involves more than mere logical coherence. It is a matter of the intelligent pursuit of circumstantially appropriate objectives.[4] Accordingly its demands are few in type but elaborate in extent:

- aligning one's beliefs with the available evidence;
- maintaining consistency within one's beliefs;

- making reasonable efforts to ensure the adequacy of the available evidence to the problems at hand.

But there is nothing personalistic or idiosyncratic about such principles. What is evidence for one will have to count as evidence for another if *evidence* it indeed is. What is an authentically cogent inference with you is cogent inference with me. And so on.

The generality of access at issue with objectivity becomes crucial here. If something makes good rational sense, it must be possible in principle for anyone and everyone to see that this is so. This matter of good reasons is not something subjective or idiosyncratic; it is objective and lies in the public domain. Cogent reason is impersonal. There is no exclusively personalized rational cogency: what is cogent for me will and must be equally cogent for anyone in the same circumstances. Robinson Crusoe may well proceed in a perfectly rational way in the context of his peculiar setting. But, he can only do so by doing what would make sense for others is similar circumstances. He must, in principle, be in a position to persuade others to adopt his course of action by an appeal to general principles to show them that his actions were appropriate in the circumstances. Rationality is by nature inherently general and objective in its operations, endeavoring to deal with issues in an objective manner—in such a way that anyone can see the sense of it.

We thus have the core of the solution to our initial question of what is wrong with an objectivity dismissive relativism. The answer is that in cognitive matters relativism violates objectivity and objectivity violations are at odds with being reasonable. For without a commitment to objectivity rational agents cannot expect to meet their needs in an efficient and effective way.

Let us consider some of the ramifications of this position, and in particular the question of what it is that reasonableness and rationality demands.

ABANDONING OBJECTIVITY IS PRAGMATICALLY SELF-DEFEATING

Proceeding with a view to objectivity in its impersonal mode is thus generally in our best interests. But is it not just advantageous but somehow obligatory? Objectivity's bonding to rationality shows that this is indeed the case. For insofar as we reason-capable agents have an obligation to exercise this capacity—as we indeed do—we are involved in a venture that carries the obligation to objectivity in its wake.[5] Let us see how this is so.

The subjectivity/objectivity contrast turns on the distinction between what is *accepted by me* as things stand—and quite possibly by me alone—over against that which is *acceptable for us* in general in suitably similar conditions. And this issue of a range of cogency pivots on the I/we contrast. The crucial contrast is that between what simply holds for oneself versus what is to be seen as holding for all of us. Objectivity is coordinate with generality: what is objectively so holds independently of the vagaries, contingencies, and idiosyncrasies of particular individuals.

Consider, in particular, the objectivity of claims and contentions. An objective truth does not hold *of* everyone; it holds *for* everyone. "Bald men have little or no hair" wears on its very sleeve the fact that it holds only *of* some—namely, of bald men. But it holds *for* everyone, is just as true for you as for me, irrespective of whether we are bald or not. The questions Who realizes it? or Who has reason to believe it? certainly arise—and may well be answered by saying that only some people do so and many others do not. But the question Whom is it true for? simply does not arise in that form. If that contention is true at all, it is true for everyone. Truths do not need to be *thematically* universal and they do not need to be *evidentially* universal; but they do need to be universal in point of *validity*. And this precisely is the basis of their objectivity. As truths they will necessarily have that for-everyone aspect.

Objectivity keeps us on the straight and narrow path of commitments that are binding on all rational beings alike. But cultivating objectivity is certainly no exercise in power-projection. It is not a matter of trying to speak for others, preempting their judgment by a high-handedness that constrains them into alignment with oneself. Quite to the contrary, it works exactly the other way around. The proper pursuit of cognitive objectivity calls for trying to put one's own judgment into alignment with what—as best one can determine it—the judgment of those others ought to be. It is not a matter of coordination by an imposition on others but the very reverse, one of a coordination by self-subordinated submission to the modus operandi of the group on granting it the benefit of the doubt in point of rationality. Such conformity is a requisite for objectivity but the matter of how it comes about is pivotal. It is—and must be—a matter of my conforming to them (the generality of sensible people) as opposed to any megalomaniacal insistence that they conform to me. Objectivity is a policy not of the dictatorial but of the cognitively gregarious who seek to be in cognitive harmony with the rest—at least insofar as they subscribe to the standards of rationality.

The impetus to rationality accordingly has important and immediate implications for our concern with objectivity. For rationality carries objectivity in its wake: the universality and impersonality of reason validates the pursuit of objectivity in direct consequence. Objectivity's insistence on cogent resolutions that prevent the course of reason from being deflected by wish and willfulness, biases and idiosyncrasies—will for this very circumstance foster and implement a commitment to the primacy of reason.

To proceed objectively is, accordingly, to render oneself perspicuous to others by doing what any reasonable and normally constituted person would do in one's place, thereby rendering one's proceedings intelligible to anyone. When the members of a group are objective, they secure great advantages thereby: they lay the groundwork for community by paving the way for mutual understanding, communication, collaboration. And in cognitive matters they also sideline sources of error. For the essence of objectivity lies in its factoring out of those of one's idiosyncratic predilections and prejudices that would stand in the way of other intelligent people's reaching the same result. Objectivity follows in rationality's wake because of its effectiveness as a means to averting both isolation and error.

Sometimes we are told: People never really know anything really and objectively: there is only what people think they know. But this contention is deeply problematic. To assert that there are no objective facts is self-contradictory. Whoever makes (asserts) this statement is presumably stating a matter of objective fact: they are not saying that they merely think it to be so.

What is at issue in an overt denial of facts is not a logical inconsistency but a practical inconsistency. For now the practice in which you are engaging becomes infeasible through the very nature of the way in which you are engaging in it. That is, you are engaging in a process of communication but doing it in with a machine that produces a monkey wrench that it inserts into its own works.

"Facts are precisely what there is not," says Nietzsche, "only interpretations."[6] But this, insofar as true, is itself nothing but an interpretation of our cognitive situation—and not actually a cogent one. For such an all-leveling perspective has its difficulties. For there are interpretations and interpretations—interpretations that resolve problems and interpretations that create difficulties. (If I interpret those grains of white sand as sugar and proceed accordingly I shall assuredly encounter distinctly unpleasant consequences.)

Thoroughly practical (functional/operations) reasons thus speak on behalf of endorsing the conception of objective truth. Specifically we require it:

1. As a presupposition to make communication possible by way of agreement and disagreement. To principle a commonality of focus.
2. As a contrast conception that enables us to acknowledge our own potential fallibility.
3. As a regulative ideal whose pursuit stops us from resting content with too little.
4. As an entryway into a communicative community in which we acknowledge that our own views are nowise decisive.

The commitment to objectivity thus affords us an effective practical instrumentality that facilitates communication and cognitive collaboration. It is certainly not something that we have to endorse only "on faith." Practical experience amply endows it with a track record of utility that constitutes a retrospective justification of a commitment to objectivity. Its substantiation lies largely in the fact that we simply could not get on without it.

To be sure, such a pragmatic/experiential validation of a particular cognitive perspective nowise *guarantees* the categorical correctness of the beliefs that ensue from it. All that it provides is that in following its strictures we will do as well as we can manage to in the circumstances. By allowing our beliefs by our experiences we do all that we can to assure their adequacy to our needs within the setting of the situation that confronts us. As indicated above, beyond that point—the point of having done all that we reasonably can—a certain element of methodological optimism is called for, a rationally warranted hope that once we have done all we can these best efforts of ours will bear fruit. In this department of human affairs—as in others—it is only rational to proceed in the hope and expectation that the best we can possibly do is good enough.[7]

Chapter 11

Rationality

Synopsis

- *Admittedly, there is no guarantee that success will ensue when we do the rationally appropriate thing—be it in theoretical or in practical matters.*
- *This is so because what is rational in the circumstances as we actually discern them can always prove mistaken due to conditions outside the range of our knowledge and control.*
- *Accordingly, practicable rationality can and often does fall short of the ideal.*
- *So why be rational? Simply because doing so offers us the best chances of getting it right.*
- *Accordingly even cognitive rationality has a pragmatic rationale.*
- *Are there alternative modes of rationality? Assuredly none that we could rationally adopt. For if something failed to offer us the best available option by our own lights then it would, ipso facto, fail to be the rational thing to do.*
- *The self-reliance of rationality is not viciously circular because the only sort of justification of rationality that is worth having is a rational one.*

Stage-Setting for the Problem

Man is a rational animal, and presumably the only one here on earth, though there is no reason of principle why extraterrestrial rational beings could not exist. But beings capable of an intuitive understanding of the likes and dislikes, attractions and aversions of others could in theory come to view one another as rational.

Historically, rationality has been bifurcated into two compartments: practical rationality that concerns itself with what we do and theoretical (or cognitive) rationality that concerns itself with what we think. This view of the matter was disseminated by the skeptics of classical antiquity, reasserted itself in the thought of David Hume and became permanently consolidated for modern philosophy in the works of Immanuel Kant. It is important to recognize this duality of rationality because it means that there is more to rationality than "epistemic competence" alone.

For better or worse, however, this view of the matter really makes no sense. For of course with Homo sapiens, and indeed any sort of intelligent being, action cannot be separated from thought. Those two departments of reason are actually aspects of one composite whole. Rationality consists in the intelligent pursuit of appropriate objectives. Intelligence indicates cognition, pursuit action, and appropriateness evaluation. With rationality these three factors are fused into a single integrated whole. There are not different forms or versions of rationality; rationality is a unified whole that unites them as inseparable constituents.

The road of rationality, like that of true love, does not always run smooth. For his very rationality can lead a rational agent into difficulty. For the rational resolution of problems is invariably contextsensitive to the information in hand, in such a way that what is a patently sensible and appropriate resolution in a given data-situation can cease to be so in the light of additional information—information that does not abrogate or correct our prior data, but simply augments and enlarges it.

It is clear that this circumstance pervades all sectors of rational deliberation, regardless of whether we are dealing with deductive information-extraction, inductive inference, evaluative reasoning, practical reasoning, or any other type. The fact is that the rationally appropriate resolution of a problem on the basis of one body of evidence or experience can always become unstuck when that body of evidence or experience is not actually revised but merely enlarged. The ramifications of this fact are pervasive and their implications merit closer attention.

Optimum-Instability

The rationally appropriate procedure in problem solving is to strive for the best resolution achievable in the light of the available data.[1] Rationality enjoins us to adopt the best available option. Having surveyed the range of alternatives, the rational thing to do is to resolve the choice between them in what is, all considered, the most favorable way. What is "favorable" will of course differ from context to context: with cognitive choices, "favorable" primarily hinges on substantiation, with practical choices, on effectiveness, with evaluative choices, on preferability. But the salient fact remains that rationality is always a matter of optimization relative to constraints, of doing the best one can in the prevailing circumstances. And this is a matter of the circumstances as best determinable from where he stands, and not of what the real ones may actually happen to be. The rational thing to do in resolving an issue is to make the best use of all the (relevant) information at our disposal. Clearly, it is not—and cannot rationally be—required that we should use information that we do not have and cannot obtain. (Even if, wholly unbeknownst to me, there will be a power failure in a few moments, it is nevertheless rational for me to take the elevator upstairs.) The rationality of our beliefs, actions, and evaluations is clearly a matter of the information that we can secure, not one of what might in theory be available to others who may be more favorably circumstanced than ourselves. Twentieth-century medicine is in no way deficient in point of rationality for failing to apply twenty-first-century remedies. Rationality, like politics, is an art of the possible—a matter of doing the best that is possible in the overall circumstances in which the agent functions—cognitive circumstances included.

If we had "complete information," and in particular if we knew how our decisions would eventuate—how matters will actually turn out when we decide one way or another—then rational decision-making and planning would of course be something very different from what they are. In this world we are constrained to decide, to operate, to plan, and to act in the light of incomplete information. When we marry, take a job, invest our money, improve existing technology, and so on, we have no clear idea what eventual consequences will ensue.

After all, if rationality were only possible in the light of complete information it would perforce become totally irrelevant for us. It lies in the inevitable nature of things that we must exercise our rationality amidst conditions of imperfect information. A mode of "rationality" capable of implementation only in ideal circumstances is pointless; in this world, the real world, there is no work for it to do. We have to be realistic in our understanding of rationality—recognizing that we must practice this virtue in real rather than ideal circumstances. A conception of "rationality" that asks no more of us than doing the best we possibly can is the only one that makes sense. (In fact, one that asks for more would not itself be rational.) Clearly, if rationality is to be something that we can actually implement then it has to be something whose demands we can meet in subideal conditions, conditions of incomplete information as we (inevitably) confront them.

The problem, then, is that rationality is "information-sensitive": exactly what qualifies as the most rational resolution of a particular problem of belief, action, or evaluation depends on the precise content of our data about the situation at issue. But, as we obtain more information we change the frame of reference relative to which optimality is determined. And then what was once the best, most favorable resolution may well no longer continue to be so in the light of these further developments. Optimality, in sum, is "context-sensitive" with respect to the informational context.

The history of the empirical sciences affords a familiar illustration. Beliefs in the luminiferous aether, the conservation of matter, and the like, were all sensible and rational in their day. Achieving a substantial enlargement of the data base on which we erect the structures of our theorizing generally produces those changes of mind characterized as "scientific revolutions." As significantly enhanced experimental information comes to hand, people are led to resolve their problems of optimal question-resolution in radically different ways.

It will not do to react to this state of affairs by saying: "Delay decision until your experience is perfected, your information altogether complete." To postpone a decision until then is tantamount to preventing its ever being made. A rationality we cannot deploy here and now, amid the realities of an imperfect world, is altogether useless.

We thus confront the *"predicament of reason"* generated by the confluence of the two considerations, that the rationality of a problem-resolution is "information-sensitive," and that in the real world our information is always incomplete. Mere additions to our information can always upset the applecart of rational decision. Rationality is inexorably circumstantial, and that is rational in the circumstances as we can discern them can always prove ineffective owning to conditions outside the range of our knowledge and control.

Ideal versus Practical Rationality: The Predicament of Reason

This predicament of reason at issue in the irresoluble tension between the demands of rationality and its practical possibilities inheres in the following aporetic situation that we face in cognitive matters as elsewhere:

1. We ought to do that which, as best we can determine it, is the rationally appropriate thing to do.
2. We ought to act as perfectly rational agents do.
3. What is appropriate as best we can determine it (in our suboptimal circumstances) will generally differ from what a perfectly rational agent would (ipso facto) determine to be best.

Since these theses are mutually incompatible, one of them must be sacrificed, and since 1 and 3 represent unavoidable "facts of life," it is clearly 2 that must be abandoned in the interests of consistency-restoration. Neither the rationality-abandonment of i-rejection nor the unrealistic perfectionism of 3-abandonment are plausible options. This circumstance that thesis 2 must be abandoned sets the stage for the predicament inherent in the fact that rationality seems to demand something of us that is, in the final analysis, not actually realizable.

Fortunately, the difficulty that arises here admits of a sensible resolution. The distinction between idealized and practicable rationality offers a way out. For we must distinguish between:

> *ideal* rationality, which is geared to resolutions that are rationally appropriate with (absolutely) everything relevant taken into account—that are optimal pure and simple, and

> *practicable* rationality, which is geared to resolutions that are rationally appropriate with everything relevant taken into account that we can effectively manage to take account of in the prevailing circumstances—that are optimal as best we can manage to tell.

This distinction softens the impact of rejecting premiss (2) of the predicament. For, while we cannot indeed achieve the demands of ideal rationality, we clearly should do all we can in the direction of practicable rationality. After all, ideal rational optimality is something merely "utopian" and "pie in the sky." For us, the only practicable optimality is that which is realistic and achievable-optimality as best we can get hold of it, which accordingly remains merely apparent optimality (optimality as best we can determine it). All we can ever secure in real-life situations of rational deliberation seem optima arrived at in the light of incomplete information. We can have no assurance that they will continue to be optimal in the light of a fuller appreciation of the circumstances. We can only do our best.

This sort of situation obtains throughout all areas of rational deliberation: cognitive, prudential, evaluative—right across the board. Even our optimally evidentiated beliefs are not necessarily true; even our optimally well-advised actions are not necessarily successful; even our optimally crafted appraisals are not necessarily correct. The fact of limitation confronts us in every direction.

The inexorable circumstance represented by the five following facts has to be faced:

1. The "ideally rational" thing to do is to do what is optimal.
2. The "rationally appropriate" or, simply, the "rational" thing is for us to do our utmost toward the ideally rational thing—to do that which as best we can tell is the ideally rational thing to do. Accordingly, rationality calls for optimization in the light of the available information.
3. What actually is optimal, however, is what is determinable as such (as optimal) in the light of complete information.
4. What is optimal in the light of incomplete information may well fail to be actually optimal and indeed may be counterproductive.
5. The most we can ever do is to act in the light of the available information, which generally is incomplete.

It follows inexorably from these premises that, in general, the rational thing for us to do is something which, as we must recognize full well, may actually impede realization of our objectives. This constitutes what might be called the predicament of reason: the circumstance that reason constantly calls on us to do that which, for ought we know, may prove totally inappropriate. Rational action in this world has to proceed in the face of the sobering recognition that while we doubtless should do the best we can, it may nevertheless eventuate that the seeming best we can do is quite the wrong thing. It is the course of reason (1) to aim at the absolutely best, but (2) to settle for the best that is realistically available. (After all, it would be unreasonable, nay irrational, to ask for more.) But the paradox lies in our clear recognition of the tension between the two.

A paradox accordingly emerges from the fact that while the ends of rationality are achieved only under the ideal conditions of global totality, nevertheless the actual practice of rationality must inevitably be conducted at the level of local and imperfect conditions. We can never rest complacently confident that in following reason's directions we are not frustrating the very purposes for whose sake we are calling on the guidance of reason. We have to recognize the "fact of life" that it is rationally advisable to do the best we can, while nevertheless realizing all the while that it may prove to be inappropriate.

"But the problem is created by mere ignorance." True enough! But true in a way that provides no comfort. Imperfect information is an inevitable fact of life. A "rationality" that could not be implemented in these circumstances would be totally pointless. Were rationality to hinge on complete information, it would thereby manifest its irrelevance for our concerns. There is nothing "mere" about ignorance.

Such deliberations lead into the terrain of philosophical anthropology. For, we here confront a fundamental aspect of the human condition. With creatures like ourselves that are of limited capacity, and whose cognitive range of reference is thus inevitably limited, no assurance can be attained that doing the best one can to follow the dictates of rational appropriateness will produce a rationally optimal resolution—that those apparent optima we can attain will actually yield real optima. There is never any assurance that we will actually succeed by following the best available advice. The cold and cruel reality is that (1) a decision based on incomplete information—no matter how intelligently made—can be totally counter-productive, and (2) in this life we always have to operate on the basis of incomplete information.

There stands before us the profound lesson of the biblical story of the Fall of Man, that in this world there simply are no guarantees—not even for a life conducted on principles of reason. It is this sobering situation—doubtless unwelcome, but inevitable—that betokens the predicament of reason: the circumstance that rationality requires us to do "what seems best" in the full and clear recognition that this may well fail to be, in actuality, anything like the best thing to do.[2]

The Problem of Validating Rationality

Why then should one be rational? In a way, this is a silly question. For, the answer is only too obvious—given that the rational thing to do is (effectively, by definition) that for which the strongest reasons speak, we ipso facto have good reason to do it. Kurt Baier has put this point in a way difficult to improve on:

> The question "Why should I follow reason?" simply does not make sense. Asking it shows complete lack of understanding of the meaning of a "why question." "Why should I do this?" is a request to be given the reason for saying that I should do this. It is normally asked when someone has already said, "You should do this" and answered by giving the reason. But since "Should I follow reason?" means "Tell me whether doing what is supported by the best reasons is doing what is supported by the best reasons," there is simply no possibility of adding "Why?" For the question now comes to this, "Tell me the reason why doing what is supported by the best reasons is doing what is supported by the best reasons." It is exactly like asking, "Why is a circle a circle?"[3]

In virtue of its nature as such, the rational resolution to an issue is the best solution we can manage—the one that we should adopt and would adopt if we were to proceed intelligently. The impetus to rationality is grounded in our commitment to proceeding intelligently—to "using our brains." ("Why be rational?" "It's the intelligent thing to do." "But why proceed intelligently?" "Come now; surely you jest!") After all, once we admit that something is the best thing to do, what further reason could we possibly want for doing it? Once it is settled that A is the rational thing to do which may itself take a lot of showing—then there is no more room for any further reason for doing A, no further point to asking: Why do A? For at this stage, the best of reasons—by hypothesis—already speak for doing A. Once rationality is established, there are no further extra- (or supra-) rational reasons to which we could sensibly appeal for validation. In this sense, then, the question Why do the rational thing? is simply foolish: it is a request for further reasons at a juncture at which, by hypothesis, all the needed reasons are already in.

But this line of response to our question, though perfectly cogent, is a bit too facile. The job that needs to be accomplished is actually more complicated.

Belief, action, and evaluation based on what really are—truly and actually—the "best of reasons" must necessarily be successful. This contention is simply circular, since those theoretically "best of reasons" are best exactly because it is they that assure realization of the best results. But, in this world, we are not in general in a position to proceed from the actual best as such, but only from the visible best that is at our disposal—"the best available (or discernible) reasons."

We have to content ourselves with doing "the apparently best thing"—the best that is determinable in the prevailing circumstances. But the fact remains that the alternatives whose adoption we ourselves sensibly and appropriately view as rational given the information at our disposal at the time are not necessarily actually optimal. The problem about doing the rational thing—doing that which we sensibly suppose to be supported by the best reasons—is that our information, being incomplete, may well point us in the wrong direction. Facing this "predicament of reason," we know the pitfalls, realizing full well the fragility of these "best laid schemes." So the problem remains: Why should we act on the most promising visible alternative, when visibility is restricted to the limited horizons of our own potentially inadequate vantage-point?

The answer here runs something like this: The rational person is, by definition, someone who uses intelligence to maximize the probability—that is, the responsibly formed subjective probability—that matters will eventuate favorably for the promotion of his real interests. It is just this that makes following the path of rationality the rational course. Rationality calls for adopting the overall best (visible) alternative—the best that is, in practice, available to us in the circumstances. And if A indeed is the rational thing to do in this sense, then we should expect to be worse off in doing something different from A. To be sure, things may not come to that; we could be lucky. But this is something we do not deserve and certainly have no grounds to expect.

After all, why do we endorse the world-descriptions of the science of the day? Why do we follow the medical recommendations of the physicians of the day, or the policy recommendations of the economists of the day? Because we know them to be correct—or at any rate highly likely to be true? Not at all! We know or believe no such thing—historical experience is too strongly counterindicative. Rather, we accept them as guides only because we see them as more promising than any of the identifiable alternatives that we are in a position to envision. We accept them because they afford us the greatest available subjective probability of success-discernibly the best bet. We do not proceed with unalloyed confidence, but rather with the resigned recognition that we can do no better at the moment. Similarly, the recommendations of reason afford not assurance of success, but merely the best overall chances of reaching our goals. We act, in short, on the basis of *faute de mieux* considerations, of "this or nothing better—as far as the eye reaches." In real-world situations reason trades in courses of action whose efficacy is a matter of hope and whose rationalization is a matter of this-or-nothing-better argumentation.

Like the drowning man, we clutch at the best available object. We recognize full well that even the most rationally laid scheme can misfire. Reality is not always and inevitably on the side of the strongest arguments. Reason affords no guarantee of success, but only the reassurance of having made the best rational bet—of having done as well as one could in the circumstances of the case. One cannot say, flatly and unqualifiedly: "You should be rational because rationality pays in rendering success if not certain then at any rate more probable." Rather, we have to content ourselves with: "You should be rational because this affords the best rationally foreseeable prospects of success—on the whole and in the long run."

And so, we follow reason because this makes good rational sense to do so, seeing that this affords us with the best visible prospect for realizing our objectives. One should be rational in general for just the same sort of reason as one should be rational in the specific case of the hungry man's choice between eating bread or sand—namely, that by all available indications this course represents the most promising prospect of attaining one's sensible goals.[4] We simply could not endorse that course of action as being "rationally advisable in the circumstances" if we were not convinced that it enhanced the chances of a successful issue. For it is just this that makes something into "the rational thing to do": its enhancing, as best we can tell, our chances of attaining success to a greater extent than any other available alternative. We may have trouble spotting "the rational thing to do" in particular circumstances. But once our mind is made up about this, then the issue of rational advisability is closed.

Yet the fact of the matter remains that we cannot prove that rationality pays—necessarily or even only over the probabilistic long run. We do not know that acting rationally in the particular case at hand will actually pay off—nor can we even claim with unalloyed assurance that it will probably do so (with real likelihood rather than subjective probability). We can only say that, as best we can judge the matter, it represents the most promising course at our disposal. We have no guarantees—no means are at our disposal for preestablishing that following rationality's counsel actually pays.

Throughout our cognitive and practical affairs we have to conduct our operations under conditions of risk. And so, when we do the rational thing but it just does not pan out, we simply have to "grin and bear it." The matter is one of calculated risks and plausibly expectable benefits. Rationality affords no guarantees. By the very nature of what is involved in rational procedure, the determinable odds are in its favor. But that may still be cold comfort when things go wrong. Then, all we have is the satisfaction of having done our best. The long and short of the matter is that nothing "obliges" us to be rational except our rationality itself.

Of course, one may somehow prefer not to be rational. With belief, I may prefer congeniality to truth. With action, I may prefer convenience to optimality. With value, I may prefer the pleasingly base to the more austere better. On all sides, I may willfully opt for "what I simply like," rather than for that which is normatively appropriate. But if I do this, I lose sight of the actual ends of the cognitive, practical, and evaluative enterprises, to the detriment of my real (as opposed to apparent) interests. It lies in the nature of things that reason is on the side of rationality. To be sure, she offers us no guarantees. Yet, if we abandon reason there: is no place better that we can (rationally) go.

The Pragmatic Turn: Even Cognitive Rationality Has a Pragmatic Rationale

To be sure, no considerations of theoretical general principle can possibly establish that what is apparently the rationally optimal course—what is so "as best one can tell"—is actually optimal. In this matter we cannot proceed by way of cogent inference in the evidential/cognitive order of reason, but must turn in another direction altogether, to inference in the practical order of reason. The best available justification of rationality is a practical inference along the following lines:

1. We want and need rationally cogent answers to our questions—answers that optimally reflect the available information.
2. Following the path of cognitive rationality (as standardly construed) is the, best available way to secure rationally cogent answers to our questions.

> *Therefore:* Following the path of standard cognitive rationality in matters of inquiry (that is, in answering our questions) is the rational thing to do: we are rationally well advised to answer our questions in line with the standard processes of cognitive rationality.

And as such a perspective indicates, it is appropriate to proceed rationally not because we know that by so doing we will (inevitably or probably) succeed, but because we realize that by doing so we will have done the very best we possibly can toward producing this outcome: we will have given the matter "our best shot."[5]

This practical turn is ultimately inevitable. We can do no more than to adopt an approach that represents the best and the most that we can do. For of course we cannot maintain: "If you form a belief rationally, then it will turn out to be true." This is simply not on the cards. The most we can do is to maintain:

> By all the relevant indications, there is good reason to think that a rationally formed belief is true. (That is exactly what we mean by "a rationally formed belief.")

The cogency of our practical argument rests on the fact that in real-life situations we simply have to do the best we can—that it would be senseless (and irrational!) to ask for more than this.

One could—to be sure—press the issue still one step further. For one might take the following line:

> Very well—let it be granted that acceptance is necessary to the project of rational inquiry or what you have called the "cognitive enterprise."

So what! Why should one seek to play this "rationality game" at all? After all, to call someone "a rational person" is just giving this individual an honorific pat on the back for comporting in intellectual affairs in an approved manner. It leaves open the questions: Why is this "rationality" really a good thing? What point is there to being rational?

We come here to yet another aspect of the skeptic's case against the utility of reason. In war, victory does not always lie on the side of the big battalions. In inquiry, truth does not always lie on the side of the stronger reasons. And here we face the skeptic's final challenge: What basis is there for the belief that the real is rational (to put it in Hegelian terms)! What assurance do we have that aligning our beliefs with the canons of the logical and the reasonable leads us any closer to the truth.

At this more ultimate stage, or course, considerations of cognitive or theoretical rationality can no longer themselves be deployed successfully. To rely on them in the defense of rationality is to move in a circle. The time is now at hand when one must go outside the whole cognitive/theoretical sphere. One clearly cannot marshal an ultimately adequate defense of rational cognition by an appeal that proceeds wholly on its own ground. It becomes necessary to seek a cognition-external rationale of justification, and now, at this final stage, the aforementioned pragmatic appeal to the conditions of effective action properly come into operation. Now the time is finally at hand for taking the pragmatic route.

Philosophers of pragmatic inclination have always stressed the ultimate inadequacy of any strictly theoretical defense of cognitive rationality. And their instincts in this regard are surely right. One cannot marshal an ultimately satisfactory defense of rational cognition by an appeal that proceeds wholly on its own grounds. In providing a viable justification the time must come for stepping outside the whole cognitive/theoretical sphere and seeking for some extracognitive support for our cognitive proceedings. It is at just this stage that a pragmatic appeal to the condition of effective action properly comes into operation.

And this pragmatic aspect of the matter has yet another side. The pivotal role of rationality as a coordination principle must also be emphasized. The human condition is such that the adequate cultivation of our individual interest requires a coordination of effort with others and imposes the need for cooperation and collaboration. But this is achievable only if we "understand" one another. And here rationality becomes critical. It is a crucial resource for mutual understanding, for rendering people comprehensible to one another, so as to make effective communication and cooperation possible.

The following three points are crucial in this regard. (1) It is a matter of life and death for us to live in a setting where we ourselves are in large measure predictable for others, because only on this basis of mutual predictability can we achieve conditions essential to our own welfare. (2) The easiest way to become predictable for others is to act in such a way that they can explain, understand,

and anticipate my actions on the basis of the question What would I do if I were in his shoes? (3) In this regard the "apparent best" is the obvious choice, not only because of its (admittedly loose) linkage to optimality per se, but also because of its "saliency." The quest for "the best available" leads one to fix on that alternative at which others too could be expected to arrive in the circumstances—so that they too can understand one's choices.

The pursuit of optimality is accordingly a determinative factor for rationality not only through its direct benefits in yielding our best apparent chances of success, but also through its providing a principle for the guidance of action that achieves the crucial requisite of social coordination in the most efficient realizable way. (But why coordinate on what is rational, why not simply on habit or fashion or "the done thing"? Partly, because these leave us in the lurch once we get off the beaten track of "the usual course of things." And partly because they are unstable and inherently unreliable.

Alternative Modes of Rationality?

Émile Durkheim was no doubt right in insisting that "all that constitutes reason, its principles and categories, is made (by particular societies operating) in the course of history."[6] But the fact that everyone's standards and criteria of rational cognition (etc.) are historically and culturally conditioned—our own of course included—certainly does not preclude their having a rationally binding stringency for those to whom they appertain, ourselves emphatically included. In conducting our affairs in this world—as in conducting our movements within it—we have no choice but to go on from where we are. If we are rational, then our standards and criteria of rationality presumably came to be ours precisely because we deem them to be (not necessarily the best possible in theory, but at any rate) the best available to us in practice on the basis of those considerations that we can conscientiously maintain.

But are the beliefs of primitive, prescientific cultures indeed less rational than ours? A resounding negative is maintained in Peter Winch's widely cited article on "Understanding a Primitive Society,"[7] which argues that Azande beliefs about witchcraft and oracles cannot be rejected as rationally inappropriate despite their clear violation of the evidential canons of modern Western scientific culture. But just here lies the problem. Are those beliefs of theirs rational? differs significantly from Are those beliefs of theirs held rationally? The answer you get depends on the question you ask. If we ask Are those beliefs they hold rational? we, of course, mean "*rational* on our understanding of the matter." And the answer here is clearly *No*, seeing that by hypothesis our sort of rationality does not figure in *their* thinking at all. The fact that they (presumably) deem their beliefs somehow "justified" by some considerations or other that they deem appropriate is going to cut no ice in *our* deliberations regarding the cogency of those beliefs. ("Are they being "rational" by *their* lights?" is one thing and "Are they being rational by *our* lights?" another.) When *we* ask about the rational acceptability of those beliefs we mean rational for us in our circumstances and not rational for them in theirs—the issue to be considered from *our* point of view, and not from somebody else's!

The problems that arise at this juncture go back to the quarrel between E. E. Evans-Pritchard and Lucien Lévy-Bruhl. In his book on *Primitive Mentality*,[8] Lévy-Bruhl maintained that primitive people have a "prelogical mentality." Against this view, Evans-Pritchard[9] argued that primitive people were perfectly "logical" all right, but simply used a logic *different* from ours. When, for example, the Nuer maintain that swamp light *is identical with* spirit, but deny that spirit *is identical with* swamp light, they are not being illogical, but simply have in view a logic of "identity" variant from that in vogue in Western cultures. But the obvious trouble with this sort of thing is that nothing apart from bafflement and confusion can result from translating Nuer talk by into *our* identity language if indeed it is the case that what is at issue in their thought and discourse nowise answers to our identity conception. Instead of translating the claim at issue as "Swamp light *is* identical with Spirit" and then going on to explain that "is identical with" does not really mean what it says because the ground rules that govern this idea are not applicable, an anthropologist would do well to reformulate or paraphrase (if need be) the claim at issue in such a way as to render intelligible what is actually going on. The fact that the Nuer have different (and to us strange-seeming) beliefs about "spirits" no more means that they have a logic different from ours than the fact that they communicate by drums mean that they have a telephone system different from ours.[10] To reemphasize: when we ask about logical acceptability it is logical acceptability by *our* lights that is at issue.

Anthropologists do sometimes say that a certain society has a conception of rationality that is different from ours. But that is literally nonsense. Those others can no more have a *conception of rationality* that addresses an object different form ours, than they can have a *conception of iron* that addresses an object different from ours, or a *conception of elephants* that addresses objects different from ours. If they are to conceive of *those* particular things at all, then their conception must substantially accord with ours. In any discussion of ours, iron objects are *by definition* what we take them to be; "elephant" is our word and *elephant* our conception. If you are not talking about *that*, then you are not talking about *elephants* at all. You have simply "changed the subject," and exited from the domain of the discussion. Similarly, if a conception of theirs (whatever it be) is not close to what *we* call rationality, then it just is not a conception of *rationality*—it does not address the topic that *we* are discussing when we put the theme of rationality on the agenda. Rationality as we ourselves see it is a matter of striving intelligently for appropriate resolutions—using *relevant* information and *cogent* principles of reasoning in the resolution of one's conjunctive and practical problems. If that is not what they are after, then it is not *rationality* that concerns them. The issues at stake in *our* deliberations have to be the issues as *we* construe them.

Of course, they may *think* that what we call pencils are chopsticks and use them as such. Or they may *think* books to be doorstops and use them as such. But that does not mean that they conceive of pencils or books differently from us, or that they have a different conception of pencils and books. "They take pencils to be something we do not (namely chopsticks)," is fine as a way of talking. But "They believe pencils to be chopsticks" is nonsense unless it is glossed as: "They believe these sorts of things called 'pencils' to be chopsticks." And when this happens, then they do not conceive of pencils differently from ourselves, they just do not conceive of *pencils* at all. They simply do not have the (one and only) available conception of pencils—namely, ours. In such cases, if they do not have *our* concept, then they just do not have *the* concept.

When social scientists say that alien cultures have a different "rationality" from ourselves what they generally mean (strictly speaking) is (1) that they have different *objectives* (e.g., that we seek to control and change our environment to suit our purposes, while they tend to reconstitute their purposes to suit their environment—to endeavor to come into "harmony" with nature), and/or (2) that they use *problem-solving techniques* that are different from ours (e.g., that we employ empirical investigation, evidence, and science, while they use divination, omens, or oracles). But if they pursue different sorts of ends by different sorts of means they, perhaps, have a different thought style and a different intellectual ethos, but not a different *rationality*. The anthropologists' talk of different *rationality* is simply an overly dramatic (and also misleading) way of making a valid point—namely, that they do their intellectual problem-solving business in a way different from ours. But those different processes of theirs do not mean that they have a different *rationality* any more than those blowguns of their mean that they have a different *rifle*.

So, it is literally nonsense to say "The *X*'s have a different conception of rationality from the one we have." For, if they do not have ours, they do not have any. It is, after all, rationality as we conceive of it that is at issue in this discussion of ours. Whatever analogue or functional equivalent there may be with which they are working, it just is not something that we, in our language, can call "a conception of rationality."

There is no difficulty with the idea that "They *implement and apply* the conception of rationality differently from ourselves." After all, we implement and apply the idea of a medication very differently from the ancient Greeks, using medications they never dreamt of. But the matter stands differently with the *conception* of a medication as such. This remains what it always has been: "a substance used as a remedy for an ailment." When one ceases to operate with what answers to that conception, then (*ex hypothesi*) one is no longer dealing with medications at all. The discussion has moved on to other topics. "Having a *different* conception" of pencils or elephants or rational actions simply means not having that conception at all. If they do not have *our* conception of scissors, they do not have *a* conception of scissors, full stop. For, when we ask about their dealings with scissors it is our own conception that defines the terms of reference. If we recognize agents as rational at all, then we ourselves can make good rational sense of what they do! This is not because *we* are so talented and versatile, nor yet because the slogan "rationality is universal" gets it quite right. It is just because we could not and would not say that they are rational (could not and would not characterize the phenomena in this particular way) if we could not make good rational sense of what they do within the setting of their situation.

The pivotal fact lies in a "questioner's prerogative." Since the question What is their mode of rationality? is *ours*, so is the "rationality" that is at stake here. Thus, on the crucial issues—What is rationality all about?, What sorts of considerations characterize the rationality at issue?, What is appropriately at stake?—it is our own position that is determinative. When the questions are ours, the concepts that figure in them are ours as well. At this stage of establishing the constitutive ground rules of appropriateness for rationality, it is *our own position* that is decisive.

Gestalt switches are certainly possible, but they are just that: unpredictable leaps. We do not reason our way into them by the use of existing standards—if it were so it would be an implementation of the old standard that is at issue and not, as per hypothesis, a Gestalt switch. They are only seen as rational *ex post facto*, from the vantage point of the *then* prevailing "established" standard—that is, the new one.

Consider the contention:

> Surely there are no historically and culturally invariant principles of rationality. People's (altogether plausible) views about what is rational change with changes in place and time.

The response here is: yes and no. Of course, different people in different places and times conduct their "rational" affairs quite differently. But, at the level of basics, of first fundamentals, there is bound to be a uniformity. For, what all modes of "rationality" have in common is precisely this—that they all qualify as "modes of rationality" under *our* conception of the matter (which, after all, is what we're talking about). At this level of deliberation, "questioner's prerogative" prevails, and our own conception of the matter becomes determinative.

It is helpful to contemplate some analogies. There are many sorts of blades for knives. But, the fundamental principle that knives have blades at all does not depend on how people choose to make knives but on *our conception* of what a "knife" is. If the given objects, whatever they might be, do not have *blades* then they are not *knives*. It would, clearly, be the height of folly to go about in another culture asking people "Must knives have blades?" The answer is a foregone conclusion. A negative response would not counterindicate the thesis at issue, but would simply betoken a failure to comprehend.

Again, there are various quite different sorts of information that people take as evidence to substantiate a claim. But "deeper" principles like "Give more credence to that for which the evidence is stronger" or "In inquiry, endeavor to expand and extend the evidence for your claims" do not depend on the evidential practices of people, but on *our conception* of what evidence is all about. If people do not proceed in ways that conform to these evidential conceptions and principles of ours, then their practices—whatever they might be—are not *evidential* practices. And the same holds good for rationality as well.

After all, it is *our conception* of rationality that fixes the "rules of the game" at issue when we pursue our deliberations about these matters. We have to play the rationality game by our ground rules because it is exactly those ground rules that define and determine what "the rationality game" is that is at issue on our deliberations. If we were not playing the game on this basis, it would not be the *rationality* game that we were engaged in—it would not be *rationality* that is the subject of our concern. It is the determinative role of our own rationality standards that makes them absolute for us.

Rationality is in this regard like communication. What communication is is the same everywhere and for everyone—inherent in the nature of the concept that is at issue. But, of course, it is only normal and natural that different people in different places and times would transact their communicative business very differently, since what is effective in one context may fail to be so in another. Similarly, *what rationality is* is one thing (and one uniform thing from person to person within the framework of a meaningful discussion of the topic); but *what is rational* is something else again—something that is by no means uniform from person to person but variable with situation and circumstance.

To be sure, what makes our own conceptions authoritatively determinative in such matters is nothing special about us. Clearly, the Aristotelian cosmos is no longer with us—we are not the center of the world. But we certainly and inevitably are at the center *of our thought world*. Our inquiries have to be conducted within our frame of reference. We have to pose *our* questions in line with *our* ideas, to frame *our* perplexities by means of *our* concepts, to consider *our* issues in *our* terminology. If we ourselves are to classify someone as rational at all (and who else's attribution is now at issue?) then we must deem him qualified under the aegis of rationality *as we understand it*. If we ask about someone Is she tall?, we are clearly asking about her height *as we conceive it*. What—if anything—she herself thinks about height is beside the point. And exactly the same situation holds with respect to rationality. A condition of "questioner's prerogative" prevails—it is the person who puts the issue on the table who sets the frame of reference for determining what that issue involves—it is, after all, *his* question. With all these questions about rationality, it is of course "rationality" *as we ourselves conceive of it* that is operative. The topic being ours, it is we who set the terms of reference for what is at issue. At this point "epistemic relativism" comes to a stop.

Paradoxically, it is precisely the inevitable relativization of *our* questions and concerns and puzzlements to *our* terms of reference that makes those particular terms of reference absolute in our own discussions. Being framed in our terminology, it is our terminology, that is decisive for the questions that *we* raise and the inquiries that *we* conduct. If *we* ask if X is being rational in believing (or doing or evaluating) a certain thing, then the issue is clearly one of its being rational on the basis of the conception of rationality as *we* understand it. The governing absoluteness of our conception inheres in "questioner's prerogative"—in the fact that the questions and issues we address in our deliberations about rationality are in fact our own and that, since the questions are ours, it is our conceptions that are determinative for what is at issue. In this regard, our commitment to our own cognitive position is (or should be) unalloyed. We ourselves are bound to see our own (rationally adopted) standards as superior to the available alternatives—and are, presumably, rationally entitled to do so on the basis of the cognitive values we ourselves endorse.

To qualify those of an alien culture as fully rational we must maintain both that they are conducting their inquiries intelligently by their own rules and also that in our sight these rules make good rational sense given their situation. It is ultimately "intelligent comportment" and "making sense" according to *our* standards of appraisal that makes what is at issue rationally invariant. The fact that *we* do (and must) apply *our* own idea of the matter is what makes for the universal element of rationality. What is universal about language use; namely, that to accredit another culture as rational at all is to accept it as being "rational" in *our* sense of the term—which may, to be sure, involve deciding whether their actions measure up to *their* standards. The absoluteness of (ideal) rationality is inherent in the very concept at issue.

THE SELF-RELIANCE OF RATIONALITY IS NOT VICIOUSLY CIRCULAR

To be sure, there is bound to be a skeptic who comes along to press the following objection: "Your proposed universalistic legitimation of objectivity pivots on the appropriateness of rationality. But your legitimation of reason conforms to the pattern: 'You should be rational just because that is the rational thing to do!' And this is clearly circular." It might seem questionable to establish the jurisdiction of reason by appeal to the judgment of reason itself. But, in act, of course, this circularity is not really vicious at all. Vicious circularity stultifies by begging the question but virtuous circularity merely coordinates related elements in their mutual interlinkage. The former presupposes what is to be proved, the latter simply shows how things are connected in a well-coordinated and mutually supportive interrelationship. The self-reliance of rationality merely exemplifies this latter circumstance of an inherent coordination among its universe components. It is not a matter of sequential validation but rather of a legitimation process that is coordinative and coherentistic.

This practical line of argumentation may still seem to leave the situation in an unsatisfactory state. It says (roughly): "You should be rational in resolving your choices because *it is rational to believe that* the best available prospects of optimality-attainment are effectively realized in this way." To be sure, one might deem it preferable if that italicized clause were wholly suppressed. A skeptic is bound to press the following objection:

> The proposed practicalistic legitimation of reason conforms to the pattern: "You should be rational just because that is the rational thing to do!" And this is clearly circular.

It might seem questionable to establish the jurisdiction of reason by appeal to the judgment of reason itself. But, in fact, of course, this circularity is not really vicious at all. Vicious circularity stultifies by begging the question; virtuous circularity merely coordinates related elements in their mutual interlinkage. The former presupposes what is to be proved, the latter simply shows how things are connected together in a well-coordinated and mutually supportive interrelationship. The self-reliance of rationality merely exemplifies this latter circumstance of an inherent coordination among its universe components.

Admittedly, the reasoning at issue has an appearance of vitiating circularity because the force of the argument itself rests on an appeal to rationality. It seems to say: "If you are going to be rational in your beliefs, then you must also act rationally, because it is rational to believe that rational action is optimal in point of goal attainment." But this sort of question begging is simply unavoidable in the circumstances. It is exactly what we want and need. Where else should we look for a rational validation of rationality but to reason itself? The only reasons for being rational that it makes sense to ask for are rational reasons. In this epistemic dispensation, we have no way of getting at the facts directly, without the epistemic detour of securing grounds and reasons for them. And it is, of course, rationally cogent grounds and reasons that we want and need. The overall justification of rationality must be reflexive and self-referential. To provide a rationale of rationality is to show that rationality stands in appropriate alignment with the principles of rationality. From the angle of justification, rationality is a cyclic process that closes in on itself, not a linear process that ultimately rests on something outside itself.

There is accordingly no basis for any rational discontent, no room for any dissatisfaction or complaint regarding a "circular" justification of rationality. We would not (should not) want it otherwise. If we bother to want an answer to the question Why be rational? at all, it is clearly a rational answer that we require.

The only sort of justification of anything—rationality included—that is worth having at all is a rational one. That presupposition of rationality is not vitiating, not viciously circular, but essential—an unavoidable consequence of the self-sufficiency of cognitive reason. There is simply no satisfactory alternative to using reason in its own defense. Already embarked on the sea of rationality, we want such assurance as can now be made available that we have done the right thing. And such reassurance can indeed be given—exactly along the lines just indicated. Given the very nature of the justificatory enterprise at issue, one just cannot avoid letting rationality sit in judgment on itself. (What is being asked for, after all, is a rational argument for rational action, a basis for rational conviction, and not persuasion by something probatively irrelevant like threats of force majeure.) One would expect, nay demand, that rationality be self-substantiating in this way—that it must emerge as the best policy on its own telling.

From the justificatory point of view, rationality is and must be autonomous. It can be subject to no external authority. Rationality in general is a matter of systematization, and the justification of rationality is correspondingly a matter of systemic self-sufficiency. Rather than indicating the defect of vicious circularity, the self-referential character of a justification of rationality is a precondition of its adequacy! It is only a rational legitimation of rationality that we would want: any other sort would avail us nothing. And if such a rational validation were not forthcoming this would indicate a grave defect.

And so, the predicament of reason gives no comfort to skepticism or irrationalism, and yields no grounds for abandoning reason or, worse yet, for turning against it. Admittedly, this self-supportive legitimation of rationality is the only cogent sort of validation that we are going to get. But in the final analysis it is the only sort that it makes sense to ask for, seeing that rationality itself enjoins us to view the best we can possibly get as good enough.[11]

A desperate objection yet remains: "So rationality speaks on its own behalf. Well and good. But why should I care for rationality? Why should I set myself to do the intelligent and appropriate thing?"

At this point there is little more to be said. If I want a reason at all, I must want a rational reason. If I care about reasons at all, I am already within the project of rationality. But once I am within the project, there is nothing further external to reason that can or need be said to validate it. At that stage rationality is already at hand to provide its own support—it wears its justification on its sleeve. (The project of trying to reason with someone who stands outside the range of rationality to convince them to come into its fold is clearly an exercise in pointlessness and futility.)

One may of course quite appropriately ask questions like: Why should I cultivate the truth; why should I cultivate my best (or true) interests? But in the very act of posing these questions I am asking for reasons—that is, I am evincing my commitment to the project of rationality. Caring for the truth and for one's best interests are simply part and parcel of this commitment. If I do not care for these things, then there is really no point in raising these questions. For in this event I have *already* taken my place outside the precincts of rationality, beyond the reach of reason. And then. Of course, there is no reason why any sensible person would want to follow me there.

Irrationality—wishful thinking and self-deception—may be convenient and even, in some degree, psychologically comforting. But it is not cognitively satisfactory. If it is a viable defense of a position that we want, it is bound to be a rational one. The only validation of rationality that can reasonably be asked for—and the only one worth having—must lie in considerations of the systemic self-sufficiency of reason. In the final analysis, Why be rational? must be answered with the only rationally appropriate response: Because rationality itself obliges us to be so. In providing a *rational* justification of objectivity—and what other kind would we want?—the best we can do is to follow the essentially circular (but *nonviciously* circular!) line of establishing that reason herself endorses taking this course. The only validation of rationality's recommendations that can reasonably be asked for—and the only one worth having—must lie in considerations of the systemic self-sufficiency of reason. Reason's self-recommendation is an important *and necessary* aspect of the legitimation of the rational enterprise. And in those matters where rationality counts, objectivity is the best policy by virtue of this very fact itself.

Yet does reason's self-reliance not open the door to skepticism? For skeptics have always insisted on just this point that we cannot prove in advance of conceding reason's cogency that we will not go wrong by trusting our reason. And this—so we have granted—is quite correct. But, of course, what one *can* do is to establish that if we reject reason we cut ourselves off from any (rationally warranted) expectation of success. There are no guarantees that our ventures in trust are going to prove successful; whether our trust is actually warranted in any given circumstances (trust in ourselves, in our cognitive faculties, in other people, and the like) is something we cannot in the nature of things ascertain in advance of events. A conjunction of trust with hope and faith is germane alike to the cognitive project, the practical project, and the evaluative project. Throughout, we have to conduct our operations under conditions of risk, without assured confidence in outcomes and without advance guarantees of success. In all these matters efficacy is a matter of trustful hope and confidence in the best available option whose rationale is a matter of this-or-nothing-better argumentation.[12]

Part III

Cognitive Progress

Chapter 12

Scientific Progress

Synopsis

- *The development of inquiry in natural science is best understood on the analogy of exploration—to be sure, not in the geographical mode but rather exploration in nature's parametric space of such physical quantities as temperature, pressure, and field strength.*
- *The technology-mediated exploration that comes into play here involves interactions between man and nature that become increasingly difficult (and expensive) as we move ever farther away from the home base of the accustomed environment of our evolutionary heritage.*
- *The course of scientific progress accordingly involves a technological escalation—an ascent to successively higher levels of technological sophistication that is unavoidably required for the production of duly informative observational data.*
- *Scientific inquiry seeks to develop a harmoniously systematized coordination of theorizing conjecture with the determinable data. However, this attempt at data/theory equilibration repeatedly sustains the destabilizing shocks of ongoingly enlarged experience. Our technologically mediated entry into new regions of parameter space confronts us with this task in ever-renewed forms.*
- *In theory, the prospect of such ongoing "scientific revolutions" is potentially unending. And there is certainly no warrant for a theory of convergence that sees the innovations of theorizing science as being of constantly diminishing scope; with discoveries in natural science, later does not mean*

> lesser. However, the increasing cost of further progress is a portentous consideration.
> - We may indeed stand at the center of our cognitive realm but we do not stand at the center of the cognitive world.
> - For the advancement of cognitive progress means that we must see the science of the future as being not just different but better.

THE EXPLORATION MODEL OF SCIENTIFIC INQUIRY

A theoretical prospect of unending scientific progress lies before us. But its practical realization is something else again. One of the most striking and important facts about scientific research is that the ongoing resolution of significant new questions faces increasingly high demands for the generation and cognitive exploitation of data. And in developing natural science, we humans began by exploring the world in our own locality, and not just our spatial neighborhood but—more far-reachingly—our *parametric* neighborhood in the space of physical variable such as temperature, pressure, and electric charge. Near the "home base" of the state of things in our accustomed natural environment, we can operate with relative ease and freedom—thanks to the evolutionary attunement of our sensory and cognitive apparatus—in scanning nature with the unassisted senses for data regarding its modes of operation. But in due course we accomplish everything that can be managed by these straightforward means. To do more, we have to extend our probes into nature more deeply, deploying increasing technical sophistication to achieve more and more demanding levels of interactive capability. We have to move ever further away from our evolutionary home base in nature toward increasingly remote observational frontiers. From the egocentric standpoint of our local region of parameter space, we journey ever more distantly outward to explore nature's various parametric dimensions in the search for cognitively significant phenomena.

The appropriate picture is not, of course, one of geographical exploration but rather of the physical exploration—and subsequent theoretical systematization—of phenomena distributed over the parametric space of the physical quantities spreading out all about us. This approach in terms of exploration provides a conception of scientific research as a prospecting search for the new phenomena demanded by significant new scientific findings. As the range of telescopes, the energy of particle accelerators, the effectiveness of low-temperature instrumentation, the potency of pressurization equipment, the power of vacuum-creating contrivances, and the accuracy of measurement apparatus increases—that is, as our capacity to move about in the parametric space of the physical world is enhanced—new phenomena come into view. After the major findings accessible via the data of a given level of technological sophistication have been achieved, further major findings become realizable only when one ascends to the next level of sophistication in data-relevant technology. Thus the key to the great progress of contemporary physics lies in the enormous strides which an ever more sophisticated scientific technology has made possible through enlarging the observational and experimental basis of our theoretical knowledge of natural processes.[1]

In cultivating scientific inquiry, we scan nature for interesting phenomena and grope about for the explanatorily useful regularities they may suggest. As a fundamentally inductive process, scientific theorizing calls for devising the least complex theory structure capable of accommodating the available data. At each stage we try to embed the phenomena and their regularities within the simplest (cognitively most efficient) explanatory structure able to answer our questions about the world and to guide our interactions in it. But step by step as the process advances, we are driven to further, ever greater demands that can be met only with a yet more powerful technology of data exploration and management.

This idea of the exploration of parametric space provides a basic model for understanding the mechanism of scientific innovation in mature natural science. New technology increases the range of access within the parametric space of physical processes. Such increased access brings new phenomena to light, and the examination and theoretical accommodation of these phenomena is the basis for growth in our scientific understanding of nature.

The Demand for Data Enhancement

Natural science is fundamentally empirical, and its advance is critically dependent not on human ingenuity alone but on the monitoring observations to which we can gain access only through interactions with nature. The days are long past when useful scientific data can be had by unaided sensory observation of the ordinary course of nature. Artifice has become an indispensable route to the acquisition and processing of scientifically useful data. The sorts of data on which scientific discovery nowadays depends can be generated only by technological means.

The pursuit of natural science as we know it embarks us on a literally endless endeavor to improve the range of effective experimental intervention, because only by operating under new and heretofore inaccessible conditions of observational or experimental systemization—attaining extreme temperature, pressure, particle velocity, field strength, and so on—can we realize situations that enable us to put knowledge-expanding hypotheses and theories to the test. The enormous power, sensitivity, and complexity deployed in present-day experimental science have not been sought for their own sake but rather because the research frontier has moved on into an area where this sophistication is the indispensable requisite of further progress. In science, as in war, the battles of the present cannot be fought effectively with the armament of the past. As one acute observer has rightly remarked: "Most critical experiment [in physics] planned today, if they had to be constrained within the technology of even ten years ago, would be seriously compromised."[2]

With the enhancement of scientific technology, the size and complexity of our body of data inevitably grows, expanding on quantity and diversifying in kind. Technological progress constantly enlarges the window through which we look upon nature's parametric space. In cultivating and developing natural science we continually enlarge our view of this space and then generalize on what we see. But what we have here is not a homogeneous lunar landscape, where once we have seen one sector we have seen it all, and where theory projections from lesser data generally remain in place when further data comes our way. Historical experience shows that there is every reason to expect that our ideas about nature are subject to constant radical changes as we explore parametric space more extensively. The technologically mediated entry into new regions of parameter space constantly destabilizes the attained equilibrium between data and theory. It is the engine that moves the frontier of progress in natural science.

Technological Escalation: An Arms Race against Nature

This situation points toward the idea of a "technological level," corresponding to a certain state-of-the-art in the technology of inquiry in regard to data generation and processing. This technology of inquiry falls into relatively distinct levels or stages in sophistication—correlatively with successively "later generations" of instrumentation and manipulative machinery. These levels are generally separated from one another by substantial (roughly, order-of-magnitude) improvements in performance in regard to such information-providing parameters as measurement exactness, data-processing volume, detection sensitivity, high voltages, high or low temperatures, and so on.

The perspective afforded by such a model of technologically mediated prospecting indicates that progress in natural science has heretofore been relatively easy because in this earlier explanation of our parametric neighborhood we have been able—thanks to the evolutionary heritage of our sensory and conceptual apparatus—to operate with relative ease and freedom in exploring our own parametric neighborhood in the space of physical variables like temperature, pressure, radiation, and so on. But scientific innovation becomes even more difficult—and expensive as we push out further from our home base toward the more remote frontiers.

No doubt, nature is in itself uniform as regards the distribution of its diverse processes across the reaches of parameter space. It does not favor us by clustering them in our accustomed parametric vicinity: significant phenomena do not dry up outside our parochial neighborhood. But scientific innovation becomes more and more difficult—and expensive—as we push our explorations even further away from our evolutionary home base toward increasingly remote frontiers.

And phenomenological novelty is seemingly inexhaustible: we can never be unalloyedly confident that we have got to the bottom of it. Nature always has fresh reserves of phenomena at her disposal, hidden away in those ever more remote regions of paramative space. Successive stages in the technological state of the art of scientific inquiry accordingly lead us to ever-different views about the nature of things and the character of their laws. Without an ever-developing technology, scientific progress would soon grind to a halt. The discoveries of today cannot be made with yesterday's equipment and techniques. To conduct new experiments, to secure new observations, and to detect new phenomena, and ever more powerful investigative technology is needed. Scientific progress depends crucially and unavoidably on our technical capability to penetrate into the increasing distant—and increasingly difficult—reaches of the spectrum of physical parameters in order to explore and to explain the ever more remote phenomena encountered there.

The salient characteristic of this situation is that, once the major findings accessible at a given level of sophistication in data-technology level have been attained, further major progress in any given problem area requires ascent to a higher level on the technological scale. Every data-technology level is subject to discovery saturation, but the exhaustion of prospects at a given level does not, of course, bring progress to a stop. Once the potential of a given state-of-the-art level has been exploited, not all our piety or wit can lure the technological frontier back to yield further significant returns at this stage. Further substantive findings become realizable only by ascending to the next level of sophistication in data-relevant technology. But the exhaustion of the prospects for data extraction at a given data-technology level does not, of course, bring progress to a stop. Rather, the need for enhanced data forces one to look further and further from man's familiar "home base" in the parametric space of nature. Thus, while scientific progress is in principle always possible—there being no absolute or intrinsic limits to significant scientific discovery—the *realization* of this ongoing prospect demands a continual improvement in the technological state-of-the-art of data extraction of exploitation. And pioneering scientific research will always operate at the technological frontier because nature inexorably exact a drastically increasing effort with respect to the acquisition and processing of data for revealing her "secrets." This accounts for the need for ever recourse to more sophisticated technology for research in natural science. The increasing technological demands that are requisite for scientific progress means that each step ahead gets more complex and more expensive as those new parametric regions grow increasingly remote. With the progress of science, nature becomes less and less yielding to the efforts of further inquiry. We are faced with the need to push nature harder and harder to achieve cognitively profitable interactions. The dialectic theory and experiment carries natural science ever deeper into the range of greater costs.

We thus arrive at the phenomenon of *technological escalation*. The need for new data forces us to look further and further from man's familiar home base in the parametric space of nature. Thus while scientific progress is in principle always possible—there being no absolute or intrinsic limits to significant scientific discovery—the realization of this ongoing prospect demands a continual enhancement in the technological state of the art of data extraction or exploitation.

The perspective afforded by such a process of escalation indicates that progress in natural science was at first relatively undemanding because we have explored nature in our own parametric neighborhood.[3] And seeing that the requisite technology was relatively crude, economic demands were minimal at this stage. But over time an ongoing escalation in the resource costs of significant scientific discovery arose from the increasing technical difficulties of realizing this objective, difficulties that are a fundamental—and an ineliminable—part of an enterprise of empirical research, for we must here contrive ever more "far out" interactions with nature, operating in a continually more difficult, accordingly, sector of parametric space.

Given that we can only learn about nature by interacting with it, Newton's third law of countervailing action and reaction becomes a fundamental principle of epistemology. Everything depends on just how and *how hard* we can push against nature in situations of observational and detectional interaction. As Bacon saw, nature will never tell us more than we can forcibly extract from her with the means of interaction at our disposal. And what we can manage to extract by successively deeper probes is bound to wear a steadily changing aspect, because we operate in new circumstances where old conditions cannot be expected to prevail and the old rules no longer apply.

Physicists often remark that the development of our understanding of nature moves through successive layers of *theoretical* sophistication.[4] But scientific progress is clearly no less dependent on continual improvements in strictly *technical* sophistication:

> Some of the most startling technological advances in our time are closely associated with basic research. As compared with 25 years ago, the highest vacuum readily achievable has improved more than a thousand-fold; materials can be manufactured that are 100 times purer; the submicroscopic world can be seen at 10 times higher magnification; the detection of trace impurities is hundred of times more sensitive; the identification of molecular species (as in various forms of chromatography) is immeasurably advanced. These examples are only a small sample. . . . Fundamental research in physics is crucially dependent on advanced technology, and is becoming more so.[5]

Without an ever-developing technology, scientific progress would grind to a halt. The discoveries of today cannot be advanced with yesterday's instrumentation and techniques. To secure new observations, to test new hypotheses, and to detect new phenomena, an ever more powerful technology of inquiry is needed. Throughout the natural sciences, technological progress is a crucial requisite for cognitive progress.

Frontier research is true *pioneering:* what counts is not just doing it but doing it *for the first time*. Apart from the initial replication of claimed results needed to establish the reproducibility of results, repetition in research is in general pointless. As one acute observer has remarked, one can follow the diffusion of scientific technology "from the research desk down to the schoolroom":

> The emanation electroscope was a device invented at the turn of the century to measure the rate at which a gas such as thorium loses its radioactivity. For a number of years it seems to have been used only in the research laboratory. It came into use in instructing graduate students in the mid-1930's, and in college courses by 1949. For the last few years a cheap commercial model has existed and is beginning to be introduced into high school courses. In a sense, this is a victory for good practice; but it also summarizes the sad state of scientific education to note that in the research laboratory itself the emanation electroscope has long since been removed from the desk to the attic.[6]

In science, as in a technological arms race, one is simply never called on to keep doing what was done before. And ever more challenging task is posed by the constantly *escalating* demands of science for the enhanced data that can only be obtained at the increasingly costly new levels of technological sophistication. One is always forced farther up the mountain, ascending to ever-higher levels of technological performance—and of expense. As science endeavors to extend its "mastery over nature," it thereby comes to be involved in a technology-intensive arms race against nature, with all of the practical and economic implications characteristic of such process.

The enormous power, sensitivity, and complexity deployed in present-day experimental science have not been sought for their own sake but rather because the research frontier has moved on into an area where this sophistication is the indispensable requisite of ongoing progress. In science, as in war, the battles of the present cannot be fought effectively with the armaments of the past.

THEORIZING AS INDUCTIVE PROJECTION

Even though neither present nor future science manages to depict reality with definitive finality, perhaps there is nevertheless a gradual *convergence* toward a definitively true account of nature at the level of scientific theorizing?

Considerations of great principles indicate that this is very implausible. Our exploration of physical parameter space is inevitably incomplete. We can never exhaust the whole range of temperatures, pressures, particle velocities, and so on. And so, we inevitably face the (very real) prospect that the regularity structure of the as yet inaccessible cases will not conform to the (generally simpler) patterns of regularity prevailing in the presently accessible cases. By and large, future data do not accommodate themselves to present theories. Newtonian calculations worked marvelously for predicting solar-system phenomenology (eclipses, planetary conjunctions, and the rest) but this does not show that classical physics has no need for fundamental revision.

Scientific theory-formation is, in general, a matter of spotting a local regularity of phenomena in parametric space and then projecting it "across-the-board," maintaining it globally. The theoretical claims of science are themselves never local—they are not spatiotemporally local and they are not parametrically local either. They stipulate—quite ambitiously—how things are always and everywhere. And it does not require a sophisticated knowledge of history of science to realize that our worst fears are usually realized—that it is seldom if ever the case that our theories survive intact in the wake of substantial extensions in our access to sectors of parametric space. The history of science is a history of episodes of leaping to the wrong conclusions. For (sensibly enough) the inductive method of science as a rational process of inquiry constructs the simplest, most economical cognitive structures to house these data comfortably. It calls for searching out the simplest pattern of regularity that can adequately accommodate our data regarding the issues at hand, and then projects them across the entire spectrum of possibilities in order to answer our general questions. Accordingly, scientific theorizing, as a fundamentally inductive process, involves the search for, or the construction of, the least complex theory-structure capable of accommodating the available body of data—proceeding under the aegis of established principles of inductive systematization: uniformity, simplicity, harmony, and such principles that implement the general idea of cognitive economy.

Simplicity and generality are the cornerstones of inductive systematization. And one very important point must be stressed in this connection. Scientific induction's basic idea of a *coordinative systematization of question-resolving conjecture with the data of experience* may sound like a very conservative process. But this impression would be quite incorrect. The drive to systematization embodies an imperative to broaden the range of our experience—to extend and to expand insofar as possible the database from which our theoretical triangulations proceed. In the design of cognitive systems, implicity/harmony and comprehensiveness/inclusiveness are two components of one whole. And the impetus to ever ampler comprehensiveness indicates why the ever-widening exploration of nature's parameter space is an indispensable part of the process.

Progress in natural science is a matter of dialogue or debate between theoreticians and experimentalists. The experimentalists probe nature to see its reactions, to seek out phenomena. And the theoreticians take the resultant data and weave a theoretical fabric about them. Seeking to devise a framework of rational understanding, they construct their explanatory models to accommodate the findings that the experimentalists put at their disposal. But once the theoreticians have had their say, the ball returns to the experimentalists' court. Employing new, more powerful means for probing nature, they bring new phenomena to view, new data for accommodation. Precisely because these data are new and inherently unpredictable, they often fail to fit the old theories. Theory extrapolations from the old data could not encompass them; the old theories do not accommodate them. And so a disequilibrium arises between existing theory and new data. And at this stage, the ball reenters the theoreticians' court. New theories must be devised to accommodate the new, nonconforming data. And so the theoreticians set about weaving a new theoretical structure to accommodate the new data. They endeavor to restore the equilibrium between theory and data once more. And when they succeed the ball returns to the experimentalists' court, and the whole process starts over again.

As this analogy indicated, physical nature can exhibit a very different aspect when viewed from the vantage point of different levels of sophistication in the technology of nature-investigator interaction. And thus possibility is in fact realized. As ample experience indicates, every new stage of investigative sophistication brings to the fore a different order or aspect of things. What we find in investigating nature always in some degree reflects the character of our technology of observation. What we can detect, or find, in nature is always something that depends on the mechanisms by which we search. And more sophisticated searches invariably engender changes of mind.

LATER NEED NOT BE LESSER

An insidiously alluring argument arises at this point. It contends that limited access to new phenomena does not matter for scientific progress because these further data would in any case yield only minor corrections located more decimal places out. The view that underlies such a position is that further changes are smaller changes: that a juncture has been reached where the additional advances of science so no more than provide further minor details and readjustments in a basically completed picture of how nature functions: that what we do not yet know does not matter all that much.

One particularly interesting and well-developed statement of this point of view is that of the biologist Gunther S. Stent, *The Coming of the Golden Age: A View of the End of Progress*:

> I want to consider what I believe to be intrinsic limits to the science, limits to the accumulation of meaningful statements about the events of the outer world. I think everyone will readily agree that there are *some* scientific disciplines which, by reason of the phenomena to which they purport to address themselves, are *bounded*. Geography, for instance, is bounded because its goal of describing the features of the Earth is clearly limited.... And, as I hope to have shown in the

preceding chapters, genetics is not only bounded, but its goal of understanding the mechanism of transmission of hereditary information *has*, in fact, been all but reached.... [To be sure] the domain of investigation of a bounded scientific discipline may well present a vast and practically inexhaustible number of events for study. But the discipline is bounded all the same because its goal is in view.... There is at least one scientific discipline, however, which appears to be *openended*, namely physics, or the science of matter.... But even to encounter limitations in practice ... [for] there are purely physical limits to physics because of man's own boundaries of time and energy. These limits render forever impossible research projects that involve observing events in regions of the universe more than ten of fifteen billion light-years distant, traveling very far beyond the domain of our solar system, or generating particles with kinetic energies approaching those of highly energetic cosmic rays.[7]

On such a view, natural science is approaching the end of its tether

Along these lines we also have the precedent of Charles Sanders Peirce's idea of *convergent approximation*.[8] This calls for envisaging a situation where, with the passage of time, the results we reach grow increasingly concordant. In the face of such a course of successive changes of ever-diminishing significance, we could proceed to maintain that the world may not really be *present* science claims it to be, but rather is as the ever more clearly emerging science-in-the-limit claims it to be. The reality of ongoing changes is now unimportant because with the passage of time those changes matter less and less. We increasingly approximate an essentially stable picture.

This prospect is certainly a theoretically possible one. But neither historical experience not considerations of general principle provide reason to think that it is a real possibility. Rather, one does well here to adopt the view that Stanley Jevons articulated more than a century ago:

> In the writings of some recent philosophers, especially of Auguste Comte, and in some degree John Stuart Mill, there is erroneous and hurtful tendency to represent our knowledge as assuming an approximately complete character. At least these and many other writers fail to impress upon their readers a truth which cannot be too constantly borne in mind, namely, that the utmost successes which our scientific method can accomplish will not enable us to comprehend more than an infinitesimal fraction of what doubtless is to comprehend.[9]

Nothing has happened in the interim to lead one to dissent from these strictures. We cannot realistically expect that our science, at *any* given stage of its actual development, will ever be in a position to afford us more than a very partial and incomplete access to the phenomena of nature. For, to reemphasize: in natural science, imperfect *physical* control is bound to mean imperfect *cognitive* control.

To evaluate this prospect, it is useful to return to the previously described "exploration model" of scientific progress as a matter of exploring our *parametric* neighborhood of physical variables like temperature, pressure, field strength, and so on. We must acknowledge here that it would be bizarre indeed if significant phenomena were to dry up as we move beyond our immediate parametric "neighborhood." Clearly nature distributes its various processes across *all* the reaches of parameter space, and does not favor us by clustering them in our parametric vicinity. Indeed all historical experience counterindicates this.

Any theory of convergence in science, however carefully crafted, will shatter under the impact of the *conceptual innovation* that becomes necessary to deal with the new phenomena encountered the wake of technical escalation. Such innovation continually brings entirely new, radically different scientific concepts to the fore, carrying in its wake an ongoing wholesale revision of "established fact." Investigators of the physical phenomena of an earlier era an earlier era not only did not *know* what the half-line of californium was, but they would not have *understood* it even if this fact had been explained to them. This aspect of the matter deserves closer attention.

"Which of the four elements (air, earth, fire, water) is the paramount *'arche,'* the fundamental type of stuff from which the whole of physical reality originates?" asked the early Milesians in pre-Socratic Greece. They contemplated just those four alternatives—together with the fifth possibility of a neutral, intermediate stuff. It did not occur to them that their whole inquiry was abortive because it was based on a misguided conception of "elements." Nor did it appear to be a realistic prospect to all those late nineteenth-century physicists who investigated the properties of the luminiferous aether that no such medium for the transmission of light and electromagnetism might exist at all.

In factual inquiry into the ways of the world we can do no better than to pose questions and canvass the currently visible alternatives. But the questions we can pose are limited by our conceptual horizons. And the answers we can envision are also limited by the cognitive "state-of-the-art." (The Greeks could not have asked about continental drift; the Romans could not have thought of explaining the tides through gravitation.) And, of course, the whole process of canvassing answers can come to grief because the very question being asked is based on untenable suppositions.

The problem with a convergentist theory of scientific progress is that with convergence later means lesser. And in the case of science, there is no reason to think that this will be so. Ongoing scientific progress is not simply a matter of increasing accuracy by extending the numbers at issue in our otherwise stable descriptions of nature out to a few more decimal places. Significant scientific progress is genuinely revolutionary in involving a *fundamental change of mind* about how things happen in the world. Progress of this caliber is generally a matter not of adding further facts—on the order of filling in a crossword puzzle. It is, rather, a matter of changing the very framework itself. And this fact blocks the theory of convergence.

Even extraordinary accuracy with respect to the entire range of *currently manageable* cases does not betoken actual correctness—it merely reflects adequacy over that limited range. And no matter how far we broaden that "limited range of 'presently accessible' cases," we still achieve no assurance (or even probability) that a theory-corpus which accommodates (perfectly well) the range of "presently achievable outcomes" will hold across-the-board. The upshot is that both as regards the observable *regularities* of nature and the discernable *constituents* of nature, very different results that project very different views of the situation can—and almost invariably do—emerge at successive levels of the observational state-of-the-art. Almost invariably we deal at every stage with a different order or aspect of things. And the reason why nature exhibits different aspects at different levels is not that nature herself is somehow stratified and has different levels of being or of operation but rather that (1) the character of the available nature-investigative interactions is variable and differs from level to level, and (2) the character of the "findings" at which one arrives will hinge on the character of these nature-investigative interactions. For—to reemphasize—what we detect or "find" in nature is always something that depends on the mechanisms by which we search. The phenomena we detect will depend not merely on nature's operations alone, but on the physical and conceptional instruments we use in probing them.

With any convergent process, *later* is *lesser*. But since scientific progress on matter of fundamental importance is generally a matter of replacement rather than mere supplementation, there is no good reason for seeing the *later* findings of science as lesser the significance of their bearing within the cognitive enterprise—to think that nature will be cooperative in always yielding its most important secrets early on and reserving nothing but the relatively insignificant for later on. (Nor does it seem plausible to think of nature as perverse, luring us ever more deeply into deception as inquiry proceeds.) A very small scale effect at the level of phenomena—even one that lies very far out along the extremes of a "range exploration" in terms of temperature, pressure, velocity, or the like—can force a far-reaching revolution and have a profound impact by way of major theoretical revisions. (Think of special relativity in relation to aether-drift experimentation, or general relativity in relation the perihelion of Mercury.)

Given that natural science progresses mainly by substitutions and replacements that involve comprehensive overall revisions of our picture of the processes at issue, it seems sensible to say that the shifts across successive scientific "revolutions" maintain the same level of overall significance when taken as a whole. At the cognitive level, a scientific innovation is simply a matter of change. Scientific progress is neither a convergent nor a divergent process.[10] Successive stages in the technological state-of-the-art of scientific inquiry lead us to different views about the nature of things and the character of their laws. And at each level of technical sophistication we get a substantially different overall story—with difference let by no means shrink to a vanishing point as the process moves along. Later is not lesser; it is just different. Theoretical progress in science is essentially a matter of substantiation. (As in politics, one cannot replace something with nothing.)

On this basis, one arrives at a view of scientific progress as confronting us with a situation in which every major successive stage in the evolution of science yields innovations, and these innovations are—on the whole—of roughly equal *overall* interest and importance. Accordingly, there is little alternative but to reject convergentism as a position that lacks the support not only of considerations of general principles but also of the actual realities of our experience in the history of science.

And this situation is critically important in our present context. For if later were indeed lesser, the ongoing cost escalation of scientific progress would not be so deeply problematic a factor. We would then simply forget about those very costly further purchases once a point is reached where their actual value becomes trivial. But this is most emphatically not the case in natural science where economic limitations confront us with a real and serious obstacle.[11]

Cognitive Copernicanism

The fallibility and corrigibility of our science means it cannot be viewed as providing definitive (let alone absolutely true) answers to its questions. We have no alternative but to see *our* science as both incomplete and incorrect in some (otherwise unidentifiable) respects. In *this* sense—that of its inability to deliver into our hands something that can be certified as the truth, the whole truth, and nothing but the truth science is certainly subject to a severe limitation. What we proudly vaunt as "scientific knowledge is a tissue of hypotheses—of tentatively adopted contentions, many or most of which we will ultimately come to regard as quite untenable and in need of serious revision or perhaps even rejection.

If there is one thing we can learn from the history of science, it is that the scientific theorizing of one day is looked on by that of the next as flawed and deficient. The clearest induction from the history of science is that science is always mistaken—that at *every* stage of its development, its practitioners, looking backward with the wisdom of hindsight, will view the work of their predecessors as seriously deficient and their theories as fundamentally mistaken. And if we adopt (as in candor we must) the modest view that we ourselves and our contemporaries do not occupy a privileged position in this respect, then we have no reasonable alternative but to suppose that much or all of what we ourselves vaunt as "scientific knowledge" is itself presumably wrong.[12]

No human generation can lay claim to scientific finality. We must acknowledge that the transience that has characterized all scientific theories of the historical past will possibly, even probably, characterize those of the present as well. Given the historical realities, the idea that science does—or sooner or later must—arrive at a changeless vision of "the truth of the matter" is not plausible. We have no alternative but to reject the egocentric claim that we ourselves occupy a pivotal position in the epistemic dispensation, and we must recognize that there is nothing inherently sacrosanct about our own present cognitive posture vis-à-vis that of other, later historical junctures. A kind of intellectual humility is in order—a diffidence that abstains from the hubris of pretensions to cognitive finality or centrality. Such a position calls for the humbling view that just as we think our predecessors of a hundred years ago had a fundamentally inadequate grasp on the furniture of the world, so our successors of a hundred years hence will take the same view of our purported knowledge of things. Realism requires us to recognize that, as concerns our scientific understanding of the world, some of what we see as secure knowledge is likely to prove to be no more than presently accepted error.

We must stand ready to acknowledge the fragility of our scientific knowledge and must temper our claims to scientific knowledge with a Cognitive Copernicanism. The original Copernican revolution made the point that there is nothing ontologically privileged about our own position in space. The doctrine now at issue effectively holds that there is nothing cognitively privileged about our own position in time. It urges that *there is nothing epistemically privileged about the present—any present*, our own prominently included. Such a perspective indicates not only the incompleteness of "our knowledge" but its presumptive incorrectness as well. The current state of "knowledge" is simply one state among others, all of which stand on an imperfect footing.

To be sure, this recognition of the fallibilism of our cognitive endeavors must emphatically *not* be construed as an open invitation to a skeptical abandonment of the cognitive enterprise. Instead, it is an incentive to do the very best we can. In human inquiry, the cognitive ideal is correlative with the striving for optimal systematization. And this is an ideal that, like other ideals, is worthy of pursuit, despite the fact that we must recognize that its full attainment lies beyond our grasp.

The crucial consideration thus remains that science is not and presumably never can be in a position to offer us anything that is definitive, incorrigible, and final. Our science is and must be developed within the confines of man's cognitive fallibility. He who looks to science for answers that are ultimate and absolute is destined to look in vain. We must recognize that "our science" is not something permanent, secured for the ages, unchangeable. Our theorizing about the nature of the real is a fallible estimation, the best that can be done at this time, in this particular "state-of-the-art." Our science is an historical phenomenon: it is one transitory state of things in an ongoing process.

This limitation cannot, of course, be justifiedly construed as a *defect* of science. It is an inevitable feature of whatever can be produced by imperfect triangulation from limited experience, and we must not complain of what cannot be helped. It is not a shortcoming of science as compared with other methods of inquiry, because other methods cannot overcome it either. (If *they* could, so could science.) But it is a limitation—an inherent limitation of the enterprise as we humans do and must conduct it. The aim or goal of science is to provide answers; but answers of certified correctness, definitive and final answers, are simply not available. The complete realization of the aims of science is something that will ever remain in the realm of aspiration and not that of achievement.

THE PROBLEM OF PROGRESS

The idea of scientific progress as the correlate of a movement through sequential stages of technological sophistication was already clearly discerned by the astute Charles Sanders Peirce around the turn of the century:

> Lamarckian evolution might, for example, take the form of perpetually modifying our opinion in the effort gradually to make that opinion represent the known facts as more and more observations come to be collected. . . . But this is not the way in which science mainly progresses. It advances by leaps; and the impulse for each leap is either some new observational resource, or some novel way of reasoning about the observations. Such a novel way of reasoning might, perhaps, also be considered as a new observational means, since it draws attention to relations between facts which would previously have been passed by unperceived.[13]

This circumstance has far-reaching implications for the perfectibility of science. The impetus to augment our science demands an unremitting and unending effort to enlarge the domain of effective experimental intervention. For only by operating under new and heretofore inaccessible conditions of observational or experimental systematization—attaining ever more extreme temperature, pressure, particle velocity, field strength, and so on—can we bring new grist to our scientific mill.

To be sure, there is, on its basis, no inherent limit to the possibility of future progress in scientific knowledge. But the exploitation of this theoretical prospect gets ever more difficult, expensive, and demanding in terms of effort and ingenuity. With scientific progress we are engaged in a situation of price-inflation where further equal-size steps become more difficult with every step we take. New findings of equal level of significance require ever-greater aggregate efforts. In the ongoing course of scientific progress, the earlier investigations in the various departments of inquiry are able to skim the cream, so to speak: they take the "easy pickings," and later achievements of comparable significance require ever deeper forays into complexity and call for an ever-increasing bodies of information. (And it is important to realize that this cost-increase is not because latter-day workers are doing *better* science, but simply because it is harder to achieve *the same level* of science: one must dig more deeply or search more widely to achieve results of the same significance as before.)

The historical course of natural science is a sequence of radical changes of mind—even about fundamentals. How, then, can science be said to *progress?* How can we speak of an advance rather than a fortuitous ebb and flow? If science neither provides nor approaches "the ultimate truth" about nature, how can we say that it has the directionality that is essential to progress, in contradistinction to mere movement?

Historians and analysts of science are often heard to complain about the absence of any clear sense of the direction of development in the structure of modern scientific work. As one recent writer puts it:

> The blackest defect in the history of science, the cause of dullest despair for the historian, lies in the virtual absence of any general historical sense of the way science has been working for the last hundred years.[14]

But any such lack of direction at once disappears when we turn from the content of scientific discovery to its tools and their mode of employment—in short, when we turn to the *technological* side of the matter. For all of recent science has a clear thrust of development—using ever more potent instruments to press ever further outward in the exploration of physical parameter-space, forging more and more powerful physical and conceptual instrumentalities for the identification and analysis of new phenomena.

Under the influence of Thomas Kuhn's book on scientific revolutions[15] some recent writers on the philosophy of science tend to stress the ideational discontinuities produced by innovation in the history of science. Scientific change, they rightly maintain, is not just a matter of marginal revisions of opinion within a fixed and stable framework of concepts. The crucial developments involve a change in the conceptual apparatus itself. When this happens, there is a replacement of the very *content* of discussion, a shift in "what's being talked about" that renders successive positions "incommensurable." The change from the Newtonian to the Einsteinian concept of time exemplifies a shift of just this sort. These discontinuities of meaning make it impossible (so it is said) to say justifiedly that the latter stages somehow represent a "better treatment of the same subject matter," since the very subject has changed.[16] The radical discontinuity of meanings also complicates the idea of scientific progress. For if the later stage of discussion is conceptually disjointed from the earlier, how could one consider the later as an improvement on the earlier. The replacement of one thing by something else of a totally different sort can hardly qualify as meliorative. (One can improve on one's car by getting a better car, but one cannot improve on it by getting a computer or a dishwashing machine.)

To draw this sort of implication from the meaning-shift thesis is, however, to be overhasty. For real progress is indeed made, through this progress does not proceed along purely theoretical but along *practical* lines. Once one sees the validation of science as lying ultimately in the sphere of its applications, one also sees that the progress of science must be taken to rest on its pragmatic improvement: the increasing success of its applications in problem solving and control, in predictive and physical mastery over nature.[17] And the control that lies at the root of progress in science is not something arcane, sophisticated, and heavily theory-laden. It turns on the fact that *any* enhancement in control—any growth of our technological mastery over nature—will have involvements that are also discernible at the *grosso modo* level of our everyday life conceptions and dealings. *How* our control is extended will generally be a very sophisticated matter, but any fool can see *that* our control has been extended in innumerable ways. Since the days of Bacon and Hobbes, it has been recognized that the conception of the application of science to human ends (*scientia propter potentiam*) provides a perfectly workable basis for taking the expanding horizons of technological capacity as an index of scientific progress.[18]

This technological dimension endows scientific change with a continuity it lacks at the level of its ideas and concepts—a continuity that finds its expression in the persistence of problem-solving tasks in the sphere of *praxis*. Despite any *semantic* or *ideational incommensurability* between a scientific theory and its latter-day replacements, there remains the factor of the pragmatic *commensurability* that can (by and large) be formulated in suitable extrascientific language.[19]

In maintaining that the successive theses of science represent not just *change* but *progress,* one does indeed stand committed to the view that in some fashion or other something that is strictly comparable is being improved on. But the items at issue need not be the theory-laden substance of scientific claims; they can revolve about the practical concerns of our commonplace affairs. The comparisons need not be made at the level of scientific theorizing but at the rudimentary level of the commonsense affairs facilitated by their technological implementation.

Issues of *technological* superiority are far less sophisticated, but also far more manageable, than issues of *theoretical* superiority. Dominance in the technological power to produce intended results tends to operate across the board. For the factors determinative of technical superiority in prediction and control operate at a grosser and more rough-and-ready level than those of theoretical content. At the level of praxis, we can operate to a relatively large degree with the *lingua franca* of everyday affairs and make our comparisons on this basis. Just as the merest novice can detect a false note in the musical performance of a master player whose activities he could not begin to emulate, so the malfunctioning of a missile or computer can be detected by a relative amateur. The superiority of modern over Galenic medicine requires few, if any, subtle distinctions.

Traditional theories of scientific progress join in stressing the capacity of the "improved" theories to accommodate new facts. Agreeing with this emphasis on "new facts," we must, however, recognize two distinct routes to this destination: the predictive (via *theory*) and the productive (via *technology*). And it would appear proper to allow both of these routes to count. To correct the overly *theoretical* bias of traditional philosophy of science requires a more ample recognition of the role of technology-cum-production. To say this is not, of course, to deny that theory and technology stand in a symbiotic and mutually supportive relationship in scientific inquiry. But the crucial fact is that the effectiveness of the technological instrumentalities of praxis can clearly be assessed without appeal to the cognitive content of the theory brought to bear in their devising.

At bottom, the progress of science manifests itself most clearly in practical rather than theoretical regards. The progressiveness of science hinges crucially on its *applications;* it resides in the pragmatic dimension of the enterprise—the increasing success of its applications in problem solving and control, in its affording not only cognitive but physical mastery over nature. A new theory need not explain the purported facts of the old one it replaces, because these "facts" need not remain facts: the new theory may revise or dismiss them. (The phenomenon of "emission of phlogiston" disappears with the arrival of Lavoisier's oxydation.) But the practical successes of the old theory in enabling us to predict occurrences or to achieve control at the unrefined level of developments describable in the language of everyday life is something relatively unproblematic, and thus relatively secure.

From where we stand in the epistemic dispensation, there is no way of attaining a higher vantage point—no God's-eye view for comparison. What we do fortunately have as common basis across the divides of scientific change is not a higher but rather a lower standpoint: the crude vantage point of ordinary everyday life. No sophisticated complexities are needed to say that one stage in the career of science is superior to another in launching rockets and curing colds and exploding bombs. These applications operate extensively at the level of the ordinary, everyday concepts of natural-language discourse. This level is relatively stable and unchanging precisely because of the crudity of its concerns; its concepts remain relatively fixed throughout the ages, and lie deep beneath the changing surface of scientific sophistication.

Its technological and applicative dimension endows science with a theory-era-transcending comparability that it lacks at the level of its ideas and concepts, The ancient Greek physician and the modern medical practitioner might talk of the problems of their patients in very different and conceptually incommensurate ways (say, an imbalance in humors to be treated by countervailing changes in diet or regimen versus a bacterial infection to be treated by administering an antibiotic). But at the pragmatic level of practical control—that is, at the level of removing those symptoms of their patients (pain, fever, dizziness, etc.) that are describable in much the same terms in antiquity as today—both are working on "the same problem."

A later scientific theory certainly need not preserve the *content* of the earlier ones; its descriptive and taxonomic innovations can make for semantical discontinuity. Nor, again, need it preserve the theoretical successes of the earlier theories—their explanatory successes at elaborating interrelations among their own elements (for example, answering questions about the operations of the luminiferous aether). But a later and progressively superior theory must preserve and improve on the practical successes of its predecessors when these practical issues are formulated in the rough-and-ready terms of everyday-life discourse.

And so, notwithstanding its instability and change ability at the level of theoretical claims, science does indeed progress not, to be sure, by way of "approaching the ultimate truth" but by providing us with increasingly powerful instrumentalities for prediction and control. Once due prominence is given to the factor of *control over nature* in the pre- or subtheoretical construction of this idea, the substantiation of imputations of scientific *progress* becomes a more manageable project than it could ever be on an "internal," content-oriented basis.

Given "the facts of life" that characterize the situation in empirical inquiry, we have neither the inclination nor the justification to claim that the world actually is just as our present science describes it to be. Nor, as we have seen, does it make sense to identify "the real truth" with "the truth as science-in-the-limit will eventually see it to be." All that can be done in this direction is to say that the world exists as ideal or perfected science describes it to be. The real, which is to say, final and definitive truth about nature at the level of scientific generality and precision is something we certainly cannot assume to be captured by our science as it stands here and now (thought, of course, this fact nowise destroys the rationality of our endorsement of current beliefs). We cannot but take the stance that scientific truth is not something in hand, but something which—so we must suppose—is attained only in the ideal or perfected state of things. With respect to scientific issues we thus arrive at the coordinating equation:

the real truth = the truth as ideal (perfected) science purports it to be.

To be sure, in espousing this conception, we intend to make "ideal science" contingent on truth, rather than the reverse: the former is the independent, the latter the dependent variable.

But what of the objection that we could not tell that we had arrived at "the definitive truth" even if we in fact had done so? Its resolution lies in the fact that the attainment of such an objective is simply not at issue. Yet while one can never lay claim to have definitely secured "the definitive truth" in matters of scientific theorizing, this notion nevertheless serves an important role in providing a contrast-conception that constitutes a useful reminder of the fragility of our cognitive endeavors. It establishes a contrast between our present science as we have it and a perfected "ideal science," which alone can properly be claimed to afford a grasp of reality, an idea that crucially regulates our view as to the nature and status of the knowledge we lay claim to and thereby productively fosters the conduct of inquiry.[20]

Chapter 13

The Law of Logarithmic Returns and the Complexification of Natural Science

Synopsis

- *Scientific progress depends crucially on securing ever enhanced information.*
- *And this means that natural science itself grows ever more complex, not withstanding its penchant for simplicity.*
- *And not only does scientific progress involve ever greater complexity, it involves ever greater costs as well.*
- *Scientific knowledge does not correlate to the brute volume of scientific information, but only to its logarithm.*
- *The rationale of this circumstance roots in the way in which significant information always lies obscured amidst a fog of insignificance. What we thus have in science is a Law of Logarithmic Returns.*
- *In consequence, the progress of knowledge involves ever escalating demands, a circumstance bears in a fundamental and ominous way on the issue of the growth of scientific knowledge over time.*
- *In consequence, progress in botanical science is a decelerating process.*
- *And so there is indeed a prospect of making some economically based predictions about the volume of future scientific innovation, despite our inability to foresee its content.*
- *To be sure, the law of logarithmic returns pivots in a critical way on assessing the quality of the work at issue. But of course anything deserving of the name of scientific knowledge will be information of the highest quality level.*

The Principle of Least Effort and the Methodological Status of Simplicity-Preference in Science

An eminent philosopher of science has maintained that "in cases of inductive simplicity it is not economy which determines our choice.... We make the assumption that the simplest theory furnishes the best predictions. This assumption cannot be justified by convenience; it has a truth character and demands a justification within the theory of probability and induction."[1] This perspective is gravely misleading. What sort of consideration could possibly justify the supposition that "the simplest theory furnishes the best prediction"? Any such belief is surely unwarranted and inappropriate. There is simply no cogent rationale for firm confidence in the simplicity of nature. To claim the ontological simplicity of the real is somewhere between hyperbolic and absurd.

The matter becomes far less problematic, however, once one approaches it from a methodological rather than a substantive point of view. For considerations of rational economy and convenience of operation obviously militate for inductive systematicity. Seeing that the simplest answer is (*eo ipso*) the most economical one to work with, rationality creates A natural pressure toward economy—toward simplicity insofar as other things are equal. Our eminent theorist has things upside down here: it is in fact methodology that is at issue rather than among any factual presumption.

A century ago, Henri Poincaré saw the issue in a more plausible light:

> [Even] those who do not believe that natural laws must be simple, are still often obliged to act as if they did believe it. They cannot entirely dispense with this necessity without making all generalization, and therefore all science, impossible. It is clear that any fact can be generalised in an infinite number of ways, and it is a question of choice. The choice can only be guided by considerations of simplicity.... To sum up, in most cases every law is held to be simple until the contrary is proved.[2]

These observations are wholly in the right spirit. As cognitive possibilities proliferate in the course of theory-building inquiry, a principle of choice and selection becomes requisite. And here economy—along with its other systematic congeners, simplicity, and uniformity, and the like—are the natural guideposts. We subscribe to the inductive presumption in favor of simplicity, uniformity, normality, not because we know or believe that matters always stand on a basis that is simple, uniform, normal, and so on—surely we know no such thing!—but because it is on this basis alone that we can conduct our cognitive business in the most advantageous, the most *economical* way. In scientific induction we exploit the information at hand so as to answer our questions in the most straightforward, the most economical way.

Suppose, for example, that we are asked to supply the next member of a series of the format 1, 2, 3, 4, ... We shall straightaway respond with 5, supposing the series to be simply that of the integers. Of course, the actual series might well be 1, 2, 3, 4, 11, 12, 13, 14, 101, 102, 103, 104, ..., with the answer thus eventuating as 11 rather than 5. But while we cannot rule such possibilities out, they do not for an instant deter our inductive proceedings. For the inductively appropriate course lies with the production rule that is the simplest issue-resolving answer—the simplest resolution that meets the conditions of the problem. And we take this line not because we know a priori that this simplest resolution will prove to be correct. (We know no such thing!) Rather we adopt this answer, provisionally at least, just exactly because it is the least cumbersome and most economical way of providing a resolution that does justice to the facts and demands of the situation. We recognize that other possibilities of resolution exist but ignore them until further notice, exactly because there is no cogent reason for giving them favorable treatment *at this stage*. (After all, once we leave the safe harbor of simplicity behind there are always multiple possibilities for complexification, and we lack any guidance in moving one way rather than another.)

Throughout inductive inquiry in general, and scientific inquiry in particular, we seek to provide a descriptive and explanatory account that provides the simplest, least complex way of accommodating the data that experience (experimentation and observation) has put at our disposal. When something simple accomplishes the cognitive tasks in hand, as well as some more complex alternative, it is foolish to adopt the latter. We certainly need not presuppose that the world somehow is systematic (simple, uniform, and the like) to validate our penchant for the systematicity of our cognitive commitments. We contrive our problem resolutions along the lines of least resistance, seeking to economize our cognitive effort by using the most direct workable means to our ends.

Our systematizing procedures in science pivot on the injunction always to adopt the most economical (simple, general, straightforward, etc.) solution that meets the demands of the situation.

Whenever possible, we analogize the present case to other similar ones, because the introduction of new patterns complicates our cognitive repertoire. And we use the least cumbersome viable formulations because they are easier to remember and more convenient to use. The rationale of the other-things-equal preferability of simpler solutions over more complex ones is obvious enough. Simpler solutions are less cumbersome to store, easier to take hold of, and less difficult to work with. It is indeed economy and convenience that determine our regulative predilection for simplicity and systematicity in general. Our prime motivation is to get by with a minimum of complication, to adopt strategies of question-resolution that enable us among other things: (1) to continue with existing solutions unless and until the epistemic circumstances compel us to introduce changes (uniformity) be; (2) to make the same processes do insofar as possible (generality), and (3) to keep to the simplest process that will do the job (simplicity). Such a perspective combines the commonsensical precept, Try the simplest thing first, with a principle of burden of proof: Maintain your cognitive commitments until there is good reason to abandon them.

When other things are anything like equal, simpler theories are bound to be operationally more advantageous. We avoid needless complications whenever possible, because this is the course of an economy of effort. It is the general practice in scientific theory construction, to give preference to

- one-dimensional rather than multidimensional modes of description,
- quantitative rather than qualitative characterizations,
- lower- rather than higher-order polynomials,
- linear rather than nonlinear differential equations.

The comparatively simpler is for this very reason easier to work with. In sum, we favor uniformity, analogy, simplicity, and the like because they ease our cognitive labor. On such a perspective, simplicity is a concept of the practical order, pivoting on being more economical to use—that is, less demanding of resources. The key principle is that of the rational economy of means for the realization of given cognitive ends, of getting the most effective answer we can with the least complication. Complexities cannot be ruled out, but they must always pay their way in terms of increased systemic adequacy! It is thus methodology and not metaphysics that grounds our commitment to simplicity and systematicity.

The rational economy of process is the crux here. And this methodological commitment to rational process does not prejudge or prejudice the substantively *ontological* issue of the complexity of nature. Natural science is emphatically *not* bound to a Principle of Simplicity in Nature. There really are no adequate grounds for supposing the "simplicity" of the world's makeup. Instead, the so-called Principle of Simplicity is really a principle of complexity-management: "Feel free to introduce complexity in your efforts to describe and explain nature's ways. But only when and where it is really needed. Insofar as possible "keep it simple!" Only introduce as much complexity as you really need for your scientific purposes of description, explanation, prediction, and control." Such an approach is eminently sensible. But of course such a principle is no more than a methodological rule of procedure for managing our cognitive affairs. Nothing entitles us to transmute this methodological precept into a descriptive/ontological claim to the effect that nature is simple—let alone of finite complexity. Accordingly, the penchant for inductive systematicity reflected in the conduct of inquiry is simply a matter of striving for rational economy. It is based on methodological considerations that are governed by an analogue of Occam's razor—a principle of parsimony to the effect that needless complexity is to be avoided *complicationes non multiplicandae sunt praeter necessitatem.* Given that the inductive method, viewed in its practical and methodological aspect, aims at the most efficient and effective means of question-resolution, it is only natural that our inductive precepts should direct us always to begin with the most systematic, the thereby economical, device that can actually do the job at hand.[3]

It clearly makes eminent sense to move onwards from the simplest (least complex) available solution to introduce further complexities when and as—but *only* when and as—they are forced on us. Simpler (more systematic) answers are more easily codified, taught, learned, used, investigated, and so on. The regulative principles of convenience and economy in learning and inquiry suffice to provide a rational basis for systematicity-preference. Our preference for simplicity, uniformity, and systematicity in general, is now not a matter of a substantive theory regarding the nature of the world, but one of search strategy—of cognitive methodology. In sum, we opt for simplicity in inquiry (and systematicity in general) not because it is truth-indicative, but because it is teleologically more effective in conducing to the efficient realization of the goals of inquiry. We look for the dropped coin in the lightest spots nearby, not because this is—in the circumstances—the most probable location but because it represents the most sensible strategy of search: if it is not there, then we just cannot find it at all.

On such a view, inductive systematicity with its penchant for simplicity comes to be seen as an aspect, not of *reality* as such, but of our procedures for its conceptualization and accordingly of *our conception* of it, or, to be more precise, of our manner of conceptualizing it. Simplicity-preference (for example) is based on the strictly method-oriented practical consideration that the simple hypotheses are the most convenient and advantageous for us to put to use in the context of our purposes. There is thus no recourse to a substantive (or descriptively constitutive) postulate of the simplicity of nature; it suffices to have recourse to a regulative (or practical) precept of economy of means. And in its turn, the pursuit of cognitive systematicity is ontologically neutral: it is a matter of conducting our question-resolving endeavors with the greatest economy. The Principle of Least Effort is in control here—the process is one of maximally economic means to the attainment of chosen ends. This amounts to a *theoretical* defense of inductive systematicity that in fact rests on *practical* considerations relating to the efficiencies of method.

Accordingly, inductive systematicity is best approached with reference, not to reality as such—or even merely our conception of it—but to the ways and means we employ in conceptualizing it. It is noncommittal on matters of substance, representing no more than a determination to conduct our question-resolving endeavors with the greatest economy. Cognitive economy, with its balance of costs and benefits, once again comes to the forefront. For in inquiry, as elsewhere, rationality enjoins us to employ the maximally economic means to the attainment of chosen ends. And our commitment to simplicity in scientific inquiry accordingly does not, in the end, prevent us from discovering whatever complexities are actually there. And the commitment to inductive systematicity in our account of the world remains a methodological desideratum regardless of how complex or untidy that world may ultimately turn out to be.

Complexification

Scientific theory-formation is, in general, a matter of spotting a local regularity of phenomena in parametric space and then projecting it "across-the-board," maintaining it globally. But the theoretical claims of science are themselves never small-scale and local—they are not spatiotemporally localized and they are not parametrically localized either. They stipulate—quite ambitiously—how things are always and everywhere. But with the enhancement of investigative technology, the "window" through which we can look out on nature's parametric space becomes constantly enlarged. In developing natural science we use this window of capability to scrutinize parametric space, continually augmenting our database and then generalizing on what we see. What we have here is not a lunar landscape where once we have seen one sector we have seen it all, and where theory-projections from lesser data generally remain in place when further data comes our way. Instead it does not require a sophisticated knowledge of history of science to realize that our worst fears are usually realized—that our theories seldom if ever survive intact in the wake of substantial extensions in our cognitive access to new sectors of the range of nature's phenomena. The history of science is a sequence of episodes of leaping to the wrong conclusions because new observational findings indicate matters are not quite so simple as heretofore thought. As ample experience indicates, our ideas about nature are subject to constant and often radical change-demanding stresses as we "explore" parametric space more extensively. The technologically mediated entry into new regions of parameter space constantly destabilizes the attained equilibrium between data and theory. Physical nature can exhibit a very different aspect when viewed from the vantage point of different levels of sophistication in the technology of nature-investigator interaction. The possibility of change is ever present. The ongoing destabilization of scientific theories is the price we pay for operating a simplicity-geared cognitive methodology in an actually complex world.

We naturally adopt throughout rational inquiry—and accordingly throughout natural science—the methodological principle of rational economy to "Try the simplest solutions first" and then make this do as long as it can. And this means that *historically* the course of inquiry moves in the direction of ever-increasing complexity. The developmental tendency of our intellectual enterprises—natural science among them—is generally in the direction of greater complication and sophistication.

In a complex world, the natural dynamics of the cognitive process exhibits an inherent tropism toward increasing complexity. Herbert Spencer argued long ago that evolution is characterized by von Baer's law of development "from the homogeneous to the heterogeneous" and thereby produces an ever-increasing definition of detail and complexity of articulation.[4] As Spencer saw it, organic species in the course of their development confront a successive series of environmental obstacles, and with each successful turning along the maze of developmental challenges the organism becomes selectively more highly specialized in its biodesign, and thereby more tightly attuned to the particular features of its ecological context.[5] Now this view of the developmental process may or may not be correct for *biological* evolution, but there can be little question about its holding for *cognitive* evolution. For rational beings will of course try simple things first and thereafter be driven step by step toward an ever-enhanced complexification. In the course of rational inquiry we try the simple solutions first, and only thereafter, if and when they cease to work—when they are ruled out by further findings (by some further influx of coordinating information)—do we move on to the more complex. Things go along smoothly until an oversimple solution becomes destabilized by enlarged experience. For a time we get by with the comparatively simpler options—until the expanding information about the world's *modus operandi* made possible by enhanced new means of observation and experimentation insists otherwise. And with the expansion of knowledge those new accessions make ever increasing demands. And so evolution, be it natural or rational—whether of animal species or of literary genres—ongoingly confronts us with products of greater and greater complexity.[6]

Man's cognitive efforts in the development of natural science manifests a Manichaean-style struggle between complexity and simplicity—between the impetus to comprehensiveness (amplitude) and the impetus to system (economy). We want our theories to be as extensive and all-encompassing as possible and at the same time to be elegant and economical. The first desideratum pulls in one direction, the second in the other. And the accommodation reached here is never actually stable. As our experience expands in the quest for greater adequacy and comprehensiveness, the old theory structures become destabilized—the old theories no longer fit the full range of available fact. And so the theoretician goes back to the old drawing board. What he comes up with here is—and in the circumstances must be—something more elaborate, more *complex* than what was able to do the job before those new complications arose (though we do, of course, sometimes achieve local simplifications within an overall global complexification). We make do with the simple, but only up to the point when the demands of adequacy force additional complications on us. An inner tropism toward increasing complexity is thus built into the very nature of the scientific project as we have it.

And the same is true also for technological evolution, with *cognitive* technology emphatically included. Be it in cognitive or in practical matters, the processes and resources of yesteryear are rarely, if ever, up to the demands of the present. In consequence, the life-environment we create for ourselves grows increasingly complex. The Occam's Razor injunction, "Never introduce complications unless and until you actually require them," accordingly represents a defining principle of practical reason that is at work within the cognitive project as well. And because we try the simplest solutions first, making simple solutions do until circumstances force one to do otherwise, it transpires that in the development of knowledge—as elsewhere in the domain of human artifice—progress is always a matter of complexification. An inherent impetus toward greater complexity pervades the entire realm of human creative effort. We find it in art; we find it in technology; and we certainly find it in the cognitive domain as well.[7]

The methodology of science thus embodies an inherent dialectic that moves steadily from the simpler to the more complex, and the developmental route of technology sails on the same course. We are driven in the direction of ever greater complexity by the principle that the potential of the simple is soon exhausted and that high capacity demands more elaborate and powerful processes and procedures. The simpler procedures of the past are but rarely adequate to the needs of the present—had they been so today's questions would have been resolved long ago and the issues at stake would not have survived to figure on the present agenda. Scientific progress is of a nature that inherently involves an inexorable tendency to complexification in both its cognitive and its ideational dimension. What we discover in investigating nature always must in some degree reflect the character of our technology of observation. It is always something that depends on the mechanisms with which we search.[8]

Induction with respect to the history of science itself—a constant series of errors of oversimplification—soon undermines our confidence that nature operates in the way we would deem the simpler. On the contrary, the history of science is an endlessly repetitive story of simple theories giving way to more complicated and sophisticated ones. The Greeks had four elements; in the nineteenth century Mendeleev had some sixty; by the 1900s this had gone to eighty, and nowadays we have a vast series of elemental stability states. Aristotle's cosmos had only spheres; Ptolemy's added epicycles; ours has a virtually endless proliferation of complex orbits that only supercomputers can approximate. Greek science was contained on a single shelf of books; that of the Newtonian age required a roomful; ours requires vast storage structures filled not only with books and journals but with photographs, tapes, floppy disks, and so on. Of the quantities currently recognized as the fundamental constants of physics, only one was contemplated in Newton's physics: the universal gravitational constant. A second was added in the nineteenth century, Avogadro's constant. The remaining six are all creatures of twentieth-century physics: the speed of light (the velocity of electromagnetic radiation in free space), the elementary charge, the rest mass of the electron, the rest mass of the proton, Planck's constant, and Boltzmann's constant.[9] It would be naive—and quite wrong—to think that the course of scientific progress is one of increasing simplicity. The very reverse is the case: scientific progress is a matter of complexification because oversimple theories invariably prove untenable in a complex world. The natural dialectic of scientific inquiry ongoingly impels us into ever deeper levels of sophistication.[10] In this regard our commitment to simplicity and systematicity, though methodologically necessary, is ontologically unavailing. And more sophisticated searches invariably engender changes of mind moving in the direction of an ever more complex picture of the world. Our methodological commitment to simplicity should not and does not preclude the substantive discovery of complexity.

The explosive growth of information of itself countervails against its exploitation for the sake of knowledge-enhancement. The problem of coping with the proliferation of printed material affords a striking example of this phenomenon. One is forced to ever higher levels of aggregation, compression, and abstraction. In seeking for the needle in the haystack we must push our search processes to ever greater depths.

And this ongoing refinement in the division of cognitive labor that an information explosion necessitates issues in a literal dis-integration of knowledge. The "progress of knowledge" is marked by an ever-continuing proliferation of ever more restructured specialties marked by the unavoidable circumstance that the any given specialty cell cannot know exactly what is going on even next door—let alone at the significant remove. Our understanding of matters outside one's immediate bailiwick is bound to become superficial. At home base one knows the details, nearby one has an understanding of generalities, but at a greater remove one can be no more than an informed amateur.

This disintegration of knowledge is also manifolded vividly in the fact that out cognitive taxonomies are bursting at the seams. Consider the example of taxonomic structure of physics. In the eleventh (1911) edition of the *Encyclopedia Britannica,* physics is described as a discipline composed of 9 constituent branches (e.g., "Acoustics" or "Electricity and Magnetism") which were themselves partitioned into 20 further specialties (e.g., "Thermo-electricity: of "Celestial Mechanics"). The fifteenth (1974) version of the *Britannica* divides physics into 12 branches whose subfields are—seemingly—too numerous for listing. (However the 14th 1960s edition carried a special article entitled "Physics, Articles on" which surveyed more than 130 special topics in the field.) When the National Science Foundation launched its inventory of physical specialties with the National Register of Scientific and Technical Personnel in 1954, it divided physics into 12 areas with 90 specialties. By 1970 these figures had increased to 16 and 210, respectively. And the process continues unabated to the point where people are increasingly reluctant to embark on this classifying project at all.

Substantially the same story can be told for every field of science. The emergence of new disciplines, branches, and specialties is manifest everywhere.

And as though to negate this tendency and maintain unity, one finds an ongoing evolution of interdisciplinary syntheses—physical chemistry, astrophysics, biochemistry, and so forth. The very attempt to counteract fragmentation produces new fragments. Indeed, the phenomenology of this domain is nowadays so complex that some writers urge that the idea of a "natural taxonomy of science" must be abandoned altogether.[11] The expansion of the scientific literature is in fact such the natural science has in recent years been disintegrating before our very eyes. An ever larger number of ever more refined specialties has made it ever more difficult for experts in a given branch of science to achieve a thorough understanding about what is going on ever in the specialty next door.

It is, of course, possible that the development of physics may eventually carry us to theoretical unification where everything that we class among the "laws of nature" belongs to one grand unified theory—one all-encompassing deductive systematization integrated even more tightly than that Newton's *Principia Mathematica*.[12] But the covers of this elegantly contrived "book of nature" will have to encompass a mass of every more elaborate diversity and variety. Like a tricky mathematical series, it will have to generate ever more dissimilar constituents which, despite their abstract linkage are concretely as different as can be. And the integration at issue at the principle of a pyramid will cover further down an endlessly expansive range and encompassing the most variegated components. It will be an abstract unity uniting a concrete mishmash of incredible variety and diversity. The "unity of science" to which many theorists aspire may indeed come to be realized at the level of concepts and theories shared between different sciences—that is, at the level of ideational overlaps. But for every conceptual commonality and shared element there will emerge a dozen differentiations. The increasing complexity of our world picture is a striking phenomenon throughout the development of modern science.

The lesson of such considerations is clear. Scientific knowledge grows not just in extent but also in complexity, so that science presents us with a scene of ever-increasing complexity. It is thus fair to say that modern science confronts us with a cognitive manifold that involves an ever more extensive specialization and division of labor. The years of apprenticeship that separate master from novice grow ever greater. A science that moves continually from an oversimple picture of the world to one that is more complex calls for ever more elaborate processes for its effective cultivation. And as the scientific enterprise itself grows more extensive, the greater elaborateness of its productions requires an ever more intricate intellectual structure for its accommodation. The complexification of scientific process and product escalate hand and hand. And the process of complexity amplification that Charles S. Peirce took to be revealed in nature through science is unquestionably manifested in the cognitive domain of scientific inquiry itself.

THE EXPANSION OF SCIENCE

The increasing complexity of the scientific enterprise itself is reflected in the fact that research and development expenditures in the United States grew exponentially after World War II, increasing at a rate of some 10 percent per annum. By the mid-1960s, America was spending more on scientific research and development than the entire Federal budget before Pearl Harbor. This growth in the costs of science has various significant ramifications and manifestations.

Take manpower, for example, where the recent growth of the scientific community is a particularly striking phenomenon. During most of the present century the number of American scientists has been increasing at 6 percent to yield an exponential growth-rate with a doubling time of roughly twelve years.[13] A startling consideration—one often but deservedly repeated—is that well over 80 percent of ever-existing scientists (in even the oldest specialties such as mathematics, physics, and medicine) are alive and active nowadays.[14]

Again, consider the growth of the scientific literature. It is by now a familiar fact that scientific information has been growing at the (reasonably constant) exponential rate over the past several centuries. Overall, the printed literature of science has been increasing at an average of some 5 percent annually throughout the last two centuries, to yield an exponential growth-rate with a doubling time of about fifteen years—an order-of-magnitude increase roughly every half century. The result is a veritable flood of scientific literature. The *Physical Review* is currently divided into six parts, each of which is larger than the whole journal was a decade or so ago. It is reliably estimated that, from the start, about 10 million scientific papers have been published and that currently some 30,000 journals publish some 600,000 new papers each year. In fact, it is readily documented that the number of books, of journals, of journal-papers has been increasing at an exponential rate over the recent period.[15] By 1960, some 300,000 different book titles were being published in the world, and the two decades from 1955 and 1975 saw the doubling of titles published in Europe from around 130,000 to over 270,000.[16] And science has had its full share of this literature explosion. The amount of scientific material in print is of a scope that puts it beyond the reach not only of individuals but also of institutions as well. No university or institute has a library vast enough to absorb or a faculty large enough to digest the relevant products of the world's printing presses.

Then too there is the massively increasing budget of science. The historic situation regarding the costs of American science was first delineated in the findings of Raymond Ewell in the 1950s.[17] His study of research and development expenditures in the United States showed that growth here has also been exponential; from 1776 to 1954 we spent close to $40 billion, and half of that was spent after 1948.[18] Projected at this rate, Ewell saw the total as amounting to what he viewed as an astronomical $6.5 billion by 1965—a figure that actually turned out to be far too conservative.

Moreover, the proliferation of scientific facilities has proceeded at an impressive pace over the past hundred years. (In the early 1870s there were only eleven physics laboratories in the British Isles; by the mid-1930s there were more than three hundred;[19] today there are several thousand.) And, of course, the scale of activities in these laboratories has also expanded vastly. It is perhaps unnecessary to dwell at length on the immense cost in resources of the research equipment of contemporary science. Radiotelescopic observatories, low-temperature physics, research hospitals, and lunar geology all involve outlays of a scale that require the funding and support of national governments—sometimes even consortia thereof. In a prophetic vein, Alvin M. Weinberg (then Director of the Oak Ridge National Laboratory) wrote: "When history looks at the twentieth century, she will see science and technology as its theme; she will find in the monuments of Big Science—the huge rockets, the high-energy accelerators, the high-flux research reactors—symbols of our time just as surely as she finds in Notre Dame a symbol of the Middle Ages."[20] Of course, exponential growth cannot of course continue indefinitely. But the fact remains that that science has become an enormous industry that has a far-flung network of *training* centers (schools, colleges, universities), and of *production* centers (laboratories and research institutes).

Natural science, in sum, has become a vast and expensive business. But what sort of relationship obtains between resource investment and returns here? Just how productive is the science enterprise?

The Law of Logarithmic Returns

Certain evaluative distinctions and classifications have played a pivotal role in epistemology and the theory of cognition, for example, the familiar distinctions between the true and the false, and between the well evidentiated (probable) and the evidentially counterindicated (improbable). But the portentous distinction between real knowledge and mere information has been generally neglected. And this is eminently unfortunate. After all, items of information were not created equal. Some claims will, even if true, be insignificant and make little or no impact on the larger cognitive scheme of things. (What difference does my personal preference for pink over purple make to anything?) And other truths will be distinctly important—the theory of relativity for example, or Avogadro's law in chemistry. Few issues regarding cognition can rival this distinction in point of its far-reaching bearing on matters of discovery, learning, understanding, and insight—alike in scientific and in everyday contexts. And so it will be this distinction that concerns us here.

En route to knowledge we must begin with information. How can one measure the volume of information generated in a field of suitable or scholarly inquiry? Various ways suggest themselves. The size of the literature of a field (as measured by the sheer bulk of publication in it) affords one possible measure. And there are various other possibilities as well. For one can also proceed by way of inputs rather than outputs, measuring, for example, the number of workdays that investigators devote to the topics at issue, or the amount of resource investment in the relevant information-engendering technology. So much for the quantitative assessment of *information*. But what about *knowledge?*

By "knowledge" in this context one must, of course, understand *putative* knowledge that is not necessarily correct but merely represents a conscientiously contrived best estimate of what the truth of the matter actually is.[21] But this is not all there is to it. Information is simply a collection of (supposedly correct) beliefs or assertions, while knowledge, by contrast, is something more select, more deeply issue-resolving. Like all usable information it must be duly evidentiated and at least *presumably* correct. But there is also an additionally evaluative aspect, seeing that knowledge is a matter of *important* information: information that is *significantly* informative. Not every insignificant smidgeon of information constitutes knowledge, and the person whose body of information consists of utter trivia really knows virtually nothing.

To provide a simple illustration for this matter of significance, let us suppose object-descriptive color taxonomy—for the sake of example, an oversimple one based merely on Blue, Red, and Other. Then that single item of *knowledge* represented by "knowing the color" of an object—namely, that it is red—is bound up with many different items of (correct) *information* on the subject (that it is not Blue, is rather similar to some shades of Other, etc.). As such information proliferates, we confront a situation of redundancy and diminished productiveness. Any knowable fact is always potentially surrounded by a vast prenumbral cloud of relevant information. And as our information grows to be ever more extensive, those really *significant* facts become more difficult to discern. Knowledge certainly increases with information, but at a far less than proportional rate.

One instructive way of measuring the volume of available information is via the opportunities for placement in a framework for describing or classifying the features of things. With n concepts, you can make n^2 two-concept combinations. With m facts, you can project m^2 fact-connecting juxtapositions, in each of which some sort of characteristic relationship is at issue.[22] A single step in the advancement of knowledge is here accompanied by a massive increase in the proliferation of information. Extending the previous example, let us also contemplate *shapes* in addition to *colors*, again supposing only three of them: Rectangular, Circular, Other. Now when we *combine* color and shape there will be $9 = 3 \times 3$ possibilities in the resultant (cross-) classification. So with that complex, dual-aspect piece of knowledge (color + shape) we also launch into a vastly amplified (i.e., multiplied) information spectrum over that increased classification-space. In moving cognitively from n to $n + 1$ cognitive parameters we enlarge our knowledge additively but expand our information field multiplicatively.

If a control mechanism can handle three items—be it in physical or in cognitive management—then we can lift ourselves to higher levels of capacity by hierarchical layering. First we can, by hypothesis, manage three base-level items; next we can, by grouping three of these into a first-level complex, manage nine base level items; and thereupon by grouping three of these first-level complexes into a second level complex we can manage twenty-seven base-level items. Hierarchical grouping is thus clearly a pathway to enhanced managerial capacity. But in cognitive contexts of information management it is clear that a few items of high-level information (= knowledge) can and will correspond to a vast range of low-level information (= mere truths). Knowing (in our illustration) where we stand within each of those three levels—that is having three pieces of *knowledge*—will position us in a vastly greater information space (viz., one of 27 compartments). And this situation is typical. The relational structure of the domain means that a small range of knowledge (by way of specifically high-grade information) can always serve to position one cognitively within a vastly greater range of low-level information.

It is instructive to view this idea from a different point of view. Knowledge commonly develops via distinctions (A vs. non-A) that are introduced with ever greater elaboration to address the problems and difficulties that one encounters with less sophisticated approaches. A situation obtains that is analogous to the "Game of Twenty Questions" with an exponentially exfoliating possibility space being traced out stepwise ($2, 4, 8, 16, \ldots, 2^n, \ldots$). With n descriptors one can specify for 2^n potential descriptions that specify exactly how, over all, a given object may be characterized. When we add a new descriptor we increase by one additional unit the amount of knowledge but double the amount of available information. The *information* at hand grows with 2^n, but the knowledge acquired merely with n. The cognitive exploitation of information is a matter of dramatically diminishing returns.

Again, consider an illustration of a somewhat different nature. The yield of knowledge from information afforded by legibility-impaired manuscripts and papyri and inscriptions in classical paleography provides an instructive instance. If we can decipher 70 percent of the letters in such a manuscript we can reconstruct the phrase at issue. If we can make out 70 percent of the phrases we can pretty well figure out the sentences. If we can read 70 percent of its sentences we can understand the message of the whole text. So some one-third of the letters suffice to carry the whole message. A vast load of information stands coordinate with a modest body of knowledge. From the standpoint of knowledge, information is highly redundant, albeit unavoidably so.

A helpful perspective on this situation comes to view through the idea of "noise," considering that expanding bodies of information encompass so much cognition-impeding redundancy and unhelpful irrelevancy that it takes successive many-fold increases in information to effect successive fixed-size increases in actual knowledge. It is supposed that internal "noise" or variability in the information-system is such that the increment needed to effect a cognitive advance of a certain fixed size or significance is proportional to the magnitude of the starting position as a whole.

Consider now the result of combining two ideas, namely that

1. Knowledge is distinguished from mere information as such by its significance. In fact: *Knowledge is simply particularly significant information*—information whose significance exceeds some threshold level (say *q*). (In principle there is room for variation here according as one sets the quality level of entry qualification and the domain higher or lower.)
2. The significance of *additional* information is determined by its impact on *preexisting* information. Significance in this sense is a matter of the relative (percentage-wise) increase that the new acquisitions effect upon the body of preexisting information (I), which may—to reiterate—be estimated in the first instance by the sheer volume of the relevant body of information: the literature of the subject, as it were. Accordingly: *The significance of incrementally new information can be measured by the ratio of the increment of new information to the volume of information already in hand*: $\Delta I / I$.

Putting the ideas of these two principles together, we have it that a new item of actual knowledge is one for which the ratio of information increments to preexisting information exceeds a fixed threshold-indicative quantity *q*:

$$\frac{\Delta I}{I} \geq q.$$

Thus knowledge-constituting *significant* information arises when the proportional extent of the change effected by a new item in the preexisting situation (independently of what that preexisting situation is) exceeds a duly fixed threshold.

On the basis of these considerations, it follows that the cumulative total amount of knowledge (*K*) encompassed in an overall body of information of size *I* is given by the *logarithm* of *I*. This is so because we have

$$K = \int \frac{dI}{I} = \log I + \text{const} = \log cI$$

where *K* represents the volume of actual *knowledge* that can be extracted from a body of bare *information* I,[23] with the constant at issue open to treatment as a unit-determinative parameter of the measuring scale, so that the equation at issue can be simplified without loss to $K = \log I$.[24] In milking additional information for significant insights it is generally the *proportion* of the increase that matters: its percentage rather than its brute amount. We accordingly arrive at the *Law of Logarithmic Returns* governing the extraction of significant *knowledge* from bodies of mere *information*.

The ramifications of such a principle for cognitive progress are not difficult to discern. Nature imposes increasing resistance barriers to intellectual as to physical penetration. Consider the analogy of extracting air for creating a vacuum. The first 90 percent comes out rather easily. The next 9 percent is effectively as difficult to extract than all that went before. The next 0.9 is just as difficult. And so on. Each successive order-of-magnitude step involved a massive cost for lesser progress; each successive fixed-size investment of effort yields a substantially diminished return. Intellectual progress is exactly the same: when we extract *knowledge* (i.e., high-grade, nature-descriptively significant information from mere information of the routine, common "garden variety," the same sort of quantity/quality relationship obtained. Initially a sizable proportion of the available is high grade—but as we press further this proportion of what is cognitively significant gets ever smaller. To double knowledge we must quadruple information.

It should thus come as no surprise that knowledge coordinates with information in multiplicative layers. With texts we have the familiar stratification: article/chapter, book, library, library system. Or pictographically: sign, scene (= ordered collection of signs), cartoon (= ordered collection of scenes to make it a story). In such layering, we have successive levels of complexity corresponding to successive levels of informational combining combinations, proportional with n, n^2, n^3, and so on. The logarithm of the levels—$\log n$, $2\log n$, $3\log n$, and so forth—reflect the amount of "knowledge" that is available through the information we obtain about the state of affairs prevailing at each level. And this is something that increases only one step at a time, despite the exponential increase in information.

The general purport of such a Law of Logarithmic Returns as regards expanding information is clear enough. Letting $K(I)$ represent the quantity of knowledge inherent in a body of information I, we begin with our fundamental relationship: $K(I) = \log I$. On this basis of this fundamental relationship, the knowledge of a two-sector domain increases additively notwithstanding a multiplicative explosion in the amount of information that is now on the scene. For if the field (F) under consideration consists of two subfields (F_1 and F_2), then because of the cross-connections obtaining within the information-domains at issue the overall information complex will take on a multiplicative size:

$$I = \inf(F) = \inf(F_1) \times \inf(F_2) = I_1 \times I_2.$$

With compilation, information is multiplied. But in view of the indicated logarithmic relationship, the knowledge associated with the body of compound information I will stand at:

$$K(I) = \log I = \log (I_1 \times I_2) = \log I_1 + \log I_2 = K(I_1) + K(I_2)$$

The knowledge obtained by compiling two information-domains (subfields) into an overall aggregates will (as one would expect) consist simply in *adding* the two bodies of knowledge at issue. While compilation increases *information* by multiplicative leaps and bounds, the increase in *knowledge* is merely additive.

THE RATIONALE AND IMPLICATIONS OF THE LAW OF LOGARITHMIC RETURNS

The Law of Logarithmic Returns present us with an epistemological analogue of the old Weber-Fechner law of psychophysics, asserting that inputs of geometrically increasing magnitude are required to yield *perceptual* outputs of arithmetically increasing intensity. The presently contemplated law envisions a parallelism of perception and conception in this regard. It stipulates that (informational) inputs of geometrically increasing magnitude are needed to provide for (cognitive and thus) *conceptual* outputs of arithmetically increasing magnitude. In searching for meaningful patterns, the ongoing proliferation of data points makes contributions of rapidly diminishing value.

It is not too difficult to come by a plausible explanation for the sort of information/knowledge relationship that is represented by $K = \log I$. The principal reason for such a K/I imbalance may lie in the efficiency of intelligence in securing a view of the *modus operandi* of a world whose law-structure is comparatively simple. For here one can learn a disproportionate amount of general fact from a modest amount of information. (Note that whenever an infinite series of 0's and 1's, as per 01010101 . . . , is generated—as this series indeed is—by a relatively *simple* law, then this circumstance can be gleaned from a comparatively short initial segment of this series.) In rational inquiry we try the simple solutions first, and only if and when they cease to work—when they are ruled out by further findings (by some further influx of coordinating information)—do we move on to the more complex. Things go along smoothly until an oversimple solution becomes destabilized by enlarged experience. We get by with the comparatively simpler options until the expanding information about the world's *modus operandi* made possible by enhanced new means of observation and experimentation demands otherwise. But with the expansion of knowledge new accessions set ever increasing demands. At bottom, then, there are thus two closely interrelated reasons for that K/I disparity:

1. Where order exists in the world, intelligence is rather efficient in finding it.

2. If the world were not orderly—were not amenable to the probes of intelligence—then intelligent beings would not and could not have emerged in it through evolutionary mechanisms.

The implications for cognitive progress of this disparity between knowledge and mere information are not difficult to discern. Nature imposes increasing resistance barriers to intellectual as to physical penetration. Consider the analogy of extracting air for creating a vacuum. The first 90 percent comes out rather easily. The next 9 percent is effectively as difficult to extract than all that went before. The next .9 is proportionally just as difficult. And so on. Each successive order-of-magnitude step involves a massive cost for lesser progress; each successive fixed-size investment of effort yields a substantially diminished return. Initially a sizable proportion of the available is high grade—but as we press further this proportion of what is cognitively significant gets ever smaller. To double knowledge we must quadruple information. As science progress, the important discoveries that represent real increases in knowledge are surrounded by an ever increasing penumbra of mere items of information. (The mathematical literature of the day yields an annual crop of over 200,000 new theorems.[25])

This state of affairs clearly has the most far-reaching implications for the progress of knowledge. This is illustrated in Table 12.1. There is an immense K/I imbalance, with ongoingly increasing information I, the corresponding increase in knowledge K, shrinks markedly. To increase knowledge by equal steps we must amplify information by successive orders of magnitude.

Extracting knowledge from information thus requires ever greater effort. And the matter can be viewed in another perspective. Nature imposes increasing resistance barriers to intellectual as to physical penetration. Extracting knowledge from information thus requires ever greater effort. For the greater a body of information, the larger the *patterns of order* that can potentially obtain and the greater the effort needed to bring particular orderings to light. With two-place combinations of the letters A and B (yielding the four pairs AA, AB, BA, and BB) we have only two possible patterns of order—namely, "The same letter all the way through" (AA and BB), and "Alternating letters" (AB and

TABLE 12.1
The Structure of the Knowledge/Information Relation

I	cI (with $c = .1$)	$\log cI\ (= K)$	$\Delta(K)$	K as % of I
100	10	1	—	1.0
1,000	100	2	1	.2
10,000	1000	3	1	.03
100,000	10,000	4	1	.004

BA). But as we add more letters, the possibilities proliferate massively. There is accordingly a law of diminishing returns in operation here: each successive fixed-size investment of effort yields a substantially diminished return. Intellectual progress is exactly the same: when we extract actual *knowledge* (i.e., high-grade, nature-descriptively significant information) from mere information of the routine, common "garden variety," the same sort of quantity/quality relationship obtained.

This situation is reflected in Max Planck's appraisal of the problems of scientific progress. He wrote that *"with every advance [in science] the difficulty of the task is increased; ever larger demands are made on the achievements of researchers*, and the need for a suitable division of labor becomes more pressing."[26] The Law of Logarithmic Returns at once characterizes and explains this circumstance of why substantial findings are thicker on the informational ground in the earlier phase of a new discipline and become ever attenuated in the natural course of progress.

This line of thought also serves to indicate, however, why the question of the rate of scientific progress is somewhat tricky. For this whole issue will turn rather delicately on fundamentally *evaluative* considerations. Thus consider the past once more. At the crudest—but also most basic—level, where progress is measured simply by the growth of the scientific literature, there has for centuries been astonishingly swift and sure progress: and exponential growth with a doubling time of roughly fifteen years. At the more sophisticated, and demanding level of high-quality results of a suitably "important" character, there has also been exponential growth, but only at the pace of a far longer doubling time, perhaps thirty years, approximating the reduplication with each successive generation envisaged by Henry Adams at the turn of this century.[27] Finally, at the maximally exacting level of the really crucial insights that fundamentally enhance our picture of nature, our analysis has it that science has been maintaining a merely constant pace of progress.

It is this last consideration that is crucial for present purposes and brings us back to the point made at the very outset. For what we have been dealing with here is an essentially seismological standard of importance. It is based on the question, If the thesis at issue were abrogated or abandoned, how large would the ramifications and implications of this circumstance be? How extensive would be the shocks and tremors felt throughout the whole range of what we (presumably) know?" And what is at issue is exactly a kind of cognitive Richter Scale based on the idea of successive orders of magnitude of impact. The crucial determinative factor for increasing importance is the extent of seismic disturbance of the cognitive terrain. Would we have to abandon and/or rewrite the entire textbook, or a whole chapter, or a section, or a paragraph, or a sentence, or a mere footnote?

And so, while the question of the rate of scientific progress does indeed involve the somewhat delicate issue of setting of an evaluative standard, our stance here can be rough and ready—and justifiably so because the details pretty much wash out. *Viewing science as a cognitive discipline*—a body of knowledge whose task is the unfolding of a rational account of the *modus operandi* of nature—stands correlative with its accession of really major discoveries: the seismically significant, cartography-revising insights into nature. And here progress has historically been sure and steady—but essentially linear. However, this perspective lays the foundation for our present analysis—that the historical situation has been one of a *constant* progress of science as a *cognitive discipline* notwithstanding its *exponential* growth as a *productive enterprise* (as measured in terms of resources, money, manpower, publications, etc).[28] If we look at the cognitive situation of science in its quantitative aspect, the Law of Logarithmic Returns pretty much says it all. In the struggle to achieve cognitive mastery over nature, we have been confronting an enterprise of ever-escalating demands, with the exponential growth in the *enterprise* associated with a merely linear growth in the *discipline*. And this situation regarding the past has important implications for the future. For while one cannot—as we have seen—hope to predict the *content* of future science, the law does actually put us into a position to make plausible estimates about its *volume*.[29]

The Growth of Knowledge

The Law of Logarithmic Returns also connects with—and is interestingly illustrated by—the idea of a cognitive life span as expounded by the sagacious Edward Gibbon in his *Memoirs of My Life*: "The proportion of a part to the whole is the only standard by which we can measure the length of our existence. At the age of twenty, one year is a tenth perhaps of the time which has elapsed within our consciousness and memory; at the age of fifty it is no more than a fortieth, and this relative value continues to decrease till the last sands are shaken out [of the hourglass measure of our life span] by the hand of death."[30] On this basis, knowledge development is a matter of adding a given percentage increment of what has gone before. Thus fresh experience superadds its additional increment ΔE to the preexisting total E in such a way that its effective import is measured by the proportion that movement bears to the total: $\Delta E/E$. And cumulatively we of course have it that this comes to the logarithm of E: $\int \Delta E/E = \log E$. On such an approach, an increment to one's lifetime has a *cognitive* value determined on strict analogy with Daniel Bernoulli's famous proposal to measure the *utility* value of incremental economic resources by means of a logarithmic yardstick.[31]

The Law of Logarithmic Returns clearly has substantial implications for the *rate* of scientific progress.[32] In particular, it stands coordinate with a principle of a swift and steady decline in the comparative cognitive yield of additions to our body of mere information. With the enhancement of scientific technology, the size and complexity of this body of data inevitably grows, expanding on quantity and diversifying in kind. Technological progress constantly enlarges the window through which we look out on nature's parametric space. In developing natural science, we use this window to scrutinize parametric space, continually augmenting our database and then generalizing on what we see. And what we have here is not a homogeneous lunar landscape, where once we have seen one sector we have seen it all, and where theory projections from lesser data generally remain in place when further data comes our way. Historical experience shows that there is every reason to expect that our ideas about nature are subject to constant radical changes as we enhance the means of broadening our database. But of course the expansion of knowledge proceeds at a far slower rate than the increase of bare information.

It is illuminating to look at the implications of this state of affairs in a historical perspective. The salient point is that the growth of knowledge over time involves ever-escalating demands. Progress is always possible—there are no absolute limits. But increments of the same cognitive magnitude have to be obtained at an ever-increased price in point of information development and thus of resource commitment as well. Here our basic formula that knowledge stands proportionate to the logarithm of information ($K \approx \log I$). And this means the increase of knowledge over time stands to the increase of information in a proportion fixed by the *inverse* of the volume of already available information:

$$\frac{d}{dt} K \approx \frac{d}{dt} \log I \approx \frac{1}{I} \frac{d}{dt} I.$$

The more knowledge we already have in hand, the slower (by vary rapid decline) will be the rate at which knowledge grows with newly acquired information. And the larger the body of information we have, the smaller will be the proportion of this information that represents real knowledge.

In developmental perspective, there is good reason to suppose that our body of bare *information* will increase more or less in proportion with our resource investment in information gathering. But then, if this investment grows exponentially over time (as has historically been the case in the recent period), we shall in consequence have it that

$$I(t) \approx c^t \text{ and correspondingly also } \frac{d}{dt} I(t) \approx c^t,$$

and accordingly,

$$K(t) \approx \log I(t) \approx \log c^t \approx t,$$

and consequently,

$$\frac{d}{dt} K(t) = \text{constant}.$$

It will follow on this basis that, since *exponential* growth in *I* is coordinated with a merely *linear* growth in *K*, the rate of progress of science in the information-exploding past has actually remained essentially constant.

On this basis, it is not all that difficult to find empirical substantiation of our law of logarithmic returns ($K \approx \log I$). The phenomenology is that while information has increased exponentially in the past (as shown by the exponential increase in journals, scientists, and outlays for the instrumentalities of research), real knowledge has expanded only linearly. Throughout modern times the number of scientists has been increasing at roughly 5 percent per annum.[33] Thus well over 80 percent of ever-existing scientists (in even the oldest specialties, e.g., mathematics, physics, and medicine) are alive and active at the present time. And scientific information has also been growing at the (reasonably constant) exponential rate over the past several centuries, so as to produce a veritable flood of scientific literature in our time. The *Physical Review* is now divided into six parts, each of which is larger than the whole journal was a decade or so ago. The total volume of scientific publication is truly staggering. It is reliably estimated that, from the start, well over 10 million scientific papers have been published and that by the mid-1970s some 30,000 journals were publishing some 600,000 new papers each year. It is readily documented that the number of books, of journals, of journals-papers has been increasing at an exponential rate over the recent period.[34] To be precise, the printed literature of science has been increasing at some 5 percent *per annum* throughout the last two centuries, to yield an exponential growth-rate with a doubling time of about fifteen years and an order-of-magnitude increase roughly every half century. And much the same story holds for the recruitment of people and the expenditures on equipment—that is for resource commitment in general.

But during this period the progress of science per se—the progress of authentic scientific knowledge as measured in the sort of first-rate discoveries of the highest level of significance has progressed at a more or less linear rate. There is much evidence to this claim. It suffices to consider the size of encyclopedias and synoptic textbooks, or again, the number of awards given for "really big contributions" (Nobel Prizes, academy memberships, honorary degrees), or the expansion of the classificatory taxonomy of branches of science and problem areas of inquiry.[35] All of these measures of preeminently *cognitive* contribution conjoin to indicate that there has in fact been a comparative constancy in year-to-year progress, the exponential growth of the scientific enterprise notwithstanding.

The Law of Logarithmic Returns reflects the fact that in any matured branch of natural science, continually greater capabilities in terms of technological capacity are required to realize further substantial results. The purchase price of significant new findings constantly increases; once all the significant findings accessible at a given state-of-the-art level of investigative technology have been realized, one must continually move on to a new, more complex (and thus more expensive) level. Ever more accurate measurements, more extreme temperatures, higher voltages, more intricate combinations, and so on, come to be required. With the passage of time, the ongoing requirement for personnel and material to sustain smooth progress has been increasing at an exponentially increasing rate. The phenomenon of cost escalation is explained through a combination of the finitude of the body of substantial results realizable at a given level of investigative technology, together with a continual and steep increase in the resource costs of pushing from one level to the next.

We are thus embarked on a literally limitless endeavor to improve the range of effective experimental intervention, since only by operating under new and heretofore inaccessible conditions of observational or experimental systematization—attaining extreme temperature, pressure, particle velocity, field strength, and so on—can we realize those circumstances that enable us to put our hypotheses and theories to the test. As one acute observer has rightly remarked: "Most critical experiments [in physics] planned today, if they had to be constrained within the technology of even ten years ago, would be seriously compromised."[36] Observational, experimental, and information processing technology is involved in an ever-continuing quest for substantial (order-of-magnitude) improvements in performance. And throughout the natural sciences, technological progress is a crucial requisite for cognitive progress.

And so, *viewing science as a cognitive discipline*—a growing body of knowledge whose task is the unfolding of a rational account of the *modus operandi* of nature—the progress of science stands correlative with its accession of really major discoveries: the seismically significant, cartography-revising insights into nature. And this perspective lays the foundation for our present analysis—that the historical situation has been one of a *constant* progress of science as a *cognitive discipline* notwithstanding its *exponential* growth as a *productive enterprise* (as measured in terms of resources, money, manpower, publications, etc.). The Law of Logarithmic Returns says it all.[37]

The Declaration of Scientific Progress

The limits of science are very real, but they are not inherently intellectual matters of human incapacity or deficient brainpower. They are fundamentally economic limits imposed by the technological character of our access to the phenomena of nature. The overoptimistic idea that we can push science ever onward to the solution of all questions that arise shatters in the awkward reality that the price of problem solving inexorably increases to a point beyond the limits of affordability.

In theory, the prospect of progress will never be exhausted: scientific innovation need never stop because ever new phenomena remain to be detected. But since ongoing progress requires us to meet ever increasing demands, the handwriting in on the wall. We have to accept the fact that it will—and inevitably must—eventually slow down. This state of affairs is particularly clear in the context of the Law of Logarithmic Returns. If, as was argued above, scientific progress has indeed been linear in the exponential growth past, than it will obviously have to become markedly slower in the zero-growth future.

And the preceding deliberations indicate the reason why this is so. For we accept the idea that our investment in science must reach a condition of (at best) steady-state stability—and that correspondingly the accumulation of mere scientific *information* will also remain constant—then we have the situation that:

$$\frac{d}{dt} I(t) = \text{constant, and accordingly: } I(t) \approx t.$$

On this basis, the volume of knowledge, which by the Law of Logarithmic Returns grows with the logarithm of the volume of information, will stand proportionate to the logarithm of elapsed time:

$$K(t) \approx \log I(t) \approx \log t.$$

And consequently the rate of increase in knowledge will stand proportionate to the *inverse* increase of the elapsed time:

$$\frac{d}{dt} K(t) \approx \frac{1}{t}.$$

The constant growth of I in a steady-state era will thus be associated with a time-inversely proportional rate of growth of K. Where information grows linearly, knowledge too will grow—but at an ever-decreasing rate. And so, the Law of Logarithmic Returns indicate that scientific progress is entering into an era of ongoing deceleration as steady-state conditions come to obtain with respect to resource investment. A zero-growth future bodes a radical and continual slowing of the pace of scientific progress.[38]

Thus if the general picture of an unremitting cost-escalation in the economics of scientific progress is even partly correct, the pace of scientific advance is destined to become markedly slower in the zero-growth era that inevitably lies ahead. The half-millennium commencing around 1650 will eventually come to be regarded as among the great characteristic developmental transformations of human history, with the age of the Science Explosion see to be a unique—and finite—in its own historical structure as the Bronze Age and the Industrial Revolution.

Predictive Implications of the Information/Knowledge Relationship

Our earlier deliberations indicated that the science of the future is inherently impredictable. But here one must be careful to keep in view the distinction between substantive issues and structural ones—between particular individual questions, theses, and theories within science, and issues regarding the overall structure of the discipline itself. Our strictures against the predictability of future science relate to the level of substantive detail: they bear on science as an investigative *discipline* with respect to substantive questions about the *modus operandi* of nature rather than as a productive *enterprise* (or "industry") that addresses such matters. For while we cannot say what the science of the future will discover, we can make some pretty shrewd estimates of the volume and the rate of its progress. Discovery, after all, depends on effort, and the amount of human energy and resource commitment is something that lies in the domain of plausible prediction.

At the level of generality, then, various plausible estimates regarding the future of science can indeed be made. For we can safely predict that it will be incomplete, that its agenda of availably open questions will be extensive, that its rate of progress will be slower, and so on. And we can confidently predict *that* future science will be more elaborate, complex, and conceptually sophisticated than the science of the present day—that it will have greater taxonomic diversification, greater substantive complexity, further high-level unification and low-level proliferation, increased taxonomic speciation of subject-matter specialties, and the like.

After all, in cognitive forecasting about science at the *volumetric*—rather than substantively *contentual*—level, a key role is played by the relationship that the amount of significant *knowledge* (K) that can be extracted from a body of *information* (of size I) is given by the relation: $K \approx \log I$.[39] This Law of Logarithmic Returns, as it might be called, constitutes a prime predictive instrument in the matter of volumetric forecasting about the cognitive domain. For this principle relates investment and returns, and investment is something that is relatively predictable and amenable to advance estimation. In particular, when our information I increases linearly over time—a situation that will have to obtain in a steady-state condition of constant resource-investment, seeing that investment and crude information are proportionate—then this relationship yields a drastic retardation that contemplates an ever longer waiting time for equal steps of significant scientific progress.[40]

But, of course, all this tells us only about the *structure* of future science, and not about its *substance*. The difficulty is that while we can predict *that* the science of the future will be more complex, cumbersome, and slower-paced in its advance than ours, we cannot foresee *how* it will be so in terms of the content of its discoveries. We can safely assert that the adequate exposition of a future science—be it physics or chemistry or whatever—will take up many more pages than ours. It is just that we do not have a clue as to what it is that those pages will say. At the substantive level, science prediction is a matter less of audacity than of foolhardiness. But shape and quantity, of course, is something else again.

The Centrality of Quality and Its Implications

The present deliberations also serve to indicate, however, why the question of the rate of scientific progress is delicate and rather tricky. For this whole issue turns rather delicately on fundamentally *evaluative* considerations. Thus consider once more the development of science in the recent historic past. As we saw above, at the crudest—but also most basic—level, where progress is measured simply by the growth of the scientific literature, there has for centuries been the astonishingly swift and sure progress of an exponential growth with a doubling time of roughly fifteen years. And at the more sophisticated, and demanding level of high quality results of a suitably "important" character, there has also been exponential growth—though only at the pace of a far longer doubling time, perhaps thirty years, approximating the reduplication with each successive generation envisaged by Henry Adams at the turn of this century.[41] Finally, at the maximally exacting level of the really crucial insights—significant scientific knowledge that fundamentally enhances our understanding of nature—our analysis has it that science has been maintaining a merely constant pace of progress.

It is this last consideration that is crucial for present purposes, carrying us back to the point made at the very outset. For what we have been dealing with that essentially seismological standard of importance based on the question "If a certain finding were abrogated or abandoned, how extensive would the ramifications and implications of this circumstance be? How great would be the shocks and tremors felt throughout the whole range of what we (presumably) know?" And what is at issue is exactly a kind of cognitive Richter Scale based on the idea of successive orders of magnitude of impact. The crucial determinative factor for increasing importance is the extent of seismic disturbance of the cognitive terrain. Would we have to abandon and/or rewrite the entire textbook, or a whole chapter, or a section, or a paragraph, or a sentence, or a mere footnote?

And so, while the question of the rate of scientific progress does indeed involve the somewhat delicate issue of the evaluative standard that is at issue, our stance here can be rough and ready—and justifiably so because the precise details do not affect the fundamental shape of the overall evaluation. *Viewing science as a cognitive discipline*—a body of knowledge whose task is the unfolding of a rational account of the modus operandi of nature—we have it that progress stands correlative with its accession of really major discoveries: the seismically significant, cartography-revising insights into nature. And this has historically been growing at a rate that sure and steady—but essentially linear. However, this perspective lays the foundation for our present analysis—that the historical situation has been one of a CONSTANT (linear) progress of science as a *cognitive discipline* notwithstanding its EXPONENTIAL growth as a *productive enterprise* (as measured in terms of resources, money, manpower, publications, etc.).

In the struggle for cognitive control over nature, we have been confronting an enterprise of ever escalating demands. And so, while one cannot hope to predict the *content* of future science, the F/I-relationship does actually put us into a position to make plausible estimates about its *volume*. To be sure, there is, on this basis, no inherent limit to the possibility of future progress in scientific knowledge. But the exploitation of this theoretical prospect gets ever more difficult, expensive, and demanding in terms of effort and ingenuity. New findings of equal significance require ever greater aggregate efforts. In the ongoing course of scientific progress, the earlier investigations in the various departments of inquiry are able to skim the cream, so to speak: they take the "easy pickings," and later achievements of comparable significance require ever deeper forays into complexity and call for an ever-increasing bodies of information. (And it is important to realize that this cost increase is not because latter-day workers are doing *better* science, but simply because it is harder to achieve *the same level* of science: one must dig deeper or search wider to achieve results of the same significance as before.)

A mixed picture emerges. On the one hand, the course of scientific progress is a history of the successive destabilization of theories. On the other hand, the increasing resource requirement for digging into ever deeper layers of complexity is such that successive triumphs in our cognitive struggles with nature are only to be gained at an increasingly greater price. The world's inherent complexity renders the task of its cognitive penetration increasingly demanding and difficult. The process at issue with the growth of scientific knowledge in our complex world is one of drastically diminishing returns. This situation means that grappling with ever vaster bodies of information in the construction of an ever more cumbersome and complex account of the natural world is the unavoidable requisite of scientific progress.

To be sure, we constantly seek to "simplify" science striving for an ever smaller basis of ever more powerful explanatory principles. But in the course of this endeavor we invariably complicate the structure of science itself. We secure greater power (and, as it were, *functional* simplicity) at the price of greater complexity in *structural* regards.[42] By the time the physicists get that grand unified theory, the physics they will have on their hands will be complex to the point to almost defying comprehension. The mathematics gets ever more elaborate and powerful, the training transit to the frontier ever longer. And so, despite its quest for greater operational simplicity (economy of principles), science itself is becoming ever more complex (in its substantive content, its reasonings, its machinery, etc.). Simplicity of process is here more than offset by complexity of product. And the phenomenon of diminishing returns reflects the price that this ongoing complexification exacts.[43]

Chapter 14

The Imperfectability of Knowledge
(Knowledge as Boundless)

Synopsis

- *Perfected science would have to realize four theoretical desiderata: (a) erotetic completeness (including explanatory completeness), (b) pragmatic completeness, (c) predictive completeness, and (d) temporal finality (the ω-condition).*
- *There is, as a matter of fundamental general principle, no practicable way for us to establish that any one of these desiderata is realized. Our science must be seen as inherently incompletable, with an ever-receding horizon separating where we are from where we would ideally like to be.*
- *For perfected science is not a realizable condition of things but an idealization that provides a helpful goal-direction as well as a useful contrast-conception to highlight the limited character of what we do and can attain.*
- *Perfection is dispensable as a goal for natural science and we certainly need not presuppose its potential attainability to validate the enterprise. The motive force of scientific progress is not the a fronte pull of an unattainable perfection but the a tergo push of removing recognized shortcomings.*

Conditions of Perfected Science

How far can the scientific enterprise advance toward a definitive understanding of reality? Might science attain a point of recognizable completion? Is the achievement of perfected science a genuine possibility, even in theory when all of the "merely practical" obstacles are put aside as somehow incidental?

What would *perfected science* be like? What sort of standards would it have to meet? Clearly, it would have to complete in full the discharge of natural science's mandate or mission. Now, the goal-structure of scientific inquiry covers a good deal of ground. It is diversified and complex, spreading across both the cognitive/theoretical and active/practical sectors. It encompasses the traditional quartet of description, explanation, prediction, and control, in line with the following picture:

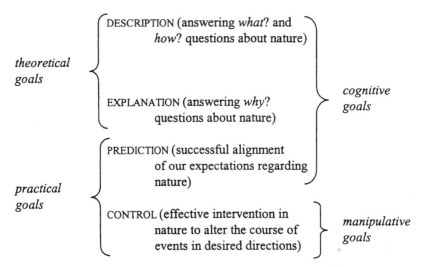

The theoretical sector concerns itself with matters of characterizing, explaining, accounting for, and rendering intelligible—with purely intellectual and informative issues, in short. By contrast, the practical sector is concerned with guiding actions, canalizing expectations, and, in general, with achieving the control over our environment that is required for the satisfactory conduct of our affairs. The former sector thus deals with what science enables us to *say*, and the latter with what it enables us to *do*. The one relates to our role as spectators of nature, the other to our role as active participants.

It thus appears that if we are to claim that our science has attained a perfected condition, it would have to satisfy (at least) the four following conditions:

1. *Erotetic completeness:* It must answer, in principle at any rate, all those descriptive and explanatory questions that it itself countenances as legitimately raisable, and must accordingly explain everything it deems explicable.
2. *Predictive completeness:* It must provide the cognitive basis for accurately predicting those eventuations that are in principle predictable (that is, those which it itself recognizes as such).

3. *Pragmatic completeness:* It must provide the requisite cognitive means for doing whatever is feasible for beings like ourselves to do in the circumstances in which we labor.
4. *Temporal finality* (the omega-condition): It must leave no room for expecting further substantial changes that destabilize the existing state of scientific knowledge.

Each of these modes of substantive completeness deserves detailed consideration. First, however, one brief preliminary remark. It is clear that any condition of science that might qualify as "perfected" would have to meet certain formal requirements of systemic unity. If, for example, there are different routes to one and the same question (for instance, if both astronomy and geology can inform us about the age of the earth), then these answers will certainly have to be consistent. Perfected science will have to meet certain requirements of structural systematicity in the manner of its articulation: it must be coherent, consistent, consonant, uniform, harmonious, and so on. Such requirements represent purely formal cognitive demands upon the architectonic of articulation of a body of science that could lay any claim to perfection. Interesting and important though they are, we shall not, however, trouble about these *formal* requirements here, our present concern being with various *substantive* matters at issue in the four aforementioned items.[1]

THEORETICAL ADEQUACY:
ISSUES OF EROTETIC COMPLETENESS

Could we ever actually achieve erotetic completeness (Q-completeness)—the condition of being able to resolve, in principle, all of our (legitimately posable) questions about the world? Could we ever find ourselves in this position?[2]

In theory, yes. A body of science certainly could be such as to provide answers to all of the questions that it allows to arise. But just how meaningful would this mode of completeness be?

It is sobering to realize that the erotetic completeness of a state of science does not necessarily betoken its comprehensiveness or sufficiency. It might reflect the paucity of the range of questions we are prepared to contemplate—a deficiency of imagination, so to speak. When the range of our knowledge is sufficiently restricted, then its question-resolving completeness will merely reflect this impoverishment rather than its intrinsic adequacy. Conceivably, if improbably, science might reach a purely fortuitous equilibrium between problems and solutions. It could eventually be "completed" in the narrow erotetic sense of providing an answer to every question that arises in the then-existing (albeit still imperfect) state of knowledge without thereby being completed in the larger sense of answering the questions that arise in probing more deeply. And so, our corpus of scientific knowledge could be erotetically complete and yet fundamentally inadequate. Thus, even if realized, this erotetic mode of completeness would not be particularly meaningful.

Any adequate theory of inquiry must recognize that the ongoing process of science is a process of *conceptual* innovation that always leaves certain theses wholly outside the cognitive range of the inquirers of any particular period. This means that there will always be facts (or plausible candidate-facts) about a thing that we do not *know* because we cannot even conceive of them. For to grasp such a fact calls for taking a perspective of consideration that we simply do not have, since the state of knowledge (or purported knowledge) is not yet advanced to a point at which its entertainment is feasible. In bringing conceptual innovation about, cognitive progress makes it possible to consider new possibilities that were heretofore conceptually inaccessible.

The language of emergence can perhaps be deployed profitably to make the point. But what is at issue is not an emergence of *the features of things* but an emergence in our *knowledge* about them. The blood circulated in the human body well before Harvey; uranium-containing substances were radioactive before Becquerel. The emergence at issue relates to our cognitive mechanisms of conceptualization, not to the objects of our consideration in and of themselves. Real-world objects are conceived of as antecedent to any cognitive interaction—as being there right along, or "pregiven," as Edmund Husserl puts it. Any cognitive changes or innovations are to be conceptualized as something that occurs on our side of the cognitive transaction, and not on the side of the objects with which we deal.[3]

And the prospect of change can never be dismissed in this domain. The properties of a thing are literally open-ended: we can always discover more of them. Even if we view the world as inherently finitistic, and espouse a Principle of Limited Variety which has it that nature can be portrayed descriptively with the materials of a finite taxonomic scheme, there can be no a priori guarantee that the progress of science will not engender an unending sequence of changes of mind regarding this finite register of descriptive materials. And this conforms exactly to our expectation in these matters. For where the real things of the world are concerned, we not only expect to learn more about them in the course of scientific inquiry, *we expect to have to change our mind about their nature and mode of comportment*. Be it elm trees, or volcanoes, or quarks that are at issue, we have every expectation that in the course of future scientific progress people will come to think differently about them than we ourselves do at this juncture.

Cognitive inexhaustibility thus emerges as a definitive feature of our conception of a real thing. For in claiming knowledge about them we are always aware that the object transcends what we know about it—that yet further and different facts concerning it can always come to light, and that all that we *do* say about it does not exhaust all that *can* be said about it.

The preceding considerations illustrate a more general circumstance. Any claim to the realization of a *theoretically* complete science of physics would be one that affords "a complete, consistent, and unified theory of physical interaction that would describe all possible observations."[4] But to check that the state of physics on hand actually meets this condition, we would need to know exactly what physical interactions are indeed *possible*. And to warrant us in using the state of physics on hand as a basis for answering *this* question, we would *already* have to be assured that its view of the possibilities is correct—and thus already have preestablished its completeness. The idea of a consolidated erotetic completeness shipwrecks on the infeasibility of finding a meaningful way to monitor its attainment.

After all, any judgment we can make about the laws of nature—any model we can contrive regarding how things work in the world—is a matter of theoretical triangulation from the data at our disposal. And we should never have unalloyed confidence in the definitiveness of our data base or in the adequacy of our exploitation of it. Observation can never settle decisively just what the laws of nature are. In principle, different law-systems can always yield the same observational output: as philosophers of science are want to insist, observations *underdetermine* laws. To be sure, this worries working scientists less than philosophers, because they deploy powerful regulative principles—simplicity, economy, uniformity, homogeneity, and so on—to constrain uniqueness. But neither these principles themselves nor the uses to which they are put are unproblematic. No matter how comprehensive our data or how great our confidence in the inductions we base on them, the potential inadequacy of our claims cannot be averted. One can never feel secure in writing *finis* on the basis of purely theoretical considerations.

We can reliably estimate the amount of gold or oil yet to be discovered, because we know a priori the earth's extent and can thus establish a proportion between what we know and what we do not. But we cannot comparably estimate the amount of knowledge yet to be discovered, because we have and can have no way of relating what we know to what we do not. At best, we can consider the proportion of available questions we can in fact resolve; and this is an unsatisfactory procedure. The very idea of cognitive limits has a paradoxical air. It suggests that we claim knowledge about something outside knowledge. But (to hark back to Hegel), with respect to the realm of knowledge, we are not in a position to draw a line between what lies inside and what lies outside—seeing that, *ex hypothesi* we have no cognitive access to that latter. One cannot make a survey of the relative extent of knowledge or ignorance about nature except by basing it on some picture of nature that is already in hand—that is, unless one is prepared to take at face value the deliverances of existing science. This process of judging the adequacy of our science on its own telling is the best we can do, but it remains an essentially circular and consequently inconclusive way of proceeding. The long and short of it is that there is no cognitively adequate basis for maintaining the completeness of science in a rationally satisfactory way.

To monitor the theoretical completeness of science, we accordingly need some theory-external control on the adequacy of our theorizing, some theory-external reality-principle to serve as a standard of adequacy. We are thus driven to abandoning the road of pure theory and proceeding along that of the practical goals of the enterprise. This gives special importance and urgency to the pragmatic sector.

Pragmatic Completeness

The arbitrament of praxis—not theoretical merit but practical capability—affords the best standard of adequacy for our scientific proceedings that is available. But could we ever be in a position to claim that science has been completed on the basis of the success of its practical applications? On this basis, the perfection of science would have to manifest itself in the perfecting of control—in achieving a perfected technology. But just how are we to proceed here? Could our natural science achieve manifest perfection on the side of control over nature? Could it ever underwrite a recognizably perfected technology?

The issue of "control over nature" involves much more complexity than may appear on first view. For just how is this conception to be understood? Clearly, in terms of bending the course of events to our will, of attaining our ends within nature. But this involvement of *"our ends"* brings to light the prominence of our own contribution. For example, if we are inordinately modest in our demands (or very unimaginative), we may even achieve "complete control over nature" in the sense of being in a position to do *whatever we want* to do, but yet attain this happy condition in a way that betokens very little real capability.

One might, to be sure, involve the idea of omnipotence, and construe a "perfected" technology as one that would enable us to do literally *anything*. But this approach would at once run into the old difficulties already familiar to the medieval scholastics. They were faced with the challenge: "If God is omnipotent, can he annihilate himself (contra his nature as a *necessary* being), or can he do evil deeds (contra his nature as a *perfect* being), or can he make triangles have four angles (contrary to *their* definitive nature)?" Sensibly enough, the scholastics inclined to solve these difficulties by maintaining that an omnipotent God need not be in a position to do literally anything but rather simply anything that it *is possible* for him to do. Similarly, we cannot explicate the idea of technological omnipotence in terms of a capacity to produce and result, wholly without qualification. We cannot ask for the production of a *perpetuum mobile,* for spaceships with "hyperdrive" enabling them to attain transluminar velocities, for devices that predict essentially stochastic processes such as the disintegrations of transuranic atoms, or for piston devices that enable us to set *independently* the values for the pressure, temperature, and volume of a body of gas. We cannot, in sum, as of a "perfected" technology that it should enable us to do anything that we might take it into our heads to do, no matter how "unrealistic" this might be.

All that we can reasonably ask of it is that perfected technology should enable us to do anything *that it is possible for us to do*—and not just what we might *think* we can do but what we really and truly can do. A perfected technology would be one that enabled us to do anything that *can possibly* be done by creatures circumstanced as we are. But how can we deal with the pivotal conception of "can" that is at issue here? Clearly, only science—real, true, correct, and in sum *perfected* science—could tell us what in fact is realistically possible and what circumstances are indeed inescapable. Whenever our "knowledge" falls short of this, we may well "ask the impossible" by way of accomplishment (e.g., spaceships in "hyperdrive"), and thus complain of incapacity to achieve control in ways that put unfair burdens on this conception.

Power is a matter of the "effecting of things possible"—of achieving control—and it is clearly cognitive state-of-the-art in science which, in teaching us about the limits of the possible, is itself the agent that must shape our conception of this issue. *Every* law of nature serves to set the boundary between what is genuinely possible and what is not, between what can be done and what cannot, between which questions we can properly ask and which we cannot. We cannot satisfactorily monitor the adequacy and completeness of our science by its ability to effect "all things possible," because science alone can inform us about what is possible. As science grows and develops, it poses new issues of power and control, reformulating and reshaping those demands whose realization represents "control over nature." For science itself brings new possibilities to light. (At a suitable stage, the idea of "splitting the atom" will no longer seem a contradiction in terms.) To see if a given state of technology meets the condition of perfection, we must *already* have a body of perfected science in hand to tell us what is indeed possible. To validate the claim that our technology is perfected, we need to *preestablish* the completeness of our science. The idea works in such a way that claims to perfected control can rest only on perfected science.

In attempting to travel the practicalist route to cognitive completeness, we are thus trapped in a circle. Short of having supposedly perfected science in hand, we could not say what a perfected technology would be like, and thus we could not possibly monitor the perfection of science in terms of the technology that it underwrites.

Moreover, even if (*per impossible*) a "pragmatic equilibrium" between what we can and what we wish to do in science were to be realized, we could not be warrantedly confident that this condition will remain unchanged. The possibility that "just around the corner things will become unstuck can never be eliminated. Even if we "achieve control," for all intents and purposes, we cannot be sure of not losing our grip on it—not because of a loss of power but because of cognitive changes that produce a broadening of the imagination and a widened apprehension as to what "having control" involves.

Accordingly, the project of achieving practical mastery can never be perfected in a satisfactory way. The point is that control hinges on what we want, and what we want is conditioned by what we think possible, and *this* is something that hinges crucially on theory—on our beliefs about how things work in this world. And so control is something deeply theory-infected. We can never safely move from apparent to real adequacy in this regard. We cannot adequately assure that seeming perfection is more than just that. We thus have no alternative but to *presume* that our knowledge (i.e., our purported knowledge) is inadequate at this and indeed at any other particular stage of the game of cognitive completeness.

One important point about control must, however, be noted with care. Our preceding negative strictures all relate to attainment of perfect control—of being in a position to do everything possible. But no such problems affect the issue of amelioration—of doing some things better and *improving* our control over what it was. It makes perfectly good sense to use its technological applications as standards of scientific advancement. (Indeed, we have no real alternative to using pragmatic standards at this level, because reliance on theory alone is, in the end, going to be circular.) While control does not help us with *perfection*, it is crucial for monitoring *progress*. Standards of assessment and evaluation are such that we can implement the idea of improvements (progress), but not that of completion (realized perfection). We can determine when we have managed to *enlarge* our technological mastery, but we cannot meaningfully say what it would be to *perfect* it. (Our conception of the *doable* keeps changing with changes in the cognitive state-of-the-art, a fact that does not, of course, alter our view of what *already has been done* in the practical sphere.)

With regard to technical perfectibility, we must recognize that (1) there is no reason to expect that its realization is possible, even in principle, and (2) it is not monitorable: even if we had achieved it, we would not be able to claim success with warranted confidence. In the final analysis, them, we cannot regard the *realization* of "completed science" as a meaningful prospect—we cannot really say what it is that we are asking for. (To be sure, what is meaningless here is not the idea of perfected science as such but the idea of *achieving* it.) These deliberations further substantiate the idea that we must always presume our knowledge to be incomplete in the domain of natural science.

Predictive Completeness

Predictive completeness rests on the idea of being able to predict everything that occurs. It represents a forlorn hope. For predictors are—of necessity!—

bound to fail even in much simpler self-predictive matters. Thus consider confronting a predictor with the problem posed by the question:

P_1: When next you answer a question, will the answer be negative?

This is a question which—for reasons of general principle—no predictor can ever answer satisfactorily.[5] For consider the available possibilities:

Answer given	Actually correct answer?	Agreement?
YES	NO	NO
NO	YES	NO
CAN'T SAY	NO	NO

On this question, there just is no way in which a predictor's response could possibly agree with the actual fact of the matter. Even the seemingly plausible response "I can't say" automatically constitutes a self-falsifying answer, since in giving this answer the predictor would automatically make "No" into the response called for by the proprieties of the situation.

Of course, the problem poses a perfectly meaningful question to which *another* predictor could give a putatively correct answer—namely, by saying: "No—that predictor cannot answer this question at all; the question will condemn a predictor to baffled silence." However, while the question posed in P_1 will be irresolvable by a *particular* computer, it could—in theory—be answered by *other* predictors and so is not predictively intractable flat-out. But there are other questions that indeed are computer insolubilia for computers-at-large. One of them is:

P_2: What is an example of a predictive question that no predictor will ever state?

In answering *this* question the predictor would have to stake a claim of the form: "Q is an example of a predictive question that no predictor will ever state." And in the very making of this claim the predictor would falsify it. It is thus automatically unable to effect a satisfactory resolution. However, the question is neither meaningless nor irresolvable. A *noncomputer* problem solver could in theory answer it correctly. All the same, its presupposition, "There is a predictive question that no predictor will ever consider" is beyond doubt true. What we thus have in P_2 is an example of an in-principle solvable—and thus "meaningful"—question which, as a matter of necessity in the logical scheme of things, no predictor can ever resolve satisfactorily.

The difficulties encountered in using physical control as a standard of "perfection" in science all also hold with respect to *prediction,* which, after all, is simply a mode of *cognitive* control.

Suppose someone asks: "Are you really still going to persist in plaints regarding the incompleteness of scientific knowledge when science can predict *everything?*" The reply is simply that science will *never* be able to predict literally everything: the very idea of predicting *everything* is simply unworkable. For then, whenever we predict something, we would have to predict also the effects of making those predictions, and then the ramification of *those* predictions, and so on *ad indefinitum.* The very most that can be asked is that science put us into a position to predict, not *everything,* but rather *anything* that we might choose to be interested in and to inquire about. And here it must be recognized that our imaginative perception of the possibilities might be much too narrow. We can only make predictions about matters that lie, at least broadly speaking, within our cognitive horizons. Newton could not have predicted findings in quantum theory any more than he could have predicted the outcome of American presidential elections. One can only make predictions about what one is cognizant of, takes note of, deems worthy of consideration. In this regard, one can be myopic either by not noting or by losing sight of significant sectors of natural phenomena.

Another important point must be made regarding this matter of unpredictability. Great care must be taken to distinguish the ontological and the epistemological dimensions, to keep the entries of these two columns apart:

unexplainable	not (yet) explained
by chance	by some cause we do not know of
spontaneous	caused in a way we cannot identify
random	lawful in ways we cannot characterize
by whim	for reasons not apparent to us

It is tempting to slide from epistemic incapacity to ontological lawfulness. But we must resist this temptation and distinguish what is inherently uncognizable from what we just do not happen to cognize. The nature of scientific change makes it inevitably problematic to slide from present to future incapacity.

Sometimes, to be sure, talk in the ontological mode is indeed warranted. The world no doubt contains situations of randomness and chance, situations in which genuinely stochastic processes are at work in a ways that "engenders unknowability." But these ontological claims must root in knowledge rather than ignorance. They can only be claimed appropriately in those cases in which (as in quantum theory) *we can explain inexplicability*—that is, in which we can account for the inability to predict/explain/control within the framework of a positive account of why the item at issue is actually unpredictable/unexplainable/unsolvable.

Accordingly, these ontological based incapacities do *not* introduce matters that "lie beyond the limits of knowledge." On the contrary, positive information is the pivot point. The only viable limits to knowability are those that root in knowledge—that is, in a model of nature that entails that certain sorts of things are unknowable. It is not a matter of an incapacity to answer appropriate questions ("We 'just don't know' why that stochastic process eventuated as it did"). Rather, in the prevailing state of knowledge, these question are improper; they just do not arise.

Science itself sets the limits to predictability—insisting that some phenomena (the stochastic processes encountered in quantum physics, for example) are inherently unpredictable. And this is always to some degree problematic. The most that science can reasonably be asked to do is to predict what it itself sees as in principle predictable—to answer every predictive question that it itself countenances as proper. Thus if quantum theory is right, the position and velocity of certain particles cannot be pinpointed conjointly. This renders the question "What will the exact position and velocity of particle X be at the exact time t?" not insoluble but illegitimate. Question-illegitimacy represents a limit that grows out of science itself—a limit on appropriate questions rather than on available solutions.

And in any case the idea that science might one day be a position to predict *everything* is simply unworkable. To achieve this, it would be necessary, whenever we predict something, to predict also the effects of making those predictions, and then the ramifications of *those* predictions, and so on *ad indefinitum*. The very most that can be asked is that science puts us into a position to predict, not *everything*, but rather *anything* that we might choose to be interested in and to inquire about.

Temporal Finality

Scientists from time to time indulge in eschatological musings and tell us that the scientific venture is approaching its end.[6] And it is, of course, entirely *conceivable* that natural science will come to a stop, and will do so not in consequence of a cessation of intelligent life but in C. S. Peirce's more interesting sense of completion of the project: of eventually reaching a condition after which even indefinitely ongoing inquiry will not—and indeed in the very nature of things *cannot*—produce any significant change, because inquiry has come to "the end of the road." The situation would be analogous to that envisaged in the apocryphal story in vogue during the middle 1800s regarding the Commissioner of the United States Patents who resigned his post because there was nothing left to invent.[7]

Such a position is in theory possible. But here, too, we can never determine that it is actual.

There is no practicable way in which the claim that science has achieved temporal finality can be validated. The question "Is the current state of science, S, final?" is one for which we can never legitimate an affirmative answer. For the prospect of future changes of S can never be precluded. One cannot plausibly move beyond "We have (in S) no good reason to think that S will ever change" to obtain "We have (in S) good reason to think that S will never change." To take this posture toward S is to *presuppose its completeness*.[8] It is not simply to take the natural and relatively unproblematic stance that that for which S vouches is to be taken as true but to go beyond this to insist that whatever is true finds a rationalization within S. This argument accordingly embeds *finality* in *completeness,* and in doing so jumps from the frying pan into the fire. For it shifts from what is difficult to what is yet more so. To hold that if something is so at all, then S affords a good reason for it is to take so blatantly ambitious (even megalomaniacal) a view of S that the issue of finality seems almost a harmless appendage.

Moreover, just as the appearance of erotetic and pragmatic equilibrium can be a product of narrowness and weakness, so can temporal finality. We may think that science is unchangeable simply because we have been unable to change it. But that's just not good enough. Were science ever to come to a seeming stop, we could never be sure that is had done so not because it is at "the end of the road" but because we are at the end of our tether. We can never ascertain that science has attained the ω-condition of final completion, since from our point of view the possibility of further change lying "just around the corner" can never be ruled out finally and decisively. No matter how final a position we *appear* to have reached, the prospects of its coming unstuck cannot be precluded. As we have seen, future science is inscrutable. We can never claim with assurance that the position we espouse in immune to change under the impact of further data—that the oscillations are dying out and we are approaching a final limit. In its very nature, science "in the limit" related to what happens in the long run, and this is something about which we *in principle* cannot gather information: any information we can actually gather inevitably pertains to the short run and not the long run. We can never achieve adequate assurance that *apparent* definitiveness is *real*. We can never consolidate the claim that science has settled into a frozen, changeless pattern. The situation in natural science is such that our knowledge of nature must ever be presumed to be incomplete.

The idea of achieving a state of recognizably completed science is totally unrealistic. Even as widely variant modes of behavior by three-dimensional objects could produce exactly the same two-dimensional shadow-projections, so very different law-systems could in principle engender exactly the same phenomena. We cannot make any definitive inferences from phenomena to the nature of the real. The prospect of perfected science is bound to elude us.

One is thus brought back to the stance of the great Idealist philosophers (Plato, Spinoza, Hegel, Bradley, Royce) that human knowledge inevitably falls short of recognizably "perfected science" (the Ideas, the Absolute), and must accordingly be looked on as incomplete. We have no alternative but to proceed on the assumption that the era of innovation is not over—that *future* science can and will prove to be *different* science.

As these deliberations indicate, the conditions of perfected science in point of description, explanation, prediction, and control are all unrealizable. Our information will inevitably prove inconclusive. We have no reasonable alternative to seeing our present-day science as suboptimal, regardless of the question of what date the calendar shows.

But what about completeness in point of discovery nature's laws? Even a system that is finitely complex both in its physical makeup and in its functional laws might nevertheless be infinitely complex in the phenomena that it manifests over time. For the operations of a structurally and lawfully finite system can yet exhibit an infinite intricacy in *productive complexity*, manifesting this limitless diversity in the working our of its processes rather than as regards its spatiostructural composition or the nomic comportment of its basic components. Even it the number of constituents of nature were small, the ways in which they can be combined to yield phenomena in space-time might yet be infinite. Think here of the examples of letters, syllables, words, sentences, paragraphs, books, genres (novels, reference books, etc.) libraries, library systems. Even a finite nature can, like a typewriter with a limited an operationally simple keyboard, yield an endlessly varied test. It can produce a steady stream of new products—"new" not necessarily in kind but in their functional interrelationships and thus in their theoretical implications. And on this basis our knowledge of nature's workings can be endlessly enhanced and deepened by contemplating an unending proliferation of phenomena.

Moreover, there is no need to assume a ceiling to such a sequence of levels of integrative complexity of phenomenal diversity. The different levels each exhibit an order of their own. The phenomena we attain in the n-th level can have features whose investigation takes us to the $(n + 1)$ level. New phenomena and new laws presumably arise at every level of integrative order. The diverse facets of nature can generate conceptually new strata of operation that yield a potentially unending sequence of levels, each giving rise to its own characteristic principles of organizations, themselves quite unpredictable from the standpoint of the other levels. In this way, even a relatively simple world as regards its basic operations can come to have an effectively infinite cognitive depth, once the notion of a natural phenomenon is broadened to include not just the processes themselves and the products they produce but also the relationship among them.

Consider, for example, some repeatedly exemplified physical feature and contemplate the sequence of 0's and 1's projected according to the rule that the i-th entry in the sequence is 1 if this feature is exemplified on occasion number

i, and 0 if not. Whenever two such feature-concepts, say C and C', generate such sequences in the manner of

C: 0100110100...
C': 1001011010...,

then we can introduce the corresponding matching sequence for C and C', 0010010001..., which is such that its i-th position is 1 if the two base sequences agree at their respective i-th positions, and 0 if they disagree. Such matching sequences will have a life of their own. For example, even if two base sequences are random, their matching sequences need not be—for example, when those base sequences simply exchange 0s and 1s. (Even random phenomena can be related by laws of coordination.)

Note that the present discussion does not propound the *ontological* theses that natural science cannot be pragmatically complete or even ω-definitive, but the *epistemological* thesis that science cannot ever be *known to be so*. The point is not that the requirements of definitive knowledge cannot in the nature of things be satisfied but that they cannot be *implemented* (i.e., be *shown* to be satisfied). The upshot is that science must always be presumed to be incomplete, not that it necessarily always is so. No doubt this is also true. It cannot, however, be demonstrated on the basis of epistemological general principles.

The prospect of change can never be dismissed in this domain. The properties of a thing are literally open-ended: we can always discover more of them. For where the real things of the world are concerned, we not only expect to learn more about them in the course of scientific inquiry, *we expect to have to change our mind about their nature and mode of comportment*. Be it elm trees, or volcanoes, or quarks that are at issue, we have every expectation that in the course of future scientific progress people will come to think differently about them than we ourselves do at this juncture. And any cognitive changes or innovations are to be conceptualized as something that occurs on our side of the cognitive transaction, and not on the side of the objects with which we deal.[9]

Cognitive inexhaustibility thus emerges as a definitive feature of our conception of a real thing. In claiming knowledge about such things, we are always aware that the object transcends what we know about it—that yet further and different facts concerning it can always come to light, and that all that we *do* say about it does not exhaust all that *can* be said about it. With science there is always more to be done, the doing of it becomes increasingly difficult. And man's material resources are limited. These limits inexorably circumscribe our cognitive access to the real world. There will always be interactions with nature of such a scale (as measured in energy, pressure, temperature, particle-velocities, etc.) that their realization would require the concurrent deployment of resources of so vast a scope that we can never realize them. If there are interactions to which we have no access, then there are (presumably) phenomena that we cannot discern. It would be very unreasonable to expect nature to confine the distribution of cognitively significant phenomena to those ranges that lie within our reach.

The world's furnishings are cognitively opaque; we cannot see to the bottom of them. Knowledge can become more extensive without thereby becoming more complete. And this view of the situation is rather supported than impeded if we abandon a cumulativist/preservationist view of knowledge or purported knowledge for the view that new discoveries need not supplement but can displace old ones.[10]

Were science ever to come to a seeming stop, we could never be sure that is had done so not because it is at "the end of the road" but because we are at the end of our tether. We can never ascertain that science has attained the ω-condition of final completion, since from our point of view the possibility of further change lying "just around the corner" can never be ruled out finally and decisively. The reality of natural science is such that our knowledge of nature must ever be presumed to be incomplete.

We have no alternative but to proceed on the assumption that the era of innovation is not over—that *future* science can and will prove to be *different* science. But of course while we can securely forecast *that* this will be so we will not(cannot(be specific regarding the details of *how* it will be so.

"Perfected Science" as an Idealization That Affords a Useful Contrast Conception

Reasons of general principle block us from ever achieving a position from which we can make good the claim that the several goals of science have actually been reached. Perfection is simply no a goal or *telos* of the scientific enterprise. It is not a realizable condition of things but at best a useful contrast conception that keeps actual science in its place and helps to sensitize us to its imperfections. The validation of this idealization lies not in its future *achievability* but in its ongoing *utility* as a regulative ideal that affords a contrast to what we do actually attain—so as to highlight its salient limitations.

With respect to the moral aspirations of man's will, Kant wrote:

> Perfection [of the moral will] is a thing of which no rational being in the world of sense is at any time capable. But since it is required [of us] as practically necessary, it can be found only in an endless progress to that complete fitness; on principles of pure practical reason, it is necessary to assume such a practical progress as the real object of our will.... Only endless progress from lower to higher stages of moral perfection is possible to a rational but finite being.[11]

Much the same story surely holds on the side of the cognitive perfecting of man's knowledge. Here, comparable regulative demands are a work governing the practical venture of inquiry, urging us to the ever fuller realization of the potentialities of the human intellect. The discontent of reason is a noble discontent. The scientific project is a venture in self-transcendence; one of the strongest motivations of scientific work is the urge to go beyond present science—to "advance the frontiers." Man's commitment to an ideal of reason in his pursuit of an unattainable systematic completeness is the epistemic counterpart of our commitment to moral ideals. It reflects a striving toward the rational that is all the more noble because it is not finally attainable. If the work of inquiring reason in the sphere of natural science were completable, this would be something utterly tragic for us.

Ideal science is not something we have in hand here and now. Nor is it something toward which we are moving along the asymptotic and approximative lines envisaged by Peirce.[12] Existing science does not and never will embody perfection. The cognitive ideals of completeness, unity, consistency, and definitive finality represent an aspiration rather than a coming reality, an idealized *telos* rather than a realizable condition of things. Perfected science lies outside history as a useful contrast-case that cannot be secured in this imperfect world.

The idea of "perfected science" is that *focus imaginarius* whose pursuit canalizes and structures out inquiry. It represents the ultimate *telos* of inquiry, the idealized destination of a journey in which we are *still* and indeed are *ever* engaged, a grail of sorts that we can pursue but not possess. The ideal of perfection thus serves a fundamentally regulative role to mark the fact that actuality falls short of our cognitive aspirations. It marks a contrast that *regulates* how we do and must view our claims, playing a rile akin to that of the functionary who reminded the Roman emperor of his mortality in reminding us that our pretensions are always vulnerable. Contemplation of this idea reminds us that the human condition is suspended between the reality of imperfect achievement and the ideal of an unattainable perfection. In abandoning this conception—in rejecting the idea of an "ideal science" that alone can properly be claimed to afford a definitive grasp of reality—we would abandon an idea that crucially regulates our view regarding the nature and status of the knowledge to which we lay claim. We would then no longer be constrained to characterize our view of things as *merely* ostensible and purported. We would be tempted to regard our picture of nature as real, authentic, and final in a manner that we at bottom realize it does not deserve.

What is being maintained here is not the "completed/perfected science" is a senseless idea as such but that the idea of *attaining* it is senseless. It represents a theoretically realizable state whose actual realization we can never achieve. What is unrealizable is not perfection as such but the *epistemic condition* of recognizing its attainment. (Even if we arrive, we can never tell that we're there!)

Does this situation not destroy "the pursuit of perfection" as a meaningful endeavor? Here it is useful to heed the distinction between a *goal* and an *ideal*.

A goal is something that we hope and expect to achieve. An ideal is merely a wistful inkling, a "wouldn't it be nice if"—something that figures in the mode of aspiration rather than expectation. A goal motivates us in striving for its attainment; an ideal stimulates and encourages. And ideal does not provide us with a destination that we have any expectation of reaching; it is something for whose actual attainment we do not even hope. It is in *this* sense that "perfected science" is an ideal.

Perfected science is not something that exists here and how, nor is it something that lies ahead at some eventual offing in the remote future. It is not a real thing to be met with in this world. It is an idealization that exists "outside time"—that is, which cannot actually come to realization at all. It lies outside history as a useful contrast-case that cannot be numbered among the achieved realities of this imperfect world. Existing science does not and presumably never will embody to perfection cognitive ideals of completeness, unity, consistency, and so on. These factors represent an aspiration rather than a coming reality: a telos, not a realizable condition of things—a hypothetical condition from which any and all of the negatives of the realized actual positions have been removed.

The concept of science perfected—of an ideal and completed science that captures the real truth of things and satisfies all of our cognitive ideals (definitiveness, completeness, unity, consistency, etc.)—is at best a useful fiction, a creature of the fictive imagination and not the secured product of inquiring reason. This "ideal science" is, as the very name suggests, an idealization, something that involves the removal in thought of limitations that obtain in fact. It involves the use of mind to move from a concern with the order of things as they are into the order of things as they ought to be.

Here, as elsewhere, we must reckon appropriately with the standard gap between aspiration and attainment. In the practical sphere—in craftsmanship, for example, or the cultivation of our health—we may *strive* for perfection, but cannot ever claim to *attain* it. And the situation in inquiry is exactly parallel with what we encounter in such other domains—ethics specifically included. The value of an ideal, even of one that is not realizable, lies not in the benefit of its attainment (obviously and *ex hypothesi!*) but in the benefits that accrue from its pursuit. The view that it is rational to pursue an aim only if we are in a position to achieve its attainment or approximation is mistaken; it can be perfectly valid (and entirely rational) if the indirect benefits of its pursuit and adoption are sufficient—if in striving after it, we realize relevant advantages to a substantial degree. An unattainable ideal can be enormously productive.

In sum, perfected science is not a realizable condition of things but an idealization that provides a useful contrast-conception to highlight the limited character of what we do and can attain. And so, the legitimation of the ideas of "perfected science" lies in its facilitation of the ongoing evolution of inquiry. In this domain, we arrive at the perhaps odd-seeming posture of an invocation of practical utility for the validation of an ideal.

THE DISPENSABILITY OF PERFECTION

The cognitive situation of natural science is such as to invite description in theological terms. For the ambiguity of the human condition is only too manifest here. We cannot expect ever to reach a position of definitive finality in this imperfect dispensation: we do have "knowledge" of all sorts, but it is manifestly imperfect. Expelled from the garden of Eden, we are deprived of access to "the God's-eye point of view." Definitive and comprehensive adequacy is denied us: we have no basis for claiming to know "the truth, the whole truth, and nothing but the truth" in scientific matters. We yearn for absolutes, but have to settle for plausiblities; we desire what is definitively correct, but have to settle for conjectures and estimates.

In this imperfect epistemic dispensation, we have to reckon with the realities of the human condition. Age disagrees with age; different states-of-the-art involve naturally discordant conceptions and incommensurate positions. The moral of the story of the Tower of Babel applies. The absolutes for which we yearn represent an ideal that lies beyond the range of practicable realizability. We simply have to do the best we can with the means at our disposal. To aspire to absolutes—for definitive comprehensiveness—is simply unrealistic.

It is sometimes maintained that such a fallibilist and imperfectionist view of science is unacceptable. To think of science as *inevitably* incomplete and to think of "the definitive answers" in scientific matters as *perpetually* unattainable is, we are told, to write science off as a meaningful project.

But in science as ins the moral life, we can operate perfectly well in the realization that perfection is unattainable. No doubt here and there some scientists nurse the secret hope of attaining some fixed and final definitive result that will stand, untouchable and changeless, through all subsequent ages. But unrealistic aspirations are surely by no means essential to the scientific enterprise as such. In science as in other domains of human endeavor, it is a matter of doing the best we can with the tools that come to hand. After all, the fact that *perfection* is unattainable does nothing to countervail against the no less real fact that *improvement* is realizable. And this undeniable prospect of realizable progress—of overcoming genuine defect and deficiencies that we find in the work of our predecessors—affords ample impetus to scientific innovation.

We can understand "progress" in two senses. On the one hand, there is *O*-progress, defined in terms of increasing distance from the starting point (the "origin"). On the other hand, there is *D*-progress, defined in terms of decreasing distance from the goal (the "destination"). Consider the picture:

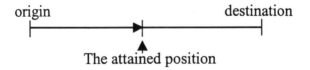

Ordinarily, the two modes of progress are entirely equivalent: we increase the distance traveled from O by exactly the same amount as we decrease the distance remained to D. But if the situation departs from the indicated picture in that there is no attainable destination—if we are engaged on a journey that, for all we know, is literally endless and has no determinable destination, or only one that is "infinitely distant"—then we just cannot manage to decrease our distance from it.

The fact of it is that scientific progress is not generated *a fronte* by the pull of an unattainable ideal; it is stimulated *a tergo* by the push of dissatisfaction with the deficiencies of achieved positions. The labors of science are not pulled forward by the mirage of (unattainable) perfection. We are pushed onward by the (perfectly realizable) wish to do better than our predecessors in the enterprise.

Seeing that in natural science we are embarked on a journey that is literally endless, it is only O-progress that can be achieved, and no D-progress. We can gauge our progress only in terms of how far we have come, and not in terms of how far we have to go. Embarked on a journey that is in principle endless, we simply cannot say that we are nearing the goal.

The upshot is straightforward. The idea of *improving* our science can be implemented without difficulty, since we can clearly improve our performance as regards its goals of prediction, control, and the rest. But the idea of *perfecting* our science cannot be implemented.[13]

Part IV

Cognitive Limits and the Quest for Truth

Chapter 15

The Rational Intelligibility of Nature

Synopsis

- *Eminent authorities have repeatedly characterized it as a miracle that nature is intelligible to us humans.*
- *But this ability of mind to come to terms with nature is a two sided one. On our side, it roots in the fact that we ourselves are a part of nature, a species evolved as intelligent beings within its ambit.*
- *And, on nature's side, it lies in the very fact of its affording a setting within which the evolution of intelligent beings is possible.*
- *Thus while nature's nature may itself be a mystery, the fact that intelligent being evolved within it can come to comprehend various aspects of its workings is only natural and to be expected.*
- *On the other hand, such argumentation certainly does not support the unrealistic conclusion that our understanding of nature is somehow complete or perfect.*

Explaining the Possibility of Natural Science

How is natural science—and, in particular, *physics*—possible at all? How is it that we insignificant humans, inhabitants of a minor satelite of a minor star in one of the worlds myriad galaxies, can manage to unlock nature's secrets and gain access to her laws? And how can our mathematics—seemingly a free creative invention of the human imagination—be used to characterize the workings of nature with such uncanny effectiveness and accuracy? Why is it that the majestic lawful order of nature is intelligible to us humans in our own conceptual terms?

This issue remained unproblematic as long as people thought of the world as the product of the creative activity of mathematicizing intelligence—as the work of a creator who proceeds on mathematical principles (*more mathematico*, in Spinoza's expression) in designing nature. For then one could take the line that God endows nature with a mathematically intelligible order and mind with a duly consonant mathematicizing intelligence. There is thus no problem about how the two get together—God simply arranged it that way. But, of course, if *this* is to be the canonical rationale for mind's grasp on nature's laws, then in foregoing explanatory recourse to God, we also—to all appearances—lose our grip on the intelligibility of nature.

Some of the deepest intellects of the day accordingly think that this possibility is gone for ever, confidently affirming that there just is no way to solve this puzzle of nature's being intelligible in a mathematically lawful manner. In classic treatise Erwin Schroedinger characterizes the circumstance that man can discover the laws of nature as "a miracle that may well be beyond human understanding."[1] Eugene Wigner asserts that "the enormous usefulness of mathematics in the natural sciences is something bordering on the mysterious, and there is no rational explanation for it"[2] and he goes on to wax surprisingly lyrical in maintaining that:

> The miracle of the appropriateness of the language of mathematics for the formulation of the laws of physics is a wonderful gift which we neither understand nor deserve.[3]

Even Albert Einstein stood in awe before this problem. In a letter written in 1952 to an old friend of his Berne days, Maurice Solovine, he wrote:

> You find it curious that I regard the intelligibility of the world (in the measure that we are authorized to speak of such an intelligibility) as a miracle or an eternal mystery. Well, *a priori* one should expect that the world can be rendered lawful only to the extent that we intervene with our ordering intelligence ... [But] the kind of order, on the contrary, created, for example, by Newton's theory of gravitation, is of an altogether different character. Even if the axioms of the theory are set by men, the success of such an endeavor presupposes in the objective world a high degree of order that we were *a priori* in no way authorized to expect. This is the "miracle" that is strengthened more and more with the development of our knowledge.... The curious thing is that we have to content ourselves with recognizing the "miracle" without having a legitimate way of going beyond it.[4]

According to all these eminent physicists we are confronted with a profound mystery. As they see it, we have to acknowledge *that* nature is intelligible, but have no prospect of understanding *why* this is so. The problem of nature's intelligibility by means of our mathematical resources is seen as intractable, unresolvable, hopeless. All three of these distinguished Nobel laureates in physics unblushingly employ the word *miracle* in this connection.

Perhaps, however, the very question is illegitimate and should not be raised at all. Perhaps the issue of nature's intelligibility is not just *intractable*, but actually *inappropriate* and improperly based on a false presupposition. For to ask for an explanation of *why* scientific inquiry is successful presupposes that there indeed *is* an explanatory rationale for this fact. But if this circumstance is something fortuitous and accidental, then of course no such rationale will exist at all. Just this position is advocated by various philosophers—for example, by Karl Popper, who wrote:

> [Traditional treatments of induction] all assume not only that our quest for [scientific] knowledge has been successful, but also that we should be able to explain why it is successful. However, even on the assumption (which I share) that our quest for knowledge has been very successful so far, and that we now know something of our universe, this success becomes [i.e., remains] miraculously improbable, and therefore inexplicable; for an appeal to an endless series of improbable accidents is not an explanation. (The best we can do, I suppose, is to investigate the almost incredible evolutionary history of these accidents . . .).[5]

Mary Hesse, too, thinks that it is inappropriate to ask for an explanation of the success of science "because science might, after all, be a miracle."[6]

And so, on this grand question of how the success of natural science is possible at all, some of the shrewdest scientific intellects of the day avow themselves baffled, and unhesitatingly enshroud the issue in mystery or miracle. On this sort of view, the question of the intelligibility of nature becomes an illegitimate pseudoproblem—a forbidden fruit at which sensible minds should not presume to nibble. We must simply rest content with the fact itself acknowledging that any attempt to explain it is foredoomed to failure because of the inappropriateness of the very project.

Surely, however, such an approach has very questionable merit. Eminent authorities to the contrary notwithstanding, the question of nature's intelligibility through natural science is not only interesting and important, but is also surely one that we should, in principle, hope and expect to answer in a more or less sensible way. Clearly, this important issue needs and deserves a strong dose of demystification.

How is it that we can make effective use of mathematical machinery to characterize the *modus operandi* of nature? The pure logical theorist seems to have a ready answer. He says: "Mathematics *must* characterize reality. Mathematical propositions are purely *abstract* truths whose validation turns on conceptual issues alone. Accordingly, they hold of *this* world because they hold of *every possible* world."

But this response misses the point of present concerns. Admittedly, the truths of *pure* mathematics obtain in and of every possible world. But they do so only in virtue of the fact that they are strictly hypothetical and descriptively empty—wholly uncommitted regarding the substantive issues of the world's operations. Their very conceptual status means that the theses of pure mathematics are beside the point of our present purposes. It is not the *a priori* truth of pure mathematics that concerns us, its ability to afford truths of reason. Rather, what is at issue is the *empirical applicability* of mathematics, its pivotal role in framing the *a posteriori*, contingent truths of lawful fact that render nature's ways amenable to reason.

After all, the circumstance that pure mathematics holds true in a world does not mean that this world's *laws* have to be characterizable in relatively straightforward mathematical terms. It does not mean that nature's operations have to be congenial to mathematics and graspable in terms of simple, neat, elegant, and rationally accessible formulas. In short, it does not mean that the world must be mathematically tractable and "mathematophile" in admitting to the sort of concise descriptive treatment it receives in mathematical physics.

How, then, are we to account for the fact that the world appears to us to be so eminently intelligible in the mathematical terms of our natural science?

The answer to this question of the cognitive accessibility of nature to mathematicizing intelligence has to lie in a somewhat complex, two-sided story in which both sides, intelligence and nature, must be expected to have a part. After all, when two things fit together, each must accommodate the other. Let us trace out this line of thought—one step at a time.

"Our" Side

Our human side of this bilateral story is relatively straightforward. After all, Homo sapiens is an integral part of nature. We are connected into nature's scheme of things as an intrinsic component thereof—courtesy of the processes of evolution. Our experience is thus inevitably an experience *of nature*. (That after all is what "experience" is—our intelligence-mediated reaction to the world's stimulating impacts on us.) So the kind of mathematics—the kind of theory of periodicity and structure—that we devise in the light of this experience is the kind that is in principle applicable to nature as we can experience it. As C. S. Peirce insisted, evolutionary pressures conform our intellectual processes to the *modus operandi* of nature. For nature not only *teaches* us (when we chose to study it) but also *forms* us (whether or not we chose to study it). And it proceeds in doing the latter in a way that is not, and cannot be, without implications for the former.

Our mathematics is destined to be attuned to nature because it itself is a natural product as a thought-instrument of ours: it fits nature because it reflects the way we ourselves are emplaced within nature as integral constituents thereof. Our intellectual mechanisms—mathematics included—fit nature because they are themselves a product of nature's operations as mediated through the cognitive processes of an intelligent creature that uses its intelligence to guide its interaction with a nature into which it is itself fitted in a particular sort of way.

The very selfsame forces that are at work in shaping the physical world are also at work in shaping our bodies and brains and in providing the stimuli that impinge on our senses and our minds. It is these interactions between thought and world that condition our sense of order and beauty—of regularity, symmetry, economy, elegance. Evolutionary pressure coordinates the mind with its environment. Even as we are destined to find healthy foods palatable and reproductively advantageous activities pleasant, so nature's inherent order and structure are bound to prove congenial to our mathematical sense of elegance and beauty.

The modes of order that attract the attention of mathematical theorists interested in structures—and that underlie their ideas of beautiful theories—are thus, unsurprisingly, also at work in the nature within which these conceptualizations arise. The mathematical mechanisms we employ for understanding the standard features of things themselves reflect the structure of our *experience*. In particular, the mathematics of an astronomically remote civilization whose experiential resources differ from ours might well be substantially different from mathematics as we ourselves know it. Their dealings with collective manifolds might be entirely anumerical—purely comparative, for example, rather than quantitative. Especially if their environment is not amply endowed with solid objects or stable structures congenial to measurement—if, for example, they were jellyfish-like creatures swimming about in a soupy sea—their "geometry" could be something rather strange, largely topological, say, and geared to flexible structures rather than fixed sizes or shapes. Digital thinking might go undeveloped while certain sorts of analogue reasoning might be highly refined. For example, if the intelligent aliens were a diffuse assemblage of units that constitute wholes in ways permitting overlap, then social concepts might become so paramount in their thinking that nature would throughout be viewed in fundamentally social categories, with those aggregates we think of as *physical* structures contemplated by them in *social* terms. Communicating by some sort of "telepathy" based on variable odors or otherwise "exotic" signals, they might, for example, devise a complex theory of empathetic thought-wave transmittal through an ideaferous aether. The processes that underlie their mathematicizing might be very different indeed.

Admittedly mathematics is not a natural science but a theory of hypothetical possibilities. Nevertheless these possibilities are possibilities as conceived by beings who do their possibility-conceiving with a nature-evolved and nature-implanted mind. It is thus not surprising that the sort of mathematics we contrive is the sort of mathematics we find applicable to the conceptualization of nature. After all, the intellectual mechanisms we devise in coming to grips with the world—in transmuting sensory interaction with nature into intelligible experience—have themselves the aspect (among many other aspects) of being nature's contrivances in adjusting to its ways a creature it holds at its mercy.

It is no more a miracle that the human mind can understand the world through its intellectual resources than that the human eye can see it through its physiological resources. The critical step is to recognize that the question "Why do our conceptual methods and mechanisms fit 'the real world' with which we interact intellectually?" is to be answered in basically the same way as the question: "Why do our bodily processes and mechanisms fit the world with which we interact physically?" In neither case can we proceed on the basis of purely theoretical grounds of general principle. Both issues are alike to be resolved in essentially evolutionary terms. It is no more surprising that our minds can grasp nature's ways than it surprising that our eyes can accommodate nature's rays or own stomachs nature's food. As we have noted from the outset, evolutionary pressure can take credit for the lot: they are part and parcel of what is mandated by attainment to our niche in nature's scheme of things. There is nothing "miraculous" or "lucky" in our possession of efficient cognitive faculties and processes—effective "hardware" and "software" for productive inquiry. If we did not, we just would not be here as inquiring creatures emplaced in nature thanks to evolutionary processes.

Nevertheless, it could perhaps be the case that we succeed in mathematicizing nature only as regards the immediate local microenvironment that defines our particular limited ecological niche. The possibility still remains open that we secure a cognitive hold on only a small and peripheral part of a large and impenetrable whole. And so, man's own one-sided contribution to the matter of nature's intelligibility cannot be the *whole* story regarding the success of science. For even if we do reasonably well in regard to our own immediate evolutionary requirements, this might still be very inadequate in the larger scheme of things. Nature's receptiveness to our cognitive efforts remains to be accounted for—the fact that nature is *substantially* amenable to reason and not just *somewhat* (and perhaps only very marginally) so.

To clarify this issue we must therefore move on to consider nature's contribution to the bilateral mind/nature relationship.

Nature's Side

What needs to be explained for present purposes is not just why mathematics is merely of *some* utility in understanding the world, but why it is actually of *very substantial* utility in that its employment can provide intelligent inquirers with an impressively adequate and accurate grasp of nature's ways. We must thus probe more deeply into the issue of nature's amenability to inquiry and its accessibility to the probes of intelligence.

To be sure, the effective applicability of mathematics to the description of nature is in no small part due to the fact that we actually devise our mathematics to fit nature through the mediation of experience. But how can one get beyond this to establish that nature simply *"must"* have a fairly straightforward law structure? Are there any fundamental reasons why the world that we investigate by the use of our mathematically informed intelligence should operate on relatively simple principles that are readily amenable to mathematical characterization?

There are indeed. For a world in which intelligence emerges by anything like standard *evolutionary* processes has to be pervaded by regularities and periodicities in the organism-nature interaction that produces and perpetuates organic species. And this means that nature must be cooperative in a certain very particular way—it must be stable enough and regular enough and structured enough for there to be appropriate responses to natural events that can be "learned" by creatures. If such "appropriate responses" are to develop, nature must provide suitable stimuli in a duly structured way. An organically viable environment—to say nothing of a *knowable* one—must incorporate experientiable structures. There must be regular patterns of occurrence in nature that even simple, single-celled creatures can embody in their makeup and reflect in their *modus operandi*. Even the humblest organisms, snails, say, and even algae, must so operate that certain *types of stimuli* (patterns of recurrently discernible impacts) call forth appropriately corresponding *types of response*—that such organisms can "detect" a structured pattern in their natural environment and react to it in a way that proves to their advantage in evolutionary terms. Even its simplest creatures can maintain themselves in existence only by swimming in a sea of detectable regularities of a sort that will be readily accessible to intelligence. Their world must encapsulate straightforwardly "learnable" patterns and periodicities of occurrence in its operations—relatively simply laws, in other words.

Let us suppse an inquiring being emplaced by evolution within nature that forms its mathematicized conceptions and beliefs about this nature on the basis of physical interaction with it. Clearly, if such a being is to achieve a reasonably appropriate grasp of the world's workings, *then* nature too must "do its part" in being duly cooperative. Given the antecedent supposition, nature must, obviously, permit the evolution of inquiring beings. And to do this, it must present them with an environment that affords sufficiently stable patterns to make coherent "experience" possible, enabling them to derive appropriate *information* from those structured interactions that prevail in nature at large. Nature's own contribution to solving the problem of its mathematical intelligibility must accordingly be the possession of a relatively simple and uniform law structure—one that deploys so uncomplicated a set of regularities that even a community of inquirers possessed of only rather moderate capabilities can be expected to achieve a fairly good grasp of the processes at work in their environment.

Accordingly, a world in which intelligence can develop by evolutionary processes also *must*—on this very basis—be a world amenable to understanding in mathematical terms.[7] It must be a world whose cognizing beings will find much grist to their mill in endeavoring to "understand" the world. Galileo long ago hit close to the mark when he wrote in his *Dialogues* that: "Nature initially arranged things her *own* way and subsequently so constructed the human intellect as to be able to understand her."[8] And of course nature's construction of mathematicizing mind has proceeded by evolutionary processes.

The development of *life* and, thereafter of *intelligence* in the world may or may not be inevitable; the emergence of intelligent creatures on the world's stage may or may not be surprising in itself and as such. But once they are there, and once we realize that they got there thanks to evolutionary processes, it can no longer be seen as surprising that their efforts at characterizing the world in mathematical terms should be substantially successful. *A world in which intelligent creatures emerge through the operation of evolutionary processes must be an intelligible world.*

On this line of deliberation, then, nature admits of mathematical depiction not just because it has laws—is a *cosmos*—but because as an evolution-permitting world it must have many *relatively simple* laws. And those relatively simple laws must be there because if they were not, then nature just would not afford the sort of environment requisite for the evolutionary development of intelligent life. An intelligence-containing world whose intelligent creatures came by this intelligence through evolutionary means must be substantially intelligible in mathematical terms.[9]

The apparent success of human mathematics in characterizing nature is thus nowise amazing. It may or may not call for wonder that intelligent creatures should evolve at all. But thereupon, once they have safely arrived on the scene through evolutionary means, it is only natural and to be expected that they should be able to achieve success in the project of understanding nature in mathematical terms. A mathematicizing intelligence *arrived at through evolution* must for this very reason prove to be substantially successful in achieving adequation to the world's ways.

The strictly hypothetical and conditional character of this general line of reasoning must be recognized. It does not maintain that by virtue of some sort of transcendental necessity the world has to be simple enough for its mode of operation to admit of elegant mathematical representation. Rather, what it maintains is the purely conditional thesis that *if* intelligent creatures are going to emerge in the world by evolutionary processes, *then* the world must be mathematophile, with various of its processes amenable to mathematical representation.

It must be stressed, however, that this merely conditional fact is quite sufficient for present purposes. For the question we face is why we intelligent creatures present on the world's stage should be able to understand its operations in terms of our mathematics. The conditional story at issue fully suffices to accomplish this particular job.

One brief digression to avert a possible misunderstanding is in order. Nothing whatever in the present argumentation can properly be construed to claim that the development of mathematics is an evolutionary requirement or desideratum as such—that creatures are somehow impelled to develop mathematics because it advantages them in the struggle for existence. (This idea would be a foolish anachronism, seeing that evolution produced us long before we produced mathematics.) To say that *intelligence*, the precondition of mathematics, is of evolutionary advantage, is not to claim that this is the case with mathematics itself. All that is being maintained is: (1) that intelligence is (in certain circumstances) of evolutionary advantage, (2) that any sufficiently intelligent creature *can* develop a mathematics (a theory of structure), and (3) that any sufficiently intelligent creature must be able to develop an *effectively applicable* "mathematics" in any world able to give rise to it through evolutionary means.

SYNTHESIS

The preceding deliberations maintains that the overall question of the intelligibility of nature has two sides:

1. Why is mind so well attuned to nature?
2. Why is nature so well attuned to mind?

And it is further maintained that the answers to these questions are not all that complicated—at least at the level of schematic essentials. The crux is simply this: Mind must be attuned to nature since intelligence is a generalized guide to conduct that has evolved as a natural product of nature's operations. And nature must be substantially accessible to mind if intelligence manages to evolve within nature by a specifically evolutionary route.

For nature to be intelligible, then, there must be an alignment that requires cooperation on both sides. The analogy of cryptanalysis is suggestive. If A is to break B's code, there must be due reciprocal alignment. If A's methods are too crude, too hit and miss, then A can get nowhere. But even if A is quite intelligent and resourceful success is precluded if B's procedures are simply beyond A's powers. (The cryptanalysts of the seventeenth century, clever though they were, could get absolutely nowhere in applying their investigative instrumentalities to a high-level naval code of World War II vintage.) Analogously, if mind and nature were too far out of alignment—if mind were too "unintelligent" for the complexities of nature or nature too complex for the capacities of mind—the two just could not get into step. It would be like trying to rewrite Shakespeare in a pidgin English with a 500-world vocabulary or like trying to monitor the workings of a system with ten degrees of freedom by using a cognitive mechanism capable of keeping track of four of them. If something like this were the case, mind could not accomplish its evolutionary mission. It would then be better to adopt an alignment process that does not take the cognitive route to the guidance of action. Just as any creature that evolves in nature must find due physical accommodation within it (a due harmonization of its bodily operations with its physical environs), so any mind that evolves in nature must find due intellectual accommodation within it (a due harmonization of its intellectual operations with its structural environs). In consequence, there must be a due equilibration between the mind's mathematizing operations and the world's mathematical structure.

The solution of our problem thus roots in the combination of two considerations: (1) a world that admits of the evolutionary emergence of intelligence must be sufficiently regular and simple—that is, it must be mathematophile, and (2) a sufficiently powerful intelligence must be able effectively to comprehend in mathematical terms any world in which it gains its foothold by evolutionary means. The possibility of a mathematical science of nature is accordingly to be explained by the fact that, in the light of evolution, intelligence and intelligibility must stand in mutual coordination.

Two points are accordingly paramount here:

1. Once intelligent creatures evolve, their cognitive efforts are bound to have some degree of adequacy because evolutionary pressures align them with nature's ways.
2. It should not be surprising that this alignment can eventually produce a substantially effective mathematical physics, because the structure of the operations of a nature that engenders intelligence by an evolutionary route is bound to be relatively accessible to this intelligence.

No doubt, this somewhat schematic account requires much amplification and concretization. A long and complex tale must be told about physical and cognitive evolution to fill in the details needed to put such an account into a properly compelling form. But there is surely good reason to hope and expect that a tale of this sort can ultimately be told.

And this is the pivotal point. Even if one has doubts about the particular outlines of the evolutionary story we have sketched, it must be acknowledged that *some such story* can provide a perfectly workable answer to the question of why nature's ways are intelligible to us humans in terms of our mathematical instrumentalities. The mere fact that such an account is in principle possible shows that the issue need not be painted in the black on black of impenetrable mystery.

There may indeed be mysteries in this general area. (Questions like Why should it be that *life* evolves in the world? and—even more fundamentally—Why should it be that the world exists at all? may plausibly be proposed as candidates.)

But be that as it may, the presently deliberated issue of why nature is intelligible to us, and why this intelligibility should incorporate a mathematically articulable physics, does not qualify as all that mysterious, let alone miraculous.

There is simply no need to join Einstein, Schroedinger, and Company in regarding the intelligibility of nature as a miracle or a mystery that passes all human understanding. If we are willing to learn from science itself how nature operates and how people go about conducting their inquiries into its workings, then we should be able increasingly to remove the shadow of incomprehension from the problem of how it is that a being of *this* particular sort, probing an environment of *that* particular type, and doing so by means of *those* particular evolutionarily developed cognitive and physical instrumentalities, manages to arrive at a relatively workable account of how things work in the world. We should eventually be able to see it as only plausible and to be expected that inquiring beings should emerge in nature and get themselves into a position to make a relatively good job of coming to comprehend it. We can thus *look to science itself* for the materials that enable us to understand how natural science is possible. And there is no good reason to expect that it will let us down in this regard.

Admittedly, any such scientifically informed account of science's ability to understand the world is in a way circular. It explains the possibility of our knowledge of nature on the basis of what we know of nature's ways. Its explanatory strategy uses the deliverances of natural science retrospectively to provide an account of how an effective natural science is possible. Such a procedure is *not*, however, a matter of vitiating circularity, but one of the healthy and virtuous self-sufficiency of our knowledge that is in fact an essential part of its claims to adequacy.[10] Any scientific world-picture that does not provide materials for explaining the success of science itself would thereby manifest a failing in its grasp of the phenomena of nature that betokens its own inadequacy.

Implications

But does such a scientific explanation of the success of science not explain too much? Will its account of the pervasiveness of mathematical exactness in science not lead to the (obviously problematic) consequence that "science gets it right"—a result that would fly in the face of our historical experience of science's fallibilism?

By no means! It is fortunate (and evolutionarily most relevant) that we are so positioned within nature that many "wrong" paths lead to the "right" destination—that flawed means often lead us to cognitively satisfactory ends. If nature were a combination lock where we simply "had to get it right"—and *exactly* right—to achieve success in implementing our beliefs, then we just would not be here. Evolution accordingly does not provide an argument that speaks unequivocally for the adequacy of our cognitive efforts. On the contrary, properly construed it is an indicator of our capacity to err and "get away with it." Admittedly, applicative success calls for *some* alignment of thought-governed action with "the real *nature* of things"—but only enough to get by without incurring overly serious penalties in failure.

The success of science should be understood somewhat on analogy with the success of the thirsty man who drank white grape juice, mistaking it for lemonade. It is not that he was roughly right—that such grape juice is "approximately" lemonade. It is just that his beliefs are not wrong in ways that lead us to his being baffled in his present purposes—that such defects as they have do not matter for the issues currently in hand. The fact is that we simply cannot, in the circumstances, help laboring under the impression that our science is highly successful, even though subsequent experience repeatedly disillusions us in this regard. It follows that intelligence and the "science" it devises must pay off in terms of applicative success—irrespective of whether it manages to get things substantially right or not. It is not that there are no mistakes, but that the mistakes at issue lie below the radar screen of current detectability.

We thus arrive at the picture of nature as an error-tolerant system. For consider the hypothetical situation of a species of behaviorally belief-guided creatures living in an environment that invariably exacts a great penalty for "getting it wrong." Whenever the creature makes the smallest mistake—the least little cognitive misstep—bang, it's dead! Our hypothesis is not viable: any such creature would long ago have become extinct. It could not even manage to survive and reproduce long enough to learn about its environment by trial and error. If the world is to be a home for intelligent beings who develop in it though evolution, then it has to be benign—it has to be error-tolerant. If *seeming* success in intellectually governed operations could not attend even substantially erroneous beliefs, then we cognizing beings who have to learn by experience—by trial and error—just could not have made our way along the corridor of time. For if nature were *not* error-forgiving, then a process of evolutionary trial and error could not work in matters of cognition, and intelligent organisms could not emerge at all.[11]

Accordingly, the applicative success of our science is not to be explained on the basis of its actually getting at the real truth, but rather in terms of its being the work of a cognitive being who operates within an error-tolerant environment—a world setting where applicative success may attend even theories that are substantially "off the mark." The applicative efficacy of science undoubtedly requires *some* degree of alignment between our world-picture and the world's actual arrangements—but only just enough to yield the particular successes at issue. No claims to finality or perfection can therefore be substantiated for our science as it stands here and now.

Evolution is indeed the guarantor of the reciprocal attunement of mind and nature envisioned by classical idealism. But it by no means requires that this attunement be of a very high grade by any absolute standard. It is one thing to be right and another to be so badly wrong that one is baffled of one's purposes. And this sort of functional adequacy is something very different from truth. Such a perspective indicates that the success of the applications of our current science does not betoken its actual truth, but merely means that those ways (whatever they be) in which it fails to be true are immaterial to the achievement of success—that in the context of the particular applications at issue its inadequacies and incorrectnesses lie beneath the penalty threshold of failure. This critical fact that evolution requires an error-tolerant environment means that we can explain the impressive successes of mathematicizing natural science without needing to stake untenable claims as to its definitive correctness.

Such evolutionary ruminations indicate that the *capacity* to develop an effective natural science, and also the *motivation* to pursue this cognitive project of inquiry, form a natural part of humanity's evolutionary developed natural heritage. But, at the same time, they suggest that the scope, scale, and nature of such a science is bound to be essentially conditioned by the cognitive resources and interests that we humans bring to bear on its development. The conceptual mechanisms we deploy in studying the world's ways are instrumentalities of our own devising, and in this case as in others, the sorts of tools we use conditions the sorts of artifacts we can create. We are constrained to depict the world in terms of concepts, categories, and schematism to whose formation we ourselves make a decisive contribution. This consideration harks back to the classical idealist theme of the formative contribution of the knower to the character of the known, and raises it important issues that deserve closer consideration.[12]

Chapter 16

Human Science as Characteristically Human

Synopsis

- *In theory, very different versions of "natural science" are possible in a universe where very different sorts of extraterrestrial beings might possibly exist.*
- *The reasoning that there must be one single science because there is one single world is gravely defective.*
- *However, quantitative analysis of the situation indicates that our particular human version of natural science may well be something unique to our particular situation.*
- *Moreover, it is quite wrong to think that potentially different versions of science are all positioned along one single developmental route.*
- *Natural science as we know it may well be—indeed presumably is—a characteristically human enterprise, specifically reflecting our particular manner of exploiting the material and intellectual resources at our disposal.*

The Potential Diversity of "Science"

To what extent does the involvement of our specifically human effort and action condition the character of our natural science? Does our science as a product reflect our particular *modus operandi?* It is instructive to consider this issue through the perspective of the questions whether an astronomically remote civilization might be scientifically more advanced than ourselves. For the seemingly straightforward question about the possibility of scientifically more advanced aliens turns out on closer inspection to involve considerable complexity. And this complexity relates not only to the actual or possible facts of the situation, but also—and crucially—to theoretical questions about the very ideas or concepts that are at issue here.

To begin we must confront the problem of just what it is for there to be another science-possessing civilization. Note that this is a question that *we* are putting—a question posed in terms of the applicability of *our* term "science." It pivots on the issue of whether *we* would be prepared to call certain of *their* activities—once we came to understood them—as engaging in scientific inquiry, and whether we would be prepared to recognize the product of these activities as constituting a state of science—or a branch thereof.

A scientific civilization is not merely one that possess intelligence and social organization, but one that puts these resources to work in a certain very particular sort of way. This consideration opens up the rather subtle issue of priority in regard to process versus product. We must decide whether what counts for a civilization's "having a science" is primarily a matter of the substantive *content* of their doctrines (their belief structures and theory complexes) or is primarily a matter of process, and thus of the aims and *purposes* with which their doctrines are formed.

As regards content, this turns on the issue of how similar their scientific beliefs are to ours. And a look at our own historical evolution indicates that this is clearly something on which we would be ill advised to put much emphasis at the very outset. After all, the speculations of the nature-theorists of pre-Socratic Greece, our ultimate ancestor in the scientific enterprise, bear precious little resemblance to our present-day sciences; neither does contemporary physics bear all that much doctrinal resemblance to that of Newton. So it emerges as clearly more appropriate to give prime emphasis to matters of process and purpose.

Accordingly, the question of these aliens "having a science" is to be regarded as turning not on the extent to which their substantive *findings* resemble ours, but on the extent to which their purposive *project* resembles ours—of determining that we are engaged in the same sort of inquiry in terms of the sorts of issues being addressed and the ways in which they are going about addressing them. The issue accordingly is at bottom not one of the substantive similarity of their scientifically formed beliefs to ours, but one of the functional equivalency of the *projects* at issue in terms of the quintessential goals that define the scientific enterprise as what it is: explanation, prediction, and control over nature. It is this issue of teleology that ultimately defines what it is for those aliens to have a science.

This perspective enjoins the pivotal question: To what extent would the *functional equivalent* of natural science built up by the inquiring intelligences of an astronomically remote civilization be bound to resemble our science in substantive content-oriented regards? In considering this issue, one soon comes to realize that there is an enormous potential for diversity here.

To begin with, the *machinery of formulation* used by an alien civilization in expressing their science might be altogether different. In particular, their mathematics might be very unlike ours. Their "arithmetic" could be anumerical—purely comparative, for example, rather than quantitative. They might thus have topology without analysis. And so, when the mathematical mechanisms at their disposal are very different from ours, it is clear that their description of nature in mathematical terms could also be very different (though not necessarily truer of falser.)

Second, the *orientation* of the science of an alien civilization might be very different. All their efforts might conceivably be directed at social intersections—to developing highly sophisticated analyses of intersecting agents and the economics of exchanges sociology, for example. Again, their approach to natural science might also be very different. Communicating by some sort of "telepathy" based on variable odors or otherwise "exotic" signals, they might devise a complex theory of thought-wave transmittal through an ideaferous aether. Electromagnetic phenomena might lie altogether outside their ken; if their environment does not afford them loadstones and electrical storms, and so on, the occasion to theorize about electromagnetic processes might never arise. The course of scientific development tends to flow in the channel of practical interests. A society of porpoises might lack crystallography but develop a very sophisticated hydrodynamics; one comprised of molelike creatures might never dream of developing optics. The science of a different civilization would presumably be closely geared to the particular pattern of their interaction with nature as funneled through the particular course of their evolutionary adjustment to their specific environment. Alien civilizations might scan nature very differently. The direct chemical analysis of environmental materials might prove highly useful to them, with bioanalytic techniques akin to our sense of taste and smell highly developed so as to provide the basis for a science of a very different sort. Acoustics might mean very little to them, while other sorts of pressure phenomena—say the theory of turbulence in gases—might be the subject of intense and exhaustive investigation. Rather than sending signals by radiowaves or heat radiation signals, they might propel gravity-waves through space. After all, a comparison of the "science" of different civilizations here on earth suggests that it is not an outlandish hypothesis to suppose that the very *topics* of an alien science might differ radically from those of ours. In our own case, for example, the fact that we live on the surface of our planet (unlike whales or porpoises), the fact we have eyes (unlike worms or moles) and thus can *see* the heavens, the fact that we are so situated that the seasonal positions of heavenly bodies are intricately connected without biological needs through the agricultural route to food supply, are all clearly connected with the development of astronomy.

Accordingly, the constitution of the alien inquirers—physical, biological, and social—emerges as a crucial element here. It serves to determine the agenda of questions and the instrumentalities for their resolution—to fix what counts as interesting, important, relevant, significant. In determining what is seen as an appropriate question and what is judged as an admissible solution, the cognitive posture of the inquirers must be expected to play a crucial role in shaping and determining the course of scientific inquiry itself.

Third, the *conceptualization* of an alien science might be very different. We must reckon with the theoretical possibility that a remote civilization might operate with a radically different system of concepts in its cognitive dealings with nature. To motivate this idea of a conceptually different science, it helps to cast the issue in temporal rather than spatial terms. The descriptive characterization of *alien* science is a project rather akin in its difficulty to that of describing our own *future* science. After all, it is effectively impossible to predict not only the answers but even the questions that lie on the agenda of *future* science, because these questions will grow out of the answers we obtain at yet unattained stages of the game. And the situation of an *alien* science could be much the same. As with the science of the remote future, the science of the remotely distant must be presumed to be of such a nature that we really could not achieve intellectual access to it on the basis of our own position in the cognitive scheme of things. Just as the technology of a another highly advanced civilization would most likely strike us as magic, so its science would most likely strike us as incomprehensible gibberish—until we had learned it "from the ground up." They might (just barely) be able to *teach* it to us, but they almost certainly could not *explain* it to us. After all, the most characteristic and significant sort of difference between variant conceptual schemes arises when the one scheme is committed to something the other does not envisage at all—something that lies outside the conceptual range of the other. The "science" of different civilizations, may well, like Galenic and Pasteurian medicine, in key respects simply *change the subject* so as no longer "to talk about the same things," but treat things (e.g., humors and bacteria, respectively) of which the other takes little or no cognizance at all. If, for example, certain intelligent aliens should prove a diffuse and complex aggregate mass of units comprising wholes in ways that allow of overlap, then the role of social concepts might become so paramount that nature as a whole comes to be viewed in these terms. The result would be something very difficult for us to grasp, seeing that they are based on a mode of "shared experience" with which we have no contact.

It is only reasonable to presume that the conceptual character of the (functionally understood) "science" of an alien civilization is so radically different from ours in substantive regards as to orient their thought about "the nature of things" in altogether different directions. Their approach to classification and structurization, explanatory mechanisms, their predictive concerns, and their modes of control over nature might all be very different. In all these regards they might have procedures and interests that depart significantly from our own.

Natural science—broadly construed as inquiry into the ways of nature—is something that is in principle almost infinitely plastic. Its development will trace out a historical course that is bound to be closely geared to the specific capacities, interests, environment, and opportunities of the creatures that develop it. We are deeply mistaken if we think of it as a process that must follow a route roughly parallel to ours and issue in a comparable product. It would be grossly unimaginative to think that either the journey or the destination must be the same—or even substantially similar.

The One World, One Science Argument

One recent writer raises the question "What can we talk about with our remote friends?," and answers it with the remark: "We have a lot in common. We have mathematics in common, and physics, and astronomy."[1] This line of thought begs some very big questions.

Our alien colleagues would have to scan nature for regularities using (at any rate to begin with) the sensors provided them by their evolutionary heritage. And they will note, record, and transmit those regularities that they found to be intellectually interesting or pragmatically useful. Their inquiries will have to develop by theoretical triangulation that proceeds from the lines indicated by these resources. Now this is clearly going to make for a course of development that closely gears their science to their particular situation—their biological endowment ("their sensors"), their cultural heritage ("what is interesting"), their environmental niche ("What is pragmatically useful"). Where these key parameters differ, there too we must expect that the course of scientific development will differ as well.

Admittedly there is only one universe and its laws, as best we can tell, are everywhere the same. And so if intelligent aliens investigate nature at all, they will investigate the same nature we ourselves do. But the sameness of the object of contemplation does nothing to guarantee the sameness of the ideas about it. It is all too familiar a fact that even where human (and thus *homogeneous*) observers are at issue, different constructions are often placed on "the same" occurrences. Primitive peoples thought the sun a god and the most sophisticated among the ancient peoples thought it a large mass of fire. We think of it as a large thermonuclear reactor, and heaven only knows how our successors will think of it in the year 3000. As the course of human history clearly shows, there need be little uniformity in the conceptions held about one selfsame object by differently situated groups of thinkers.

It would certainly be naive to think that because one selfsame object is in question, its description must issue in one selfsame result. This view ignores the crucial matter of one's intellectual orientation. One selfsame piece of driftwood is viewed very differently indeed by the botanist, the painter, the interior decorator, the chemist, the woodcarver, and so forth. The critical issue is that of the particular "aspect" of the item being focused on as important or interesting. With science, as with any productive enterprise, it is not only the raw material but also the mode of productive processing that serves to determine the nature of the outcome. Minds with different sorts of concerns and interests and different backgrounds of information can deal with mutually common items in ways the yield wholly disjoint and disparate results because altogether different features of the thing are being addressed. It is notorious that observers are "prisoners" (so to speak) of their cognitive preparation, interests, and predispositions—seeing only what their preestablished cognitive resources enable them to see and blind to that for which they are cognitively unprepared.

Accordingly, the sameness of nature and its laws by no means settles the issue of scientific uniformity. For a science is always the result of *inquiry* into some sector of nature and this is inevitably a matter of a *transaction* or *interaction* in which nature is but one party and the inquiring beings another. The result of such an interaction depends crucially on the contribution form both sides—from nature and from the intelligences that interact with it. A kind of "chemistry" is at issue, where nature provides only one input and the inquirers themselves provide another—one that can massively and dramatically affect the outcome in such a way that we cannot disentangle the respective contributions of the two parties, nature and the inquirer.

Each inquiring civilization must be thought of as producing its own, itself ever-changing cognitive product—all more of less adequate in their own ways—but with little if any actual overlap in conceptual content. Human organisms are essentially similar, but there is not much similarity between the medicine of the ancient Hindus and that of the ancient Greeks. There is every reason to think that the natural science of different astronomically remote civilizations should be highly diversified. Even as different creatures can have a vast variety of lifestyles for adjustment within one selfsame physical environment like this earth, so they can have a vast variety of thought-styles for cognitive adjustment within one selfsame world.

After all, throughout the earlier stages of man's intellectual history, different human civilizations have developed their "natural sciences" in a substantially different way. And the shift to an extraterrestrial perspective is bound to amplify such cultural differences. Perhaps reluctantly, we must face the fact that on a cosmic scale the "hard" physical sciences have something of the same cultural relativity to which we are accustomed with the material of the "softer" social sciences.

It seems reasonable to argue: "Common problems constrain common solutions. Intelligent alien civilizations have in common with us the problem of cognitive accommodation to a shared world. Natural science as we know it is our solution of this problem. Ergo it is likely to be theirs as well." But this tempting argument founders on its second premise. Their problem is *not* common with ours because their situation must be presumed substantially different, seeing that they live in a significantly different environment and come equipped with significantly different resources. To presuppose a common problem is in fact to beg the question.

There is no quarrel here with "the principle of the uniformity of nature." But this principle merely tells us that when exactly the same question is put to nature, exactly the same answer will be forthcoming. However, the development of a science hinges crucially on this matter of questions—to the sorts of issues that are addressed and the sequential order in which they are posed. And here the prospect of variation arises: We must expect alien beings to question nature in ways very different from our own. On the basis of an *interactionist* model, there is no reason to think that the sciences of different civilizations will exhibit anything more than the roughest sorts of family resemblance.

Our human science reflects not only our interests but also our capacities. It addresses a range of issues that are correlative with our specific modes of physical interaction with nature, the specific ways in which we monitor its processes. It is highly selective—the science of a being that secures its most crucial information through sight, monitoring developments along the spectrum of electromagnetic radiation, rather than, say, monitoring variations of pressure or temperature. A different sort of creature would have different interests and concerns. Ours is certainly not a phenomenalistic science geared to the feel of things or the taste of things. All the same, the science we have developed reflects our capacities and needs, our evolutionary heritage as a being inserted into the orbit or natural phenomena in a certain particular way.

The fact is that all such factors as capacities, requirements, interests, and course of development affect the shape and substance of the science and technology of any particular place and time. Unless we narrow our intellectual horizons in a parochially anthropomorphic way, we must be prepared to recognize the great likelihood that the "science" and "technology" of another civilization will be something *very* different from science and technology as we know it. We are led to view that our human sort of natural science may well be *sui generis*, adjusted to and coordinate with a being of our physical constitution, inserted into the orbit of the world's processes and history in our sort of way. It seems that in science as in other areas of human endeavor we are emplaced within the thought-world that our biological and social and intellectual heritage affords us.

A Quantitative Perspective

Let us attempt to give some quantitative structure to the preceding qualitative deliberations by bringing some rough order-of-magnitude estimates on the scene.

First, there is have the problem of estimating H, the number of habitable planets in the universe. This assessment can be formed by means of the following quantities, themselves represented merely as order-of-magnitude specifications:

n_1 = number of galaxies in the observable universe (10^{11}).
n_2 = average number of star systems per galaxy (10^{11}).
x_1 = fraction of star systems having suitably large and stable planets (1/10).
n_3 = average number of such planets in the temperature zone of a suitably benign solar system, where it is neither too hot nor too cold for life (1).
x_2 = fraction of temperature planets equipped with a surface chemistry capable of supporting life (1/10).

These figures—borrowed in the main from Dole and Sagan—can be subject to skepticism.[2] They, and those that are to follow, must be viewed realistically. They are not graven in stone for all the ages, but represent conjectural "best estimates" in the present state of the art. The important point, as will emerge below, is that the overall tendency of our discussion is not acutely sensitive to precision in this respect. Accordingly one should look on the calculations that are to follow as rather suggestive than in any sense conclusive. Their function is to indicate a general line of thought, and not to establish a definitive conclusion.

Given the preceding estimates, the sought-for number of habitable planets will be the product of these quantities:

$$H = 10^{20}.$$

This, of course, is a prodigiously large number, providing for some thousand million habitable planets per galaxy. We here confront a truly impressive magnitude. But it is only the start of the story.

A planet capable of supporting life might well have no life to support, let alone *intelligent* life. The point is that the physics, chemistry, and biology must all work out just right. The physical, chemical, and biological environments must all be duly auspicious and exactly the right course of triggering processes must unfold for the evolution of intelligence to run a successful course. Our next task is thus to estimate I, the number of planets on which intelligent life evolves. Let us proceed here via the following (again admittedly rough and ready) quantities:

r_1 = fraction of habitable planets on which life—that is, some sort of self-reproducing biological system—actually arises (1/100).

r_2 = fraction of these on which highly complex life-forms evolve, possessed of something akin to a central nervous system and thus capable of complex (though yet instinctively programmed) behavior forms (1/100).

r_3 = fraction of these on which intelligent and sociable beings evolve—beings who can acquire, process, and exchange factual information with relative sophistication—who can observe, remember, reason, and communicate (1/100).

As these fractions indicate, the evolutionary process that begins with the inauguration of life and moves on to the development of intelligence is certainly not an inexorable sequence, but one which could, given suitably inauspicious conditions, abort in a stabilization that freezes the whole course of development at some plateau along the way. Note that r_3 in particular involves problems. Conscious and indeed even intelligent creatures are readily conceivable who yet lack that orientation toward their environment needed to acquire, store, transmit, and process the factual information necessary to science. Where such conceptions as space, time, process, unit, function, and order are missing, it is difficult to see how anything deserving of the name "science" could exist. An intelligence unswervingly directed at the aesthetic appreciation of particular phenomena rather than their generally lawful structure is going to miss out on the scientific dimension.

And so, when we put the fractions of the preceding series to work, we arrive at

$$I = 10^{14}.$$

This unquestionably still indicates an impressively large number of intelligence-bearing planets. It would, in fact, yield a quota of some thousand per galaxy (a figure which, if, correct, would cast a shadow over the prospect of our ever establishing contact with extraterrestrial intelligence, since it would indicate its nearest locale to be some 1,000 light-years away).

As regards this figure, one can say that it would certainly be possible to take a more rosy view of the matter. One could suppose that nature has penchant for life—that a kind of Bergsonian *élan vital* is operative, so that life springs forth wherever it can possibly get a foothold. Something of this attitude certainly underlies J. P. T. Pearman's contention that the probability is 1 that life will develop on a planet with a suitable environment[3]—a stance in which most recent writers on the subject are concerned. One theorist cuts the Gordian Knot with a curious bit of reasoning:

> Biological evolution proceeds by the purely random process of mutation.... Since the process is a random one, the laws of probability

suggest that the time-scale of evolution on earth should resemble the average time-scale for the development of higher forms of life anywhere.[4]

But this blatantly ignores the crucially differentiating role of initial conditions in determining the outcome of random processes. After all, the terrain through which a random walk proceeds is going to make a lot of difference to its destination. The transition from habitability to habitation—from the possibility of life to its actuality—is surely not all that simple. Sir Arthur Eddington did well to remind us in this context of the prodigality of nature when he asked how many acorns are scattered for any one that grows into an oak.[5]

One could perhaps go on to suppose that nature incorporated a predisposition for intelligence—that there is a Teilhard-de-Chardin-reminiscent impetus toward *nous,* so that intelligence develops wherever there is life. Indeed the suggestion is sometimes made in this vein that "the adaptive value of intelligence ... is so great ... that if it is genetically feasible, natural selection seems likely to bring it forth."[6] But this argument from utility to evolutionary probability clearly has its limitation. ("[T]here are no organisms on Earth which have developed tractor treads for locomotion, despite the usefulness of tractor treads in some environments."[7]) Moreover, this suggestion seems implausibly anthropocentric. To all appearances the termite has a securer foothold on the evolutionary ladder than man; and the coelacanth can afford to smile when the survival advantages of intelligence are touted by a johnny-come-lately creature whose self-inflicted threats to long-term survival are a cause of general concern. As J. P. T. Pearman has rightly noted, "the successful persistence of a multitude of simpler organisms from ancient times argues that intelligence may confer no unique benefits for survival in an environment similar to that of earth."[8] After all, it will prove survival-conductive mainly for a being of a particularly restless disposition, a creature like man, who refuses to settle down in a secured ecological niche, but shifts restlessly from environment to environment needing continually to readjust to self-imposed changes. The value of intelligence, one might say, is not absolute but remedial—as an aid to offsetting the problems of a particular sort of lifestyle. We would do well to think of the emergence of intelligence as a long series of fortuitous twists and turnings rather than an inexorable push toward a foreordained result. It would be glib in the extreme to assume that once life arises its subsequent development would proceed in much the same way as here on earth.[9]

The indicated figures accordingly seem plausibly middle-roadish between undue pessimism and an intelligence-favoring optimism that seems unwarranted at this particular stage of the scientific game. Even so, it is clear that the proposed specification of *I* represents a strikingly substantial magnitude—one which contemplates many thousand of millions of planets equipped with intelligent creatures scattered throughout the universe.

Intelligence, however, is not yet the end of the line. (After all, dolphins and apes are presumably intelligent, but they do not have, and are unlikely to develop, a "science.") Many further steps are needed to estimate S, the number of planets throughout the universe in which scientific civilizations arise. The developmental path from intelligence to science is a road strewn with substantial obstacles. Here matters must be propitious not just as regards the physics and chemistry and biochemistry and evolutionary biology and cognitive psychology of the situation. The social-science requisites for the evolution of science as the cultural artifact of a multifocal civilization must also be met. Economic conditions, social organization, and cultural orientation must all be properly adjusted before the move from intelligence to science can be accomplished. For scientific inquiry to evolve and flourish there must, in the first place, be cultural institutions whose development requires economic conditions and a favorable social organization. And terrestrial experience suggests that such conditions for the social evolution of a developed culture are by no means always present where intelligence is. We do well to recall that of the myriad human civilizations evolved here on earth only one, the Mediterranean/European, managed to develop natural science in a form that is nowadays unproblematically recognizable as such by all advanced civilizations. The successful transit from intelligence to science is certainly not a sure thing.

Let us once more look at the matter quantitatively:

p_1 = probability that intelligent beings will (unlike dolphins) also possess developed manipulative abilities and will (unlike the higher apes) combine intelligence with manipulative ability to as to develop a technology that can be passed on as a social heritage across the generations. (.01)

p_2 = probability that technologically competent intelligent beings will group themselves in organized societies of substantial complexity—a transition that stone-age man, for example, never managed to make. (.1)

p_3 = probability that an organized society will not only acquire the means for transmitting across successive generation the political and pragmatic know-how indispensable to an "organized society" as such, but will also (unlike the ancient Egyptians) develop institutions of learning and culture for accumulating, refining, systematizing, and perpetuating factual information. (.1)

p_4 = probability that society with cultural institutions will develop an unstable (i.e., continually developing and dynamic) technology—in the way the ancient Greeks and the old Chinese mandarins, for example, never did—so as to create a technologically progressive civilization. (.01)

p_5 = probability that a technologically progressive civilization will develop and maintain an articulated "science" and concern itself with the theoretical study of nature at a level of high generality and precision. (.1)

However firm the physical quantities with which we began, we are by now skating on very thin ice indeed. These latter issues of sociology and cognitive psychology can only be quantified in the most tentative and cautious way. But the one thing that is clear is that a good many conditions of this sort have to be met and that each involves a likelihood of relatively modest proportions.

The issue of technology reflected in p_1 and p_4 is particularly critical here. The urge to an ever aggressive technological extension of self is certainly not felt by every intelligent life-form. It is a part of Western man's peculiar lifestyle to impatiently cultivate the active modification of nature in the pursuit of human convenience so as to create an artificial environment of ultra-low entropy. Even in human terms this is not a uniquely constrained solution to the problem of evolutionary adaptation. Many human societies seem to have remained perfectly content with the *status quo* for countless generations and very sophisticated cultural projects—literary criticism, for example—have developed in directions very different from the scientific. After all, a culture can easily settle comfortably into frozen traditionary pattern with respect to technology. (If their attention span is long enough, our aliens might cultivate scholastic theology *ad indefinitum*.) Moreover, unless their oral lore is something very different from ours, it is hard to see how an alien civilization could develop science without writing—a skill which even many human communities did not manage to develop. The salient point is that for "science" to emerge in a distant planet it is not enough for there to be life and intelligence; there must also be culture and progressive technology and explanatory interest and theorizing competency.

The product of the preceding sequence of probability estimates is 10^{-7}. Multiplying this by I we would obtain the following expected-value estimate of the number of "science"-possessing planetary civilizations:

$$S = 10^7 = 10,000,000.$$

This, of course, is still a large number, albeit now one that is rather modest on a cosmic scale, implying a chance of only some .01% that a given galaxy actually provides the home for a "science." (Note too that we ignore the temporal dimension—the scientific civilizations at issue may have been destroyed long ago, or perhaps simply have lost interest in doing science.) A more conservative appraisal of the sociological parameters has thus led us to a figure that is more modest by many orders of magnitude than the estimate by Shklovskii and Sagan (op. cit.) that some $10^{5\pm1}$ scientifically sophisticated civilizations exist in our galaxy alone. Nevertheless even this modest ten million is still a sizable number.

COMPARABILITY AND JUDGMENTS OF RELATIVE ADVANCEMENT OR BACKWARDNESS

Let us now come to grips with the crux of our present concerns—the issue of scientific advancement. Earlier we defined "science" in terms of a rather generic sort of *functional* equivalency. The question, however, from which we began was not whether a remote civilization has a "science" of some sort, but wether it is *scientifically more advanced* than ourselves. But if *advancement* is to be at issue, and the question is to be one of relative aspects of intelligence as such—with data processing in volumetric terms—but with the quality of the orientation of intelligence to substantive issues. If another science is to represent an advance over ours, we must clearly construe it as *our sort* of science in much more particularized and substantive terms. And given the *immense* diversity to be expected among the various modes of "science" and "technology," the number of extraterrestrial civilizations possessing a science and technology that is duly consonant and contiguous with ours—and in particular, heavily geared toward the mathematical laws of the electromagnetic spectrum—must be judged to be very small indeed.

We have come to recognize that sciences can vary (1) in their formal mechanisms of *formulation*—their "mathematics," (2) in their *conceptualization*, that is, in the kinds of explanatory and descriptive concepts they bring to bear, and (3) in their *orientation* toward the manifold pressures of nature, reflecting the varying "interest"-directions of their developers. While "science" as such is clearly not anthropocentric, science *as we have it is*—the only "science" that we ourselves know—is a specifically human artifact that must be expected to reflect in significant degree the particular characteristics of its makers. Consequently, the prospect that an alien "science"-possessing civilization has a *science* that we would acknowledge (if sufficiently informed) as representing the same general line of inquiry as that in which we ourselves are engaged seems extremely implausible. The possibility that *their* science-and-technology is "sufficiently similar" in orientation and character to substantively proximate to *ours* must be viewed as extremely remote. We clearly cannot estimate this as representing something other than a very long shot indeed—certainly no better than one in many thousands.

Just such comparability with "our sort of science" is, however, the indispensable precondition for judgments of relative advancement or backwardness vis-à-vis ourselves. The idea of their being scientifically "more advanced" is predicated on the uniformity of the enterprises—doing better and more effectively the kinds of things that *we* want science and technology to do. Any talk of advancement and progress is predicated on the sameness of the direction of movement: only if others are traveling along the same route as we can they be said to be ahead of behind us. The issue of relative advancement is linked inseparably to the idea of doing the same sort of thing better or more fully. And this falls apart when "this sort of thing" is not substantially the same. One can say that a child's expository writing is more primitive than an adult's, or that the novice's performance at arithmetic or piano playing is less developed than that of the expert. But we can scarcely say that Chinese cookery is more or less advanced than Roman, or Greek pottery than Renaissance glassblowing. The salient point for present purposes is simply that where the enterprises are sufficiently diverse, the ideas of comparative advancement and progress are inapplicable for lack of a *sine qua non* condition.

Claiming scientific superiority is not as simple as may seem at first sight. To begin with, it would not automatically emerge from their capacity to make many splendidly successful predictions. For this could be the result of precognition or empathetic attunement to nature or such-like. Again, what is wanted is not just a matter of *correct,* but of cognitively underwritten, and thus *science-guided* prediction—predictions guided by insight based on understanding and not mere lucky guesswork. And that is just exactly what is to be proved.

It clearly is not enough for establishing their being scientifically more advanced than ourselves that the aliens should perform "technological wonders"—that they should be able to do all sorts of things we would like to do but cannot. After all, bees can do that. The technology at issue must clear be the product of intelligent contrivance rather than evolutionary trial and error. What is needed for advancement is that their performatory wonders issue from superior theoretical knowledge—that is, from superior science. And then we are back in the circle.

Nor would the matter be settled by the consideration that an extraterrestrial species might be more "intelligent" then we in having a greater capacity for the timely and comprehensive monitoring and processing of information. After all, whale or porpoises, with their larger brains, may (for all we know) have to manipulate relatively larger quantities of sheer data than ourselves to maintain effective adaptation within the highly changeable environment. What clearly counts for scientific knowledge is not the *quantity* of intelligence in sheer volumetric terms but its *quality* in substantive, issue-oriented terms. Information handling does not assure scientific development. Libraries of information (or misinformation) can be generated about trivia—or dedicated to matters very different from science as we know it.

It is perhaps too tempting for humans to reckon cognitive superiority by the law of the jungle—judging as superior those who do or would come out on top in outright conflict. But surely the Mongols were not possessors of a civilization superior to that of the New Eastern cultures they overran. Again, we earthlings might easily be eliminated by not very knowledgeable creatures able to produce at will—perhaps by using natural secretions—a biological or chemical agent capable to killing us off.

The key point then, is that if they are to effect an *advance* on our science, they must both (1) be engaged in doing roughly our sort of thing in roughly our sort of way, and (2) do it significantly better. In speaking of the "science" of another civilization as "more advanced" than our own, we contemplate the prospect that they have developed *science* (*our* sort of science—"science" as we know it) further than we have ourselves. And this is implausible. Even assuming that "they" develop a "science" at all—that is, a *functional equivalent* of our science—it seems unduly parochial to suppose that they are at work constructing *our* sort of science in substantive, content-oriented terms. Diverse life-modes have diverse interests; diverse interests engender diverse technologies; diverse technologies make for diverse modes of science. And where the parties concerned are going in different directions, it makes no sense to say that one is ahead of or behind the other.

If a civilization of intelligent aliens develops a science at all, it seem plausible to expect that they will develop it in another direction altogether and produce something that we, if we could come to understand it at all, would regard as simply detached in a content-orientation—though perhaps not in intent—from the scientific enterprise as we ourselves cultivate it. (Think of the attitude of orthodox sciences to "exotic" phenomena like hypnotism, or acupuncture, let alone to parapsychology.)

The crucial consideration is that there just is not single-track itinerary of scientific/technological development that different civilizations travel in common with mere differences in speed or in staying power (not withstanding the penchant of astrophysics for the neat plotting of numerical "degrees of development" against time in the evolution of planetary civilizations—cf. J.A. Ball, *American Scientist*, vol. 68, 1980, p. 58). In cognition and even in "scientific" evolution we are not dealing with a single-track railway, but with a complex network leading to many mutually remote destinations. Even as cosmic evolution involves a red shift that carries different star systems ever farther from each other in space, so cognitive evolution may well involve a red shift that carries different civilizations even farther from each other into mutually remote thought-worlds.

The prospect that an alien civilization is going about the job of doing *our* science—a "science" that reflects the sorts of interests and involvement that *we* have in nature—better than we do ourselves must accordingly be adjudged as extremely far-fetched. Specifically two conditions would have to be met for the science of an intelligent civilization to be in a position to count as comparable to ours.

1. That, *given* that they have a "science" and a developing "technology," they have managed to couple the two and have proceeded to develop (unlike the ancient Greeks and Chinese and Byzantines) a *science-guided* technology. (Probability p_6.)
2. That their science-guided technology is sufficiently oriented toward issues regarding natural processes sufficiently close to those at which our science-guided technology is oriented that a comparison can reasonably be made between them. (Probability p_7.)

To judge by terrestrial experience, it seems rather optimistic to estimate p_6 to be even so large as one in a thousand (with $p_6 = .001$). And p_7 must also be adjudged as quite small. As we have seen science-guided technology could be oriented in very different directions. The potential diversity of different modes of "science" is encrnomous, so that there is little choice but to see p_7 as an eventuation whose chances are no better than, say, one in ten thousand (so $p_7 = .0001$). If our alien scientists are differently constructed (if they are silicon-based creatures, for example), of it their natural environment is very different, their practical interests and its accordant technology will be oriented in very different directions from ours. For example, their technology might be wholly independent of "hardware," oriented not toward physical machinery, but toward the software of mind-state manipulation, telepathy, hypnotism, autosuggestion, or the like. (Ray Bradbury's Martians destroy an expedition from earth armed with atomic weapons by thought control.) And we must not keep our imagination on a short leash in this regard. Given the diversity of different modes of "science" and the enormous spectrum of possible issues and purposes in principle available to extraterrestrial aliens, the prospect must be recognized that the direction of their science-guided technology might be vastly different from ours.

Accordingly we have it that

$$p_6 \times p_7 = 10^{-7}.$$

Now the product of this quantity with the previously estimated quantity S, the number of civilizations that possess a technologized science as we comprehend it, is clearly not going to be very substantial—it is, in fact, going to be strikingly close to 1.

If "being there" in scientific regards means having *our* sort of scientifically guided technology and our sort of technologically channeled science, then it does not seem all that far-fetched to suppose that as regards science as we have it *we might be there alone*—even in a universe amply furnished with other intelligent civilizations. The prospect that somebody else could do "our sort of thing" in the scientific sphere better than we can do it ourselves seems very remote.

Basic Principles

The overall structure of our analysis thus emerges in the picture of Table 16.1. Its figures interestingly embody the familiar situation that as one moves along a nested hierarchy of increasing complexity one encounters a greater scope for diversity—that the further layers of system complexity provide for an ever-widening spectrum of possible state and conditions. (The more fundamental that system, and narrow its correlative range of alternatives, the more complex, the wider.) If each unit ("letter," "cell," "atom") can be configured in ten ways, then each ordered group of ten such units ("word," "organ," "molecule") can be configured in 10^{10}, and each complex of ten such groups ("sentences," "organisms," "objects") in $(10^{10})^{10} = 10^{100}$ ways. Thus even if only a small fraction of what is realizable in theory is realizable in nature, any increase in organizational complexity will nevertheless be accompanied by an enormous amplification of possibilities.

To be sure, the numerical particulars that constitute the quantitative thread of the discussion cannot be given much credence. But their general tendency nevertheless conveys an important lesson. For people frequently seem inclined to reason as follows:

> There are after all, an immense number of planetary objects running about in the heavens. And proper humility requires us to recognize that there is nothing at all that special about the Earth. If it can evolve life and intelligence and civilization and science, then so can other planets, and given that there are so many other runners in the race we must assume that—even though we cannot see them in the cosmic darkness—some of them have got ahead of us in the race.

As one recent writer formulates this familiar argumentation, "Since man's existence on the earth occupies but an instant in cosmic time, surely intelligent life has progressed far beyond our level on some of these 100,000,000 (habitable) planets (in our galaxy)."[10] But such plausible-sounding argumentation overlooks the numerical complexities. Even though there are an

TABLE 16.1
Conditions for the Development of Science

Planets of sufficient size for potential habitation	10^{22}
fraction thereof with affording:	
temperate location for	10^{-1}
chemistry for life-support	10^{-1}
biochemistry for the actual emergence of life	10^{-2}
biology and psychology for the evolution of intelligence	10^{-4}
sociology for developing a culture with a "technology" and a "science"	10^{-7}
epistemology for developing science as we know it	10^{-7}

immense number of solar systems, and thus a staggering number of planets (some 10^{22} in our estimate), nevertheless, a substantial number of conditions must be met for "science" (as we understand it) to arise. The astrophysical, physical, chemical, biological, psychological, sociological, and epistemological parameters must all be in proper adjustment. There must be habitability, and life, and intelligence, and culture, and technology, and a "science" coupled to technology, and an appropriate subject-matter orientation of this intellectual product, and so on. A great many turnings must go aright en route to science of a quality comparable to ours. Each step along the way is one of finite (and often smallish) probability. And to reach the final destination, all these probabilities must be multiplied together, yielding a quantity that might be very small indeed. Even if there were only twelve turning points along this developmental route, each involving a chance of successful eventuation that is, on average, no worse than a one-in-a-hundred, the chance of an overall success would be immensely small, corresponding to an aggregate success-probability of merely 10^{-24}.

It is tempting to say "The Universe is a big place; surely we must expect that what happens in one locality will be repeated someplace else." But this overlooks the issue of probability. Admittedly cosmic locales are very numerous. But probabilities can get to be very small: No matter how massive N may be, there is that diminutive $1/N$ that can countervail against it.

The workings of evolution—be it of life or intelligence or culture of technology or science—are always the product of a great number of individually unlikely events. Things can eventuate very differently at many junctures. The unfolding of developments involves putting to nature a series of questions whose successive resolution produces a process reminiscent of the game Twenty Questions, sweeping over a possibility-spectrum of awesomely large proportions. The result eventually reached lies along a route that traces our one particular contingent path within a space of alternatives that provides for an ever divergent fanning out of alternative as each step opens up yet further possibilities. And evolutionary process is a very iffy proposition—a complex labyrinth where a great many twists and turns in the road must be taken aright for matters to end up as they do.

Of course, it all looks easy with the wisdom of hindsight. If things had not turned out appropriately at every stage, we would not be here to tell the tale. The many contingencies on the long route of cosmic, galactic, solar-systemic, biochemical, biological, social, cultural, and cognitive evolution have all turned out aright—the innumerable obstacles have all been surmounted. In retrospect it all looks easy and inevitable. The innumerable possibilities of variation along the way are easily kept out of sight and out of mind. The wisdom of hindsight makes it look very easy. It is so easy, so tempting to say that a planet on which there is life will of course evolve a species with the technical capacity for interstellar communication.[11] It is tempting, but it is also nonsense.

The ancient Greek atomists' theory of possibility afford an interesting object-lesson in this connection. Adopting a Euclideanly infinitistic view of space, the atomist taught that every (suitably general) possibility is realized in fact someplace or other. Confronting the question of Why do dogs not have horns: just why is the theoretical possibility that dogs be horned not actually realized? the atomists replied that it indeed is realized by just elsewhere—*in another region of space.* Somewhere within infinite space there is another world just like ours in every respect save one, that its dogs have horns. For the circumstance that dogs lack horns is simply a parochial idiosyncrasy of the particular local world in which we interlocutors happen to find ourselves. Reality accommodates all possibilities of worlds alternative to this through spatial distribution: as the atomists saw it, *all* alternative possibilities are in fact actualized in the various subworlds embraced within one spatially infinite superworld.

This theory of virtually open-ended possibilities was shut off by they closed cosmos of the Aristotelian world picture, which dominated European cosmological thought for almost two millennia. The breakup of the Aristotelian model in the Renaissance and its replacement by the "Newtonian" model is one of the great turning points of the intellectual tradition of the West—elegantly portrayed in Alexandre Koyré's book of the splendid title "From the Closed World to the Infinite Universe" (New York, 1957). Strangely enough, the refinitization of the universe effected by Einstein's general relativity in one of its principal interpretations produced scarcely a ripple in philosophical or theoretical circles, despite the immense stir caused by other aspects of the Einstein revolution. (Einsteinian space-time is after all, even more radically finitistic than the Aristotelian world-picture, which left open at any rate the prospect of an infinite future, with respect to time.)

To be sure, it might well seem that the finitude in question is not terribly significant because the distances and times involved in modern cosmology are so enormous. But this view is rather naive. The difference between the finite the the infinite is as big as differences can get to be. and it represents a difference that is—in this present context—of the most far-reaching significance. For this means that we have no alternative to supposing that a highly improbable set of eventuations is not going to be realized in very many places, and that something sufficiently improbable may well not be realized at all. The decisive *philosophical* importance of the Einsteinean finitization of space-time is that it means that an eventuation that is sufficiently improbable may well not be realized at all. A finite universe must "make up its mind" about its contents in a far more radical sense than an infinite one. And this is particularly manifest in the context of low-probability possibilities. In a finite world—unlike an infinite one—we cannot avoid supposing that a prospect that is sufficiently unlikely is simply not going to be realized at all, that in piling improbability on improbability we eventually outrun the reach of the actual. It is, accordingly, quite conceivable that our science represents a solution of the problem of cognitive accommodation that is terrestrially locale-specific.

Here lies a deep question. Is the mission of intelligence uniform or diversified? Two fundamentally opposed philosophical positions are possible with respect to cognitive evolution in its cosmic perspective. The one is a uniformitarian *monism* that sees the universal mission of intelligence in terms of a certain shared destination, a common cosmic "position of reason as such." The other is a particularistic *pluralism* that allows each solar civilization to forge its own characteristic cognitive destiny, and sees the mission of intelligence as such in terms of spanning a wide spectrum of alternatives and realizing a vastly diversified variety of possibilities, with each thought-form realizing its own peculiar destiny in separation from all the rest. The conflict between these doctrines must in the final analysis be settled not by armchair speculation for general principles, but by rational triangulation from the empirical data. This said, it must be observed that the whole tendency of these present deliberation is toward the pluralistic side.

In many mind there is, no doubt, a certain charm to the idea of companionship. It would be comforting to think that however estranged we are in other ways, those alien minds and ourselves share *science* at any rate—that we are fellow travelers on a common journey of inquiry. Mythology and scientific speculation alike manifest our yearning for companionship and contact. (Pascal was not the only one frightened by the eternal silence of infinite spaces.) It would be pleasant to think ourselves not only colleagues but junior collaborators whom other, wiser minds might be able to help along the way. Even as many in sixteenth-century Europe looked to those strange pure men of the Indies (East or West) who might serve as moral exemplars for sinful European man, so we are tempted to look to alien inquirers who surpass us in scientific wisdom and might assist us in overcoming our cognitive deficiencies. The idea is appealing, but it is also, alas, very unrealistic.

In the late 1600s Christian Huygens wrote:

> For 'tis a very ridiculous opinion that the common people have got among them, that it is impossible a rational Soul should dwell in any other shape than ours... This can proceed from nothing but the Weakness, Ignorance, and Prejudice of Men, as well as the humane Figure being the handsomest and most excellent of all others, when indeed it's nothing but a being accustomed to that figure that makes me think so, and a conceit... that no shape or color can be so good as our own.[12]

What is here said about people's tendency to emplace all rational minds into a physical structure akin to their own familiar one is paralleled by a tendency to emplace all rational knowledge into a cognitive structure akin to their own familiar one.

With respect to biological evolution it seem perfectly sensible to reason as follows:

> What can we say about the forms of life evolving on these other worlds? ... [I]t is clear that subsequent evolution by natural selection would lead to an immense variety of organisms; compared to them, all organisms on Earth, from molds to men, are very close relations. (Shklovskii and Sagan, op. cit., p. 350)

It is plausible that much the same situation should obtain with respect to cognitive evolution: that the "sciences" produced by different civilizations here on earth—the ancient Chinese, Indians, and Greeks, for example—should exhibit immensely greater points of similarity than obtains between our present-day science and anything devised by astronomically remote civilizations. And where movement in altogether different directions is at issue, the idea of a comparison in terms of "advance" or "backwardness" would simply be inapplicable.

The chapter's deliberations accordingly convey two principal lessons. The first is that the prospect that some astronomically remote civilization is "scientifically more advanced" than ourselves—that somebody else is doing "our sort of science" *better* than we ourselves—requires in the first instance that they be doing our sort of science at all. And this deeply anthropomorphic supposition is extremely unlikely.

Moreover, a second main lesson follows from the consideration that natural science *as we know it* is to all visible intents and purposes a characteristically human enterprise—a circumstance that endows science with an inexorably economic dimension. For this means that the sorts of results of scientific inquiry that we are able to achieve will hinge crucially on *the way* in which we deploy resources in cultivating our scientific work as well as on *the extent* to which we do so. Accordingly, our knowledge is limited by the simple fact that it is ours.[13]

Chapter 17

On Ignorance, Insolubilia, and the Limits of Knowledge

SYNOPSIS

- *Knowledge about our own ignorance is bound to be very limited.*
- *We can identify questions that we cannot answer but never facts that we do not know.*
- *Paradoxical though it may seem, even an omniscient God may fail to know some sorts of temporally perspectival knowledge.*
- *The temporal constitution of our knowledge is one of its crucial aspects.*
- *It means that later knowledge develops out of earlier knowledge doing so in ways envisioned long ago by Kant.*
- *Its developmental nature can block the way to knowledge about knowledge.*
- *The idea of unanswerable questions—insolubilia—is old and well-established.*
- *And there is good reason to think that our cognitive limits can readily engender such insolubilia.*
- *Indeed some of this can actually be identified.*
- *However, the extent of our ignorance is something that cannot be reliably assessed.*
- *The impredictability of knowledge means that the universe too is to some extent impredictable.*

Concrete versus Generic Knowledge and Ignorance

One of the most critical but yet problematic areas of inquiry relates to knowledge regarding our own cognitive shortcomings. It is next to impossible to get a clear fix on our own ignorance, because in order to know that there is a certain fact that we do not know, we would have to know the item at issue to be a fact, and just this is, by hypothesis, something we do not know.[1] And it is even difficult to obtain a taxonomy of ignorance. For the realm of ignorance is every bit as vast, complex, and many-faceted as that of knowledge itself. Whatever someone can know that they can also be ignorant about—arguably exempting with a handful of Cartesian exceptions such as the fact that knowers are pretty much bound *ex officio* realize that they themselves exist and can think.

In this connection it is instructive to note some relatively simple but nevertheless far-reaching considerations regarding the project of rational inquiry and the limits of knowledge. Let Kxp as usual abbreviate "x knows that p." And now note the contrast between the contentions:

"x knows that something has the property F": $Kx(\exists u)Fy$

and

"x knows of something that *it* has the property F": $(\exists y)KxFu$.

The variant placement of the quantifier means that there is a crucial different here, since in the second case, unlike the first, the knower in question is in a position specifically to identify the item at issue. Here in this second case our knower not merely knows generally and indefinitely that *something* has F, but knows concretely and specifically *what it is* that has F. The two cognitive situations are clearly very different. To know that someone is currently in the Library of Congress is one thing and to know who is there is quite another.

And this has wider ramifications. For the reality of it is that there is a world of difference between saying "I don't know whether p is a fact" and saying "p is a fact that I don't know." The former comes down to maintaining:

$$\sim Kip \ \& \sim Ki\sim p \qquad (i = \text{oneself}).$$

No problem there: I neither know that p nor that not-p. However, the second statement, to the effect that p is a fact that one does not know to be so, comes down to maintaining: $p \ \& \sim Kip$. But in claiming to know this, namely

$$Ki(p \ \& \sim Kip).$$

I claim to know (among other things) both that p is true and that I do not know this. And such a claim is self-contradictory.[2]

And correspondingly we must recognize that there is a crucial difference between the indefinite "I know that there is some fact that I do not know" and the specific "Such and such is a fact of which I know that I do not know it." The first is unproblematic but the second not, seeing that to know of something that it is a fact I must know it as such so that what is at issue is effectively a contradiction in terms. I can know about my ignorance only *sub ratione generalitatis* at the level of indefiniteness, but I cannot know it in concrete detail. I can meaningfully hold that two and two equals four is a *claim* (or a *purported* fact) that I do not know to be the case, but cannot meaningfully maintain that two and two equals four is an *actual* fact that I do not know to be the case. To maintain a fact as fact is to assert knowledge of it: in maintaining p as a fact one claims to know that p.

And so it lies in the nature of things that my ignorance about facts is something regarding what one can have only generic and not concrete knowledge. One can know *that* that one does not know various truths, but I am not in a position to *identify* any of the specific truths I do not know. In sum, I can have general but not specific knowledge about my ignorance, although my knowledge about *your* ignorance is unproblematic in this regard.[3]

Erotetic Incapacity

It is helpful to take an erotetic—that is, question-oriented—view of knowledge and ignorance. After all, x knows that p iff x can *cogently* give a correct answer to the question: Is p the case? Here an answer is cogently given if the giver has a satisfactory rationale for giving it.

The general distinction between the concrete and the generic also applies in the case of questions. Thus let AxQ abbreviate "x can (correctly) answer the question Q." Then we can contrast two possibilities for ignorance, the one general and nonspecific

- *Generic question-resolving incapacity*

    "There is a question Q that I cannot answer" $(\exists Q)(\sim AiQ)$

and the other particular and specific:

- *Concrete question-resolving incapacity*

    "Q_0 is a (specific) question that I cannot answer" $\sim AiQ_0$.

Here we encounter the same dichotomy between generic and specific incapacity as before. The former does not imply the latter since the second case—unlike the first—is a matter of concretely identifying the item at issue. Even as there is concrete and indefinite knowledge there is concrete and generic ignorance as well.

There is, however, a crucial difference here. For in the case of questions—unlike factual knowledge—we can be concretely specific regarding about our incapacity. We cannot coherently say "p is a specific truth (fact) I do not know." But saying "Q is a specific question I cannot answer" is perfectly unproblematic. When we look at cognition from the angle of questions rather than that of knowledge, ignorance becomes identifiable. Erotetic ignorance—inability to answer questions—is accordingly something quite different from propositional ignorance: the failure to know truths. For with erotetic ignorance we can hope to get beyond generalities to identify questions that we cannot answer. But this sort of specificity is something that we cannot manage in the realm of propositional knowledge.

Divine versus Mundane Knowledge

Let us begin here with a somewhat extreme case. A knower is *unrestrictedly* omniscient iff whenever there is something to be known, this knower knows it. In other words, whenever p is a true matter of fact, the knower knows that it is so. Thus x is omniscient in this sense if we have:

$$(\forall p)(p \Rightarrow Kxp)$$

Such a knower knows everything that is knowable. This knower's knowledge is literally unlimited: something is a truth if, and only if, our omniscient being x knows it.

By contrast, a knower is *restrictedly* omniscient iff this knower knows everything that is known. That is, whenever *anyone* knows something, this knower knows it as well: Thus x is omniscient in this weaker sense if we have:

$$(\forall p)(\exists y Kyp \Rightarrow Kxp)$$

The difference between the two modes of omniscience can come into operation only when there are unknowns, that is, truths which nobody knows at all. For when we have

$$p \ \& \ {\sim}(\exists y) Kxp$$

then the antecedent at issue with unrestricted omniscience is satisfied, while that at issue with restricted omniscience is not.

How do I know that I am not omniscient? Certainly not because I can specify particular facts that I do not know. Rather, it is because there are questions I cannot answer. And specifically because I realize full well at the level of generality that there are truth-determinate propositions whose truth-status I cannot decide, perfectly meaningful propositions about authentic matters of fact that I neither know to be true nor to be false, even though I do know that they have to be one or the other. That George Washington wondered if Martha was suitably dressed for the occasion of his first inauguration would seem to be a good example. But what of others?

Of course, we know from Kurt Gödel's work that in systems of axiomatic arithmetic there will be (formal) truths that are unprovable (indemonstrable). Accordingly, one can identify undecidable questions in formal systems of arithmetic. But—are there specifiable *substantive* questions about the world's contingent facts that are comparably undecidable? Are there also factual rather than merely formal insolubilia—questions with respect to matters of fact that are not answerable thanks not merely to some fortuitous lack of information but for deeper and more fundamental reasons?

Issues of Temporalized Knowledge

Let us consider why—and how—the answer to this question is affirmative.

The philosophical theologians of the middle ages, who loved puzzles, were wont to exercise their ingenuity regarding the question: If he is omniscient, does God know what is happening *now*? And they inclined to answer this question with the response: yes and no. Clearly an unrestrictedly omniscient God will know everything that happens in the world. And this means that he knows whatever is happening concurrently with the calendar's reading 13 January 2001 and the clock's reading 3:15 P.M. But this is B-series knowledge in McTaggert's terminology—knowledge of events in the manner of before/concurrent/after. However, as a being who does not occupy a place *within* the manifold of space and time—who, being extramundane, lacks the world-internal perspective required for indexicals like here and now—God cannot operate with the correlative concepts and so in *that* sense, the sense natural to us as mundane world-emplaced creatures who occupy a spatiotemporally qualified position in the world's scheme of things, God does *not* know what is going on *now*. He does not have temporal knowledge in the English philosopher J. M. E. McTaggart's A-series mode of matters in the range of past/present/future. In this regard (as in many others) God is quite unlike ourselves. We finite world-emplaced being who exist within space-time can ask and answer questions about temporal matters from a time-internal perspective.

And this has significant implications for us because *our* knowledge—unlike God's—is something that both has to arise within and be concerned about the temporal domain in the manner of the time-interval perspective of *A*-series temporality. It is, in fact, here that the root source of our cognitive imperfection lies.

This temporal and developmental aspect of knowledge has portentous ramifications. For one thing, it means that we are evidentially incapacitated in comparison with other knowers. Thus consider the yes-or-no question: "When next you yourself (Jones) answer a questions, will you do so in the negative?" Whichever way poor Jones replies, he is plunged into error. The best he can do is to plead incapacity and respond *"can't say."* But, of course, third parties are differently circumstanced. *Another* knower—one different from Jones himself—can answer the question by saying "No, Jones won't do so," and manage to be entirely correct. But this is something that Jones cannot coherently say on his own account.

And so one thing that follows here is that not only are finite knowers not unrestrictedly omniscient, they are not *restrictedly* omniscient either—in the sense of being able to answer correctly every question that another knower can so answer.

But the temporal aspect of the knowledge of finite beings has other, even more portentous aspects.

In particular it means that our knowledge is developmental in nature: that it admits of learning and of discoveries: that there are things (facts) that we do not and cannot know at one temporal juncture that we can and do manage to get to know at another. Knowledge does not come to us from on high, perfected and completed like Athena springing from the head of Zeus. It is the product of a process of inquiry unfolding over time—a process from which the possibilities of error of omission and commission can never be excluded. And this means that problems are bound to arise as our thought contemplates the future.

Our "picture" of the world—our worldview, as one usually calls it—is an epistemic construction built up from our personal and vicarious experience-based knowledge. And like any construction it is made over time from preexisting materials—in this case the information at our disposal. We can select these materials—but only to a limited extent. In the main they force themselves on us through the channels of our experience. The puzzle question that inevitably arises in this context is that of the accuracy or correctness of our world picture. And it is here that the temporally emplaced aspect of things comes into play.

Perhaps the best way to get a good grip on this issue is by asking how we are to relate the following four items:

1. *presently purported truth:* the truth as we ourselves see it, here and now.
2. *futured truth:* the truth as we will come to see it when we push our inquiries further and deeper.

3. *completed truth:* the truth as we would see it when and if—per impossible—our inquiries were pushed through to the point of ultimate completion.
4. *perfected truth:* the truth as we ought to come to see it if we conducted our inquiries in a manner definitively appropriate and correct.

Presumably we can identify 3 and 4 and regard the result of achieving ideal correctness and completeness as being substantially one and the same. We could thus simply speak here of "the real truth" as realized through the perfecting and completing of inquiry.

However, the real problem is how this authentic truth as per 3 and 4 is to be related to that which we actually have as per 1: our purported truth as it stands here and now.

The few things that can be said here must be said with care and caution. Indeed, we must here proceed by way of a *via negativa* exactly because we have no possible way of telling *now* exactly what sort of improvements in our presently purported knowledge the *future* progress of inquiry will demand.

KANT'S PRINCIPLE OF QUESTION EXFOLIATION

New knowledge that emerges from the progress of science can have a very different bearing on the matter of questions. Specifically, we can discover:

1. New (that is, *different*) answers to old questions.
2. New questions.
3. The inappropriateness or illegitimacy of old questions.

With (1) we learn that the wrong answer has been given to an old question: We uncover an error of commission in our previous question-answering endeavors. With (2) we discover that there are certain questions that have not heretofore been posed at all: We uncover an error of omission in our former question-asking endeavors. Finally, with (3) we find that one has asked the wrong question altogether: We uncover an error of commission in our former question-asking endeavors, which are now seen to rest on incorrect presuppositions (and are thus generally bound up with type (1) discoveries). Three rather different sorts of cognitive progress are thus involved here—different from one another and from the traditional view of cognitive progress in terms of a straightforward "accretion of further knowledge."

The coming to be and passing away of questions is a phenomenon that can be mooted on this basis. A question *arises* at the time t if it then can meaningfully be posed because all its presuppositions are then taken to be true. And a question *dissolves* at t if one or another of its previously accepted presuppositions is no longer accepted. Any state of science will remove certain questions from the agenda and dismiss them as inappropriate. Newtonian dynamics dismissed the question What cause is operative to keep a body in movement (with a uniform velocity in a straight line) once an impressed force has set it into motion? Modern quantum theory does not allow us to ask What caused this atom on californium to disintegrate after exactly 32.53 days, rather than, say, a day or two later? Scientific questions should thus be regarded as arising in an *historical* setting. They arise at some juncture and not at others; they can be born and then die away.

A change of mind about the appropriate answer to some question will unravel the entire fabric of questions that presupposed this earlier answer. For if we change our mind regarding the correct answer to one member of a chain of questions, then the whole of a subsequent course of questioning may well collapse. If we abandon the luminiferous ether as a vehicle for electromagnetic radiation, then we lose at one stroke the whole host of questions about its composition, structure, mode of operation, origin, and so on.

Epistemic change over time thus relates not only to what is *"known"* but also to what can be *asked*. The accession of "new knowledge" opens up new questions. And when the epistemic status of a presupposition changes from acceptance to abandonment or rejection, we witness the disappearance of various old ones through dissolution. Questions regarding the *modus operandi* of phlogiston, the behavior of caloric fluid, the structure of the luminiferous ether, and the character of faster-than-light transmissions are all questions that have become lost to modern science because they involve presuppositions that have been abandoned.

And this brings us to the theme of fallibilism once more. A body of knowledge may well answer a question only provisionally, in a tone of voice so tentative or indecisive as to indicate that further information is actually needed to enable us settle the matter with confidence. But even if it does firmly and unqualifiedly support a certain resolution, this circumstance can never be viewed as absolutely final. What is seen as the correct answer to a question at one stage of the cognitive venture, may, of course, cease to be so regarded at another, later stage.[4] Given that a particular state of science S sees a certain answer as appropriate to a question Q, we can never preclude the prospect that some superior successor to S will eventually come about that endorses some different answer—one that is actually *inconsistent* with the earlier one.

The second of these modes of erotetic discovery is particularly significant. The phenomenon of the ever-continuing "birth" of new questions was first emphasized by Immanuel Kant, who saw the development of natural science in terms of a continually evolving cycle of questions and answers, where, "*every answer given on principles of experience begets a fresh question, which likewise requires its answer* and thereby clearly shows the insufficiency of all scientific modes of explanation to satisfy reason."[5] This claim suggests the following

Principle of Question Propagation—Kant's Principle, as we shall call it: "The answering of our factual (scientific) questions always paves the way to further as yet unanswered questions."

Cognitive Incapacity

What we have here is, in effect, a sort of Cognitive Heracliteanism. Heraclitus said that the world is ever changing that we cannot step into the same river twice. And epistemic counterpart: the world of knowledge is ever changing. In the course of cognitive progress we do not—cannot—confront the same question agenda twice.

Thus one way in which the question-resolving capacity of our knowledge can be limited is by way of the mode of the situation described in the following thesis:

Weak-Limitation (The Permanence of Unsolved Questions). There are *always*, at every temporal stage,[6] questions to which no answer is in hand. At every juncture of cognitive history there exist then-unanswerable questions for whose resolution then-current science is inadequate (yet which may well be answerable at some later state).

Now if Immanuel Kant was right, and every state of knowledge generates further new and yet unanswered questions, then we will clearly never reach a position where all questions are resolved. Thus given Kant's Principle of Question Propagation, such a condition of weak limitation inexorably characterizes our knowledge, seeing that of the permanence of unsolved questions is at once assured.

However, while Kant's principle assures us *that* new questions will emerge from the answers we presently give to our questions it provides no detailed information about *what* these questions will be—nor about *when* they will arise. Accordingly, we realize at the level of nonspecific generality that *various questions will arise tomorrow that we cannot as yet identify today*. But since one cannot possibly *identify* the question that will arise tomorrow, it follow that one cannot possibly say whether all of the questions that will arise belong to the family of those for which one can provide satisfactory answers.

But now consider the proposition:

(P) A new question that I cannot answer within one year will arise tomorrow.

This thesis—somewhat reminiscent of a halting problem in computation theory—is clearly a proposition whose truth I am unable to determine one way or the other. Accordingly the question "Is (*P*) true or not?" is for all intents and purposes an undecidable question: is as firm a fact as can be that I am unable to determine the truth-status of (*P*) one way or the other. We therefore now have before us a specific, example—namely, Is (*P*) true?—that instances a concrete and perfectly meaningful question I cannot answer. But, of course, all this only bears on the issue of what I myself can or cannot do and does not address that of what can or cannot be done within the unbounded community of inquirers at large—now or ever. To address this issue we must dig deeper.

Insolubilia Then and Now

A medieval insolubilium was represented by a question that cannot be answered satisfactorily one way or another because every possible answer is unavailable on grounds of *a logical insufficiency of inherent coherence.* Such an insolubilium poses a paradox. By contrast, a modern insolubilium poses a puzzle. It is represented by a question that cannot be answered satisfactorily one way or anther because every possible answer is unavailable on grounds of *an evidential insufficiency of accessible information.*

An example of the former (medieval) sort of logical insolubilium is posed by the self-referential statement: "This sentence is false." Is this statement true or not? Whatever answer we give, be it yes or no, we are in deep trouble either way.[7]

But what about factual insolubilia of the modern type—informatively unanswerable questions?

Consider some possible examples of this phenomenon. In 1880 the German physiologist, philosopher, and historian of science Emil de Bois-Reymond published a widely discussed lecture of *The Seven Riddles of the Universe* (*Die sieben Welträtsel*),[8] in which he maintained that some of the most fundamental problems regarding the workings of the world were irresolvable. Reymond was a rigorous mechanist. On his view, nonmechanical modes of inquiry cannot produce adequate results, and the limit of our secure knowledge of the world is confined to the range where purely mechanical principles can be applied. As for all else, we not only *do not* have but *cannot* in principle obtain reliable knowledge. Under the banner of the slogan *Ignoramus et ignorabimus* (We *do not* know and *shall never* know), Reymond maintained a skeptically agnostic position with respect to basic issues in physics (the nature of matter and of force, and the ultimate source of motion) and psychology (the origin of sensation and of consciousness). These issues are simply *insolubilia* which transcend man's scientific capabilities. Certain fundamental biological problems he regarded as unsolved, but perhaps in principle soluble (though very difficult): the origin of life, the adaptiveness of organisms, and the development of language and reason. And as regards the seventh riddle—the problem of freedom of the will—he was undecided.

The position of du Bois-Reymond was swiftly and sharply contested by the zoologist Ernest Haeckel in a book *Die Welträtsel* published in 1889,[9] which soon attained a great popularity. Far from being intractable or even insoluble—so Haeckel maintained—the riddles of du Bois-Reymond had all virtually been solved. Dismissing the problem of free will as a pseudoproblem—since free will "is a pure dogma [which] rests on mere illusion and in reality does not exist at all"—Haeckel turned with relish to the remaining riddles. Problems of the origin of life, of sensation, and of consciousness Haeckel regarded as solved—or solvable—by appeal to the theory of evolution. Questions of the nature of matter and force, he regarded as solved by modern physics except for one residue: the problem (perhaps less scientific than metaphysical) of the ultimate origin of matter and its laws. This "problem of substance" was the only remaining riddle recognized by Haeckel, and it was not really a problem of science: in discovering the "fundamental law of the conservation of matter and force" science had done pretty much what it could do with respect to this problem—the rest that remained was metaphysics with which the scientist had no proper concern. Haeckel summarized his position as follows:

> The number of world-riddles has been continually diminishing in the course of the nineteenth century through the aforesaid progress of a true knowledge of nature. Only one comprehensive riddle of the universe now remains—the problem of substance.... [But now] we have the great, comprehensive "law of substance," the fundamental law of the constancy of matter and force. The fact that substance is everywhere subject to eternal movement and transformation gives it the character also of the universal law of evolution. As this supreme law has been firmly established, and all others are subordinate to it, we arrive at a conviction of the universal unity of nature and the eternal validity of its laws. From the gloomy *problem* of substance we have evolved the clear *law* of substance.[10]

The basic structure of Haeckel's teaching is clear: science is rapidly nearing a state where all the big problems have been solved. What remains unresolved is not so much a *scientific* as a *metaphysical* problem. In science itself, the big battle is virtually at an end, and the work that remains to be done is pretty much a matter of mopping-up operations.

But is this rather optimistic position tenable? Can we really dismiss the prospect of factual insolubilia? Let us explore this issue more closely.

Cognitive Limits

To begin with there is the prospect of what might be called the *weak limitation* inherent in the circumstance that there are certain issues on its agenda that science cannot resolve *now*. However, this condition of weak limitation is perfectly compatible with the circumstance that *every* question raisable at this stage will *eventually* be answered at such future juncture. And so, a contrasting way in which the question-resolving capacity of our knowledge may be limited can envisage the following, more drastic situation:

> *Strong-Limitation* (*The Existence of Insolubilia*). There will (as of some juncture) be then-posable questions which will *never* obtain answer, meaningful questions whose resolution lies beyond the reach of science altogether—questions that will remain ever unsolved on the cognitive agenda.

Such strong limitation the existence of immoral questions—permanently unanswerable questions (general insolubilia) that admit of no resolution within any cognitive corpus we are able to bring to realization.

However, for there to be *insolubilia* it is certainly not necessary that anything be said about the current *availability* of the insoluble question. The prospect of its actual identification *at this or indeed any other particular prespecified historical juncture is wholly untouched*. Even a position that holds that there indeed *are* insolubilia certainly need not regard them as being identifiable at the present state-of-the-are of scientific development. One can accordingly also move beyond the two preceding theses to the yet stronger principle of

> *Hyperlimitation* (*The Existence of IDENTIFIABLE insolubilia*). Our present-day cognitive agenda includes certain here-and-now specifiable and scientifically meaningful questions whose resolution lies beyond the reach of science altogether.

Awkwardly, however, a claim to identify insolubilia by pinpointing here and now issues that future inquiry will never resolve can readily go awry. Charles S. Peirce has put the key point trenchantly:

> For my part, I cannot admit the proposition of Kant—that there are certain impassable bounds to human knowledge.... The history of science affords illustrations enough of the folly of saying that this, that, or the other can never be found out. Auguste Comte said that it was clearly impossible for man ever to learn anything of the chemical constitution of the fixed stars, but before his book had reached its readers the discovery which he had announced as impossible had been made. Legendre said of a certain proposition in the theory of numbers that, while it appeared to be true, it was most likely beyond the powers of the human mind to prove it; yet the next writer on the subject gave six independent demonstrations of the theorem.[11]

To identify an insoluble problem, we would have to show that a certain inherently appropriate question is such that its resolution lies beyond every (possible or imaginable) state of future science. This task is clearly a rather tall order. Its realization is clearly difficult. But not in principle impossible.

Observe, to begin with, that even if we agree with Peirce that science is en route to a completion we may well always—at *any* given time—remain at a remove from ultimacy. For as long as the body of knowledge continues to grow there will still remain scope for the possibility of insolubilia. Even an asymptotically completeable science can accommodate a fixed region of unresolvability, as long as the scope of that science itself is growing. That is, even if the *fraction* of unresolved questions converges asymptotically to zero, the *number* of unresolved questions may be ever-growing in the context of an expanding science. For consider

No. of questions on the agenda	100	1000	10,000	10^k
Fraction of unresolved questions	$\dfrac{1}{2}$	$\dfrac{1}{4}$	$\dfrac{1}{8}$	$\left(\dfrac{1}{2}\right)^{k-1}$
No. of unresolved questions	50	250	1250	$10^k \times \left(\dfrac{1}{2}\right)^{k-1}$

These figures indicate that there is room for insolubilia even within a science that is ever-improving so as to approach asymptotic completeness. And this points toward a prospect that is well worth exploring.

Identifying Insolubilia

To elucidate the prospect of identifying scientific insolubilia, let us resume the theme of the progressive nature of knowledge, and continue the earlier considerations of second-order questions about future knowledge. Specifically, let us focus even more closely on the historicity of knowledge development.

It lies in the very nature of the situation that the detailed nature of our ignorance is—for us at least—hidden away in an impenetrable fog of obscurity. The limits of one's information set unavoidable limits to one's predictive capacities. In particular, we cannot foresee what we cannot conceive. Our questions—let alone answers—cannot outreach the limited horizons of our concepts. Having never contemplated electronic computing machines as such, the ancient Romans could also venture no predictions about their impact on the social and economic life of the twenty-first century. Clever though he unquestionably was, Aristotle could not have pondered the issues of quantum electrodynamics. The scientific questions of the future are—at least in part—

bound to be conceptually inaccessible to the inquirers of the present. The question of just how the cognitive agenda of some future date will be constituted is clearly irresolvable for us now. Not only can we not anticipate future discoveries now, we cannot even prediscern the questions that will arise as time moves on and cognitive progress with it.[12] We are cognitively myopic with respect to future knowledge. It is in principle infeasible for us to say now what questions will figure in the erotetic agenda of the future, let alone what answers they will engender.

But, of course, all of these are, by hypothesis, issues that will resolve themselves in the fullness of time. We have not as yet identified an insolubilia that can never be satisfactorily resolved.

To address this question, consider, however, the thesis:

(T) It will always be the case that there will come a time when all of the ever-resolved questions then on the agenda will be resolved within 100 years.

And now let Q^* be the question: Is T true or not? It is clear that to answer this question Q^* one way or the other we would need to have cognitive mastery over the question agenda of all future times. And, as emphasized above, just this is something that we cannot manage to achieve. By their very nature as such, the discoveries of the future are unavailable at present. Thus Q^* illustrates the sort of case we are looking for: it affords an example of a specific and perfectly meaningful question that we are in effect always and ever unable to resolve convincingly—irrespective of what the date on the calendar happens to read.

And we can move even further in this direction. For, after all, scientific inquiry is a venture in innovation. Present science can never speak decisively for future science, and present science cannot predict the specific discoveries of future inquiry. Accordingly, claims about what someone will achieve over all—and thus just where it will be going in the long run—are beyond the reach of attainable knowledge at this or any other particular stage of the scientific "state of the art." And on this basis the thesis "There are non-decidable questions that science will never resolve—even were it to continue *ad indefinitum*"—the Insolubilia Thesis as we may call it—is something whose truth-status can never be settled in a decisive way. And since this is so we have it that this question itself is self-instantiating: it is a question regarding an aspect of reality (of which of course science itself is a part) that scientific inquiry will never—at any specific state of the art—be in a position to settle decisively.

It should be noted that this issue cannot be settled by supposing a mad scientists who explodes the superbomb that blows the earth to smithereens and extinguishes all organic life as we know it. For the prospect cannot be precluded that intelligent life will evolve elsewhere. And even if we contemplate the prospect of a "big crunch" that is a reverse "big bang" and implodes our universe into an end, the project can never be precluded that at the other end of the big crunch, so to speak, another era of cosmic development awaits.

Of course, someone may possibly be of a mind to complain as follows:

> You are not giving me what I want. For let us distinguish between a base-level question in which no (essential) inference to questions and question agendas is made and a meta-level question in which there is an uneliminable reference to questions and question agendas. What I want is an example—a definitively specified instance—of an insolubilium at the base-level of substantive questions about the real world.

To such a complainer one can respond as follows:

> In its own way, your complaint is well taken; and indeed it seems to be pretty much in the spirit of Peirce's telling observation just quoted above. But it is worthwhile to look in a somewhat different light at this very question that you have just raised, viz. "Are there any base-level factual insolubilia." The reality of it is that it is somewhat beyond difficult and impossible to imagine that this is an issue that could be settled convincingly one way or the other in any state of actually available information. And so this question itself is a pretty good candidate for an insolubilium—though, to be sure, not at the base level.

Clearly that complaint cannot accomplish its intended mission.

Relating Knowledge to Ignorance

In any event, however, while there indeed are scientific insolubilia—and we can actually identify some of them—the fact remains that detailed knowledge about the *extent* of our ignorance is unavailable to us. For what is at stake with this issue of extent is the ratio of the manifold of what one does know to the manifold of that what one does not. And it is impossible in the nature of things for me to get a clear fix on the latter. For the actual situation is not that of a crossword puzzle—or of geographic exploration—where the size of the *terra incognita* can be somehow measured in advance of securing the details that are going to be filled in. We can form no sensible estimate of the imponderable domain of what can be known but is not known. To be sure, we can manage to compare what one person or group knows with what some other person or group knows. But mapping the realm of what is knowable as such is something that inevitably reaches beyond our powers. And for this reason any questions about the cognitive completeness of our present knowledge is and will remain inexorably unresolvable.

There are, of course, finite fields of knowledge. There is only so much you can know (nonrelationally, at least) about the content of Boston's 1995 telephone directory, namely the totality of what is in its pages. But that is only the case because here "what *can be* known" and "what is known" actually coincide. But this sort of thing is the case only in very special circumstances and never with respect to areas of natural science such as medicine or physics that deal with the products of nature at a level of generic generality.

Yet although ignorance lies at the core of the present discussion, it is not an exercise in radical skepticism. It does not propose to take the pessimistic line of a cognitive negativism to the effect that knowledge about the world is unavailable to us. Instead, what is being contemplated here is (1) that despite whatever we may come to know there are some matters on which we are destined to remain ignorant, and (2) that among the things that we can get to know about are various facts about the nature and extent of our own ignorance.

That our knowledge is sufficient for our immediate purposes—specifically by enabling us to answer the questions we then and there have before us—is something that is in principle readily determinable. But that it is *theoretically* adequate to answer not just our present questions but those that will grow out of them the future is something we can never manage to establish. For it is clear that the sensible management of ignorance is something that requires us to operate in the realm of practical considerations exactly because the knowledge required for theoretical adequacy on this subject is—by hypothesis—not at our disposal. We have no cogently rational alternative to proceed, here as elsewhere, subject to the basic pragmatic principle of having to accept the best that we can do as good enough.

And so we return to the point made at the very outset: the ironic, though in some ways fortunate fact is that one of the things about which we are most decidedly ignorant is the detailed nature of our ignorance itself. We simply cannot make a reliable assessment of the extent of our ignorance.

Postscript: A Cognitively Indeterminate Universe

One further consequence of these deliberations warrants being noted by way of postscript, as it were, to our discussion of the unpredictability of future knowledge. It is that our cognitive imperfection means that the universe itself is unpredictable. For a world in which it transpires that the future knowledge and thereby the future thoughts of intelligent beings are impredictable (at least in part) as a matter of fundamental principle is one in which the correlative physical phenomena are impredictable as well. As long as intelligent agents continue to exists within it a world is—and is bound to be—in part unpredictable since what intelligent beings do will always in some ways reflect the state of their knowledge, alike in its positive and its negative composition. For since mind and its operations are themselves an integral component of nature, a mental unpredictability cannot but constitute a form of natural or physical unpredictability as well. The extinction of intelligent agents would be required to make the world predictable: as long as intelligent life continues the world will continue to be impredictable.

Put in a nutshell, the position of affairs is this. Since all of the doings—actual and potential—of intelligent agents are themselves part or aspect of the constitution of the universe, then insofar as our cognitive doings are inherently impredictable, so also will this be the case with the correlative aspects of physical nature itself. To be sure, if our cognitive efforts stood outside nature, things might be different in this regard, since physical predictability might then be combined with a "merely epistemic" mental unpredictability. But this, of course, is a prospect that is implausible in the extreme.[13]

Chapter 18

Cognitive Realism

Synopsis

- *To exist is to play a role in some realm of identifiable items.*
- *Actual identification by us humans is not crucial here; our own powers are not determinative.*
- *Indeed our knowledge regarding actual existence is bound to be incomplete.*
- *As far as we are concerned, the world's real things are cognitively inexhaustible.*
- *Our conception of things are always corrigible,*
- *incomplete, and*
- *changeable.*
- *The reality of things outruns our conceptions of them.*
- *This hidden depth of real things is not a basis for slepticism but rather a prime impetus of realism.*
- *Such a realism has a pragmatic basis as serving to override the essential conceptual underpinnings for communication and discourse.*
- *Such a view of the matter gives an idealistic cast to realism.*
- *In particular it does not permit an actual-science realism holding that reality is as the science of our day describes it to be, but rather an aspirational science-realism that sees depicting reality as a regulative goal that science aspires to actualize.*

Existence

To exist (in the broadest sense of this term) is to function as a constituent of a realm, to play a role in a domain of identifiable items of some sort. In principle there are thus as many modes of existence as there are types of interrelated items, and to exist is to exist as an item of the correlative kind. There is physical existence in space and time, mathematical existence in the realm of quantities or structures, sensory existence in the spectrum of colors or the catalog of odors, and so on. This means that existence is realm-correlative and thus contextualized. Strictly considered, we should not speak of existence categorically and without qualifications. It is always a matter of existence-as: as a physical object, as a number, as a character in a Shakespeare play, or the like. Existence is accordingly not homogenous but categorially differentiated: different kinds of existing things exist in their own characteristic way. To attribute to numbers the same kind of existence that colors or that mammals have is to commit a serious category mistake.

All the same, when philosophers talk of existence they generally mean *physical existence* in the natural world. And here the term admits both a narrower and a broader construction. In its narrower construction to exist physically is to be an object in the space and time: to occupying a place here in the manner in which cats and trees and water molecules do. But to exist physically in the broader sense of the term is to play a role in the causal commerce of such things—to exist in the manner in which droughts and headaches or human desires do, and thereby to figure as part of the world's processual development. It is such actual, real-world existence—narrow and broad alike—that will be our principal object of concern here. And the overall range of such existences can be specified in an essentially recursive manner as follows:

1. the things we experience with our internal and external senses exist.
2. the things whose existence we need to postulate in order to realize an adequate causal explanation of things that exist also exist.

This view really takes the approach of a *causal* realism, a theory maintaining that to be a real existent is to be part of the world's causal commerce. Such a definition is essentially recursive—ordinary material objects are existentially real, and so is anything whatsoever that is bound up with them by linkages of cause and effect.

Accordingly, "to exist" in the physical mode is to feature as a component or aspect of the causal commerce real world. And some jargon-expression such as "to subsist" needs to be coined for contextualized existence within a framework of supposition at issue with fictions or hypotheses. Thus merely possible objects—or *possibilia*, for short—are things that merely "exist" in the sense of subsistence within a hypothetical realm on a fictional make-believe world. They are not part of the real world's actual furnishings.

Homo Mensura?

But how is this causal view of reality related to the issue of knowledge? Clearly, whatever can be *known* by us humans to be real must of course, for that very reason, actually be real. But does the converse hold? Is the real for that very reason also automatically knowable? Is it appropriate to join C. S. Peirce who, in rejecting "incognizables," insisted that whatever is real must be accessible to cognition—and indeed must ultimately become known?

Is humanly cognizable reality the only sort of reality there is? Some philosophers certainly say so, maintaining that there actually is a fact of the matter only when "we [humans] could in finite time bring ourselves into a position in which we were justified either in asserting or in denying [it]."[1] On such a view all reality is inevitably *our* reality. What we humans are not in a position to domesticate cognitively—what cannot be brought home to us by (finite!) cognitive effort—simply does not exist as a part of reality at all. Where we have no cognitive access, there just is nothing to be accessed. On such a perspective we are led back to the *homo mensura* doctrine of Protagoras: "Man is the measure of all things, of what is, that it is, of what is not, that it is not."

However, in reflecting on the issue in a modest mood, one is tempted to ask: "Just who has appointed us to this exalted role? How is it that *we humans* qualify as the ultimate arbiters of reality as such?"

Regarding this doctrine that what is real must be knowable, traditional realism takes an appropriately modest line. It insists on preserving, insofar as possible, a boundary-line of separation between ontology and epistemology; between fact and knowledge of fact, between truth-status possession and truth-status decidability with respect to propositions and between entity and observability with respect to individual things. As the realist sees it, reality can safely be presumed to have depths that cognition may well be unable to plumb.

To be sure, it is possible to reduce the gap between personal and objective cognition by liberalizing the idea of what is at issue with cognizers. Consider the following series of metaphysical theses: *For something to be real in the mode of cognitive accessibility it is necessary for it to be experientiable by:*

1. *Oneself.*
2. *One's contemporary (human) fellow inquirers.*
3. *Us humans (at large and in the long run).*
4. *Some actual species of intelligent creatures.*
5. *Some physically realizable (though not necessarily actual) type of intelligent being—creatures conceivably endowed with cognitive resources far beyond our feeble human powers.*
6. *An omniscient being (i.e., God).*

This ladder of potential knowers is critically important for construing the idea that to be is to be knowable. For here the question "By whom?" cannot really be evaded.

The idea of an experiential idealism that equates reality with experientiality is one that can accordingly be operated on rather different levels. Specifically, the "*i*-th level" idealist maintains—and the "*i*-th level" realist denies—such a thesis at stage number *i* of the preceding six-entry series. On this approach, the idealist emerges as the exponent of a experientiability theory of reality, equating truth and reality with what is experientially accessible to by "us"—with different, and potentially increasingly liberal, constructions of just who is to figure in that "us group" of qualified cognizers. But of course no *sensible* idealist maintains a position as strong as the egocentrism of the first entry on the list. Equally it is presumably the case that no *sensible* realist denies a position as weak as the deocentrism of the last. The salient question is just where to draw the line in determining what is a viable "realistic/idealistic" position.

Let us focus for the moment on the third entry of the above listing, the "man is the measure," *homo mensura* doctrine. By *this* standard, both Peirce and the Dummett of the preceding quotation are clearly *homo mensura* realists, seeing that both confine the real to what we humans can come to know. But this is strange stuff. Of course, what people can *know* to be real constitutes (*ex hypothesi*) a part or aspect of reality-at-large. That much is not in question. But the bone of contention between *homo mensura* realism and a sensible idealism is the question of a surplus—of whether reality may have parts or aspects that outrun altogether the reach of human cognition. And on this basis the *homo mensura* doctrine is surely problematic. For in the end, what we humans can know cannot plausibly be seen as decisive for what can (unqualifiedly) be known.

Undoubtedly, a mind that evolves in the world by natural selection has a link to reality sufficiently close to enable it to secure *some* knowledge of the real. But the converse is decidedly questionable. It is a dubious proposition that the linkage should be so close that *only* what is knowable for some species of actual being should be real—that reality has no hidden reserves of fact that are not domesticable within the cognitive resources of existing creatures—let alone one particular species thereof. Accordingly, it seems sensible to adopt the "idealistic" line only at the penultimate level of the above listing and to be a realist short of that. Essentially this is the position of the casual commerce realism espoused at the outset of the present discussion. As such a position sees it, the most plausible form of idealism is geared to that next-to-last position which takes the line that "to be real is to be causally active—to be a part of the world's causal commerce." For since one can always hypothesize a creature that detects a given sort of causal process, we need not hesitate to equate reality with experientiability in principle. We thus arrive at an idealism which achieves its viability and plausibility through its comparative weakness in operating at the next-to-last level, while at all of the earlier, more substantive levels our position is effectively realistic. The result is a doctrinal position that is a halfway-house compromise that combines an idealism of sorts with a realism of sorts.

A conservative idealism of this description holds that what is so as a "matter of fact" is not necessarily cognizable by "us" no matter how far—short of God!—we extend the boundaries of that "us-community" of inquiring intelligences. On the other hand, one cannot make plausible sense of "such-and-such a feature of nature is real but no possible sort of intelligent being could possibly discern it." To be real is to be in a position to make an impact somewhere on something of such a sort that a suitably equipped mind-endowed intelligent creature could detect it. What is real in the world must make some difference to it, that is *in principle* detectable. Existence-in-this-world is coordinated with perceivability-in-principle. And so, at this point, there is a concession to idealism—albeit one that is relatively weak. Since to be physically real is to be part of the world's causal commerce, it is always in principle possible for an intelligent sentient being of a suitable sort to enter into this causal situation so as to be able to monitor what is going on. Accordingly *being* and *being knowable-in-principle* can plausibly be identified.

But in any case, traditional *homo mensura* realism is untenable. There is no good reason to indulge a hubris that sees our human reality as definitive on grounds of being the only one there is. Neither astronomically nor otherwise are we the center around which all things revolve. After all, humans have the capacity not only for knowledge but also for imagination. And it is simply too easy for us to imagine a realm of things and states of things of which we can obtain no knowledge because "we have no way to get there from here," lacking the essential means for securing information in such a case.

REALISM AND INCAPACITY

Charles Sanders Peirce located the impetus to realism in the limitations of man's will—in the fact that we can exert no control over our experience and, try as we will, cannot affect what we see and sense. Peirce's celebrated "Harvard Experiment" of the Lowell Lectures of 1903 makes the point forcibly:

> I know that this stone will fall if it is let go, because experience has convinced me that objects of this kind always do fall; and if anyone has any doubt on the subject, I should be happy to try the experiment, and I will bet him a hundred to one on the result ... [I know this because of an unshakable conviction that] the uniformity with which stones have fallen has been due to some *active general principle* [of nature].... Of course, every sane man will adopt the latter hypothesis. If he could doubt it in the case of the stone—which he can't—and I may as well drop the stone once and for all—I told you so!—if anybody doubt this still, a thousand other such inductive predictions are getting

verified every day, and he will have to suppose every one of them to be merely fortuitous in order reasonably to escape the conclusion that *general principles are really operative in nature*. That is the doctrine of scholastic realism.²

In this context, however, it is important to distinguish between mental *dependency* and mental *control*. Peirce is clearly right in saying that we cannot *control* our conviction that the stone will fall that do what we will, it will remain. Nevertheless, this circumstance could conceivably still be something that *depends on us*—exactly as with the fearsomeness of heights for the man with vertigo. If the *unconscious* sphere of mind actually dictates how I *must* "see" something (as, for example, in an optical illusion of the Mueller-Leyer variety), then I evidently have no *control*. But that does not *in itself* refute mind-dependency—even of a very strong sort. There is always the prospect that we are deluding ourselves in these matters—that the limitations at issue appertain only to our *conscious* powers, and not to our mental powers as such.³

This prospect blocks Peirce's argument in the way already foreseen by Descartes in the *Meditations*:

> I found by experience that these [sensory] ideas presented themselves to me without my consent being requisite, so that I could not perceive any object, however desirous I might be, unless it were present . . . But although the ideas which I receive by the senses do not depend on my will I do not think that one should for that reason conclude that they proceed from things different from myself, since possible some facility might be discovered in me—though different from those yet known to me—which produced them.⁴

The fact of it is that we may simply delude ourselves about the range of the mind's powers: lack of control notwithstanding, dependency may yet lie with the "unconscious" sector of mind. The traditional case for realism based on the limits of causal control through human *agency* thus fails to provide a really powerful argument for mind-independence.

However, a far more effective impetus to realism lies in the limitations of man's *intellect*, pivoting on the circumstances that the features of real things inevitably outrun our cognitive reach. In placing some crucial aspects of the real together outside the effective range of mind, it speaks for a position that sees mind-independence as a salient feature of the real. The very fact of fallibilism and limitedness—of our absolute confidence that our putative knowledge does not and cannot do full justice to the real truth of what reality is actually like—is surely one of the best arguments for a realism that turns on the basic idea that there is more to reality than we humans do or can know. Traditional scientific realists see the basis for realism in the accuracy and extent of out scientific knowledge; the present metaphysical realism, by contrast, sees its basis in our realization of the inevitable *shortcomings* of our knowledge—scientific knowledge included.

Such an epistemic approach accordingly preempts the preceding sort of objection. If we are mistaken about the reach of our cognitive powers—if they do not adequately grasp, "the way things really are"—then this very circumstance itself clearly *bolsters* the case for the sort of realism now at issue. The cognitive intractability of things is something about which, in principle, we cannot delude ourselves altogether, since such delusion would illustrate rather than abrogate the fact of a reality independent of ourselves. The very inadequacy of our knowledge is one of the most salient tokens there is of a reality out there that lies beyond the inadequate gropings of mind. It is the very limitation of our knowledge of things—our recognition that reality extends beyond the horizons of what we can possible know or even conjecture about it—that betokens the mind-independence of the real.

But a qualification is in order here. One must be careful about what the presently contemplated sort of argument for realism actually manages to establish. For it does *not* establish outright that a stone—be it Peirce's or Dr. Johnson's or the geologist's—is something mind-independently real. Rather, what it shows is that our conception of a "stone"—indeed our conception of any physical object—is the conception of something that is mind-independently real, possessed of a nature extending beyond the realm of our minds. And so the realism underwritten by these deliberations is not in fact a squarely *ontological* doctrine, but a realism nevertheless geared to our conceptual scheme for thinking about things. As indicated above, the present position is a halfway-house compromise: a metaphysical realism that is unproblematically compatible with an idealism of sorts.

THE COGNITIVE OPACITY OF REAL THINGS

It is worthwhile to examine somewhat more closely the considerations that indicate the inherent imperfection of our knowledge of things.[5]

To begin with, it is clear that, as we standardly think about things within the conceptual framework of our fact-oriented thought and discourse, any real physical object has more facets than it will ever actually manifest in experience. For every objective property of a real thing has consequences of a dispositional character and these are never surveyable in tote because the dispositions which particular concrete things inevitably have endow them with an infinitistic aspect that cannot be comprehended within experience.[6] This desk, for example, has a limitless manifold of phenomenal features of the type: "having a certain appearance from a particular point of view." It is perfectly clear that most of these will never be actualized in experience. Moreover, a thing is what it does: entity and lawfulness are coordinated correlates—a good Kantian point. And this fact that things demand lawful comportment means that the finitude of experience precludes any prospect of the exhaustive manifestation of the descriptive facets of any real things.[7]

Physical things not only have more properties than they ever will overtly manifest, but they have more than they can possibly ever actually manifest. This is so because the dispositional properties of things always involve what might be characterized as mutually preemptive conditions of realization. This cube of sugar, for example, has the dispositional property of reacting in a particular way if subjected to a temperature of 10 000°C and of reacting in a certain way if emplaced for one hundred hours in a large, turbulent body of water. But if either of these conditions is ever realized, it will destroy the lump of sugar as a lump of sugar, and thus block the prospect of its ever bringing the other property to manifestation. The perfectly possible realization of various dispositions may fail to be mutually compossible, and so the dispositional properties of a thing cannot ever be manifested completely—not just in practice, but in principle. Our objective claims about real things always commit us to more than we can actually ever determine about them.

The existence of this latent (hidden, occult) sector is a crucial feature of our conception of a real thing. Neither in fact nor in thought can we ever simply put it away. To say of the apple that its only features are those it actually manifests is to run afoul of our conception of an apple. To deny—or even merely to refuse to be committed to the claim—that would manifest particular features if certain conditions came about (e.g., that it would have such-and-such a taste if eaten) is to be driven to withdrawing the claim that it is an apple. The process (corroborating the implicit contents of our objective factual claims about something real is potentially endless, and such judgments are the "nonterminating" in C. I. Lewis' sense.[8] This cognitive depth of objective factual claims inherent in the fact that their content will always outrun the evidence for making them means that the endorsement; any such claim always involves some element of evidence-transcending conjecture.

The concepts at issue (viz., "experience" and "manifestation") are such that we can only ever experience those features of a real thing that it actually manifests. But the preceding considerations show that real things always have more experientially manifestable properties that they can ever actually manifest in experience. The experienced portion of a thing is similar to the part of the iceberg that shows above water. All real things are necessarily thought of as having hidden depths that extend beyond the limits, not only of experience, but also of experientiability. To say of something that it is an apple or a stone or a tree is to become committed to claims about it that go beyond the data we have—and even beyond those which we can, in the nature of things, eve actually acquire. The "meaning" inherent in the assertoric commitments of our factual statements is never exhausted by its verification real things are cognitively opaque—we cannot see to the bottom o them. Our knowledge of such things can thus become more extensive without thereby becoming more complete.

In this regard, however, real things differ in an interesting and important way from their fictional cousins. To make this difference plain, it is useful to distinguish between two types of information about a thing, namely that which is generic and that which is not. Generic information tells about those features of a thing that it has in common with everything else of its kind or type. For example, a particular snowflake will share with all others certain facts about its structure, its hexagonal form, its chemical composition, its melting point, and so forth. On the other hand, it will also have various properties that it does not share with other members of its own "lowest species" in the classificatory order—its particular shape, for example, or the angular momentum of its descent. These are its nongeneric features.

Now a key fact about fictional particulars is that they are of finite cognitive depth. In discoursing about them we shall ultimately run out of steam as regards their nongeneric features. A point will always be reached when one cannot say anything further that is characteristically new about them—presenting non-generic information that is not inferentially implicit in what has already been said. New generic information can, of course, always be forthcoming through the progress of science. When we learn more about coal-in-general then we know more about the coal in Sherlock Holmes's grate. But the finiteness of their cognitive depth means that the presentation of ampliatively novel nongeneric information must by the very nature of the case come to a stop when fictional things are at issue.

With *real* things, on the other hand, there is no reason of principle why the availability of new, nongenerically idiosyncratic information need ever come to an end. On the contrary, we have every reason to presume these things to be cognitively inexhaustible. A precommitment to description-transcending features—no matter how far description is pushed—is essential to our conception of a real thing. Something whose character was exhaustible by linguistic characterization would thereby be marked as fictional rather than real.[9]

THE COGNITIVE INEXHAUSTIBILITY OF THINGS

As the preceding deliberations indicate, one of the most fundamental aspects of our concept of a real thing is that our knowledge of it is inevitably imperfect—that reality is such as to transcend what we can know of it. The number of true descriptive remarks that can be made about a thing—about any actual physical object—is theoretically inexhaustible. For example, take a stone. Consider its physical features: its shape, its surface texture, its chemistry, and so on. And then consider its causal background: its subsequent genesis and history. Then consider its functional aspects as relevant to its uses by the stonemason, or the architect, or the landscape decorator, and so forth. There is, in principle, no theoretical limit to the different lines of consideration available for articulating descriptive information about a thing, so that the totality of potentially available facts about a thing—about any real thing whatever—is in principle inexhaustible.

Our thought about the real things of this world pushes them beyond any finite limits. From finitely many axioms, reason can generate a potential infinity of theorems; from finitely many words, thought can exfoliate a potential infinity of sentences; from finitely many data, reflection can extract a potential infinity of items of information. Even with respect to a world of finitely many objects, the process of reflecting on these objects can, in principle, go on unendingly. One can inquire about their features, the features of these features, and so on. Or again, one can consider their relations, the relations among those relations, and so forth. Thought—abstraction, reflection, analysis—is an inherently ampliative process. As in physical reflection mirror images can reflect one another indefinitely, so mental reflection can go on and on. Given a start, however modest, thought can advance *ad indefinitum* into new conceptual domains. The circumstance of its starting out from a finite basis does *not* mean that it need ever run out of impetus (as the example of Shakespearean scholarship seems to illustrate).

It is helpful to introduce a distinction at this stage. On anything like the standard conception of the matter, a "truth" is something to be understood in *linguistic* terms—the representation of a fact through its statement in some actual language. Any correct statement in some actual language formulates a truth. (And the converse obtains as well: a *truth* must be encapsulated in a statement, and cannot exist without linguistic embodiment.) A *fact*, on the other hand, is not a linguistic entity at all, but an actual circumstance or state of affairs. Anything that is correctly statable in some *possible* language will present a fact.[10]

Every truth must state a fact, but in principle it is possible that there will be facts that cannot be stated in any actually available language and thus fail to be captured as truths. Facts afford *potential* truths whose actualization as such hinges on there being given a linguistic formulation. Truths involve a one-parameter possibilization: they include whatever *can* be stated truly in some *actual* language. Facts, on the other hand, involve a two-parameter possibilization—they include whatever *can* be stated truly in some *possible* language. Truths are *actualistically* language-correlative, while facts are *possibilistically* language-correlative.[11] Accordingly, it must be presumed that there are facts that will never be formulated as truths, though it will obviously be impossible to give concrete examples of this phenomenon.[12]

Now propositional *knowledge* regarding matters of actual fact—and even belief and conjecture regarding *supposed* fact—is always a matter of truth-recognition and thereby of linguistically formulable information. But we have no alternative to supposing that the realm of facts regarding this world is larger than the attainable body of knowledge about them—regardless of whether that "we" is construed distributively or collectively. And is not very difficult to see why this is and must be so.

As long as we are concerned with information formulated in languages of the standard (recursively developed) sort, the number of actually *articulated* items of information (truths or purported truths) about a thing is always, at any historical juncture, finite. And it remains *denumerably* infinite even over a theoretically endless long run.[13] The domain of truth is therefore enumerable: but that of fact is not comparably manageable because *no finite list* of truths about a real object exhausts the totality of true facts about it since old facts always combine to give rise to new ones.

THE CORRIGIBILITY OF CONCEPTIONS

It must be stressed that these deliberations regarding cognitive inadequacy are less concerned with the correctness of our particular claims about real things than with our characterizing conceptions of them. And in this connection it deserves stressing that there is a significant and substantial difference between a true or correct statement or contention, on the one hand, and a true or correct conception, on the other hand. To make a true contention *about* a thing, on the one hand, we merely need to get one particular fact about it straight. To have a true conception *of* the thing, on the other hand, we must get all of the important facts about it straight. And it is clear that this involves a certain normative element—namely, what the "important" or "essential" facets of something are.

Anaximander of Miletus presumably made many correct contentions about the sun in the fifth century B.C.—for example, that its light is brighter than that of the moon. But Anaximander's conception of the sun (as the flaming spoke of a great wheel of fire encircling the earth) was totally wrong.

To assure the correctness of our conception of a thing we would have to be sure—as we very seldom are—that nothing further can possibly come along to upset our view of just what its important features are and just what their character is. Thus, the qualifying conditions for true conceptions are far more demanding than those for true claims. With a correct contention about a thing, all is well if we get the single relevant aspect of it right, but with a correct conception of it we must gel the essentials right—we must have an overall picture that is basically correct. And this is something we generally cannot ascertain, if only because we cannot say with secure confidence what actually is really important or essential before the end of the proverbial day.

With conceptions—unlike propositions or contentions—incompleteness means incorrectness, or at any rate presumptive incorrectness. Having a correct or adequate conception of something as the object it is requires that we have all the important facts about it right. But since the prospect of discovering further important facts can never be eliminated, the possibility can never be eliminated that matters may so eventuate that we may ultimately (with the wisdom of hindsight) acknowledge the insufficiency or even inappropriateness of our earlier conceptions. A conception based on incomplete data must be assumed to be at least partially incorrect. If we can decipher only half an inscription, our conception of its overall content must be largely conjectural—

and thus must be presumed to contain an admixture of error. When our information about something is incomplete, obtaining an overall picture of the thing at issue becomes a matter of theorizing, or guesswork, however sophisticatedly executed. And then we have no alternative but to suppose that this overall picture falls short of being wholly correct in various (unspecifiable) ways. With conceptions, falsity can thus emerge from errors of omission as well as those of commission, resulting from the circumstance that the information at our disposal is merely incomplete, rather than actually false (as would have to be the case with contentions).

To be sure, an inadequate or incomplete *description* of something is not thereby false—the statements we make about it may be perfectly true as far as they go. But an inadequate or incomplete *conception* of a thing is ipso facto one that we have no choice but to presume to be incorrect as well,[14] seeing that where there is incompleteness we cannot justifiably take the stance that it relates only to inconsequential matters and touches nothing important. Accordingly, our conceptions of particular things are always to be viewed not just as cognitively *open-ended* but as *corrigible* as well.

We are led back to the thesis of the great idealist philosophers (Spinoza, Hegel, Bradley, Royce) that human knowledge inevitably falls short of "perfected science" (the Idea, the Absolute), and must be presumed deficient both in its completeness and its correctness.[15]

Cognitive Progress

From finitely many axioms, reason can generate a potential infinity of theorems; from finitely many words, thought can exfoliate a potential infinity of sentences; from finitely many data, reflection can extract a potential infinity of items of information. Even with respect to a world of finitely many objects, the process of reflecting on these objects can, in principle, go on unendingly. One can inquire about their features, the features of these features, and so on. Or again, one can consider their relations, the relations among those relations, and so on. Thought—abstraction, reflection, analysis—is an inherently ampliative process. As in physical reflection mirror-images can reflect one another indefinitely, so mental reflection can go on and on. Given a start, however modest, thought can advance ad indefinitum into new conceptual domains. The circumstance of its starting out from a finite basis does not mean that it need ever run out of impetus (as the example of Shakespearean scholarship seems to illustrate).

The number of true descriptive remarks that can be made about a thing—about any actual physical object—is theoretically inexhaustible. For example, take a stone. Consider its physical features: its shape, its surface texture, its chemistry, and so on. And then consider its causal background: its subsequent genesis and history. Then consider its functional aspects as relevant to its uses by the stonemason, or the architect, or the landscape decorator, and so forth. There is, in principle, no theoretical limit to the different lines of consideration available to yield descriptive truths about a thing.

It follows from such lines of thought that we cannot possibly articulate, and thus come to know explicitly, "the whole truth" about a thing. The domain of fact inevitably transcends the limits of our capacity to express it, and a fortiori those of our capacity to canvass it in overt detail. There are always bound to be more facts than we are able to capture in our linguistic terminology.

It might be possible, however, to have latent or implicit knowledge of an infinite domain through deductive systematization. After all, the finite set of axioms of a formal system will yield infinitely many theorems. And so, it might seem that when we shift from overt or explicit to implicit or tacit knowledge, we secure the prospect of capturing an infinitely diverse implicit knowledge-content within a finite, explicit linguistic basis through recourse to deductive systematization.

The matter is not, however, quite so convenient. The totality of the deductive consequences that can be obtained from any finite set of axioms is itself always denumerable. The most we can ever hope to encompass by any sort of deductively implicit containment within a finite basis of truths is a denumerably infinite manifold of truths. And thus as long as implicit containment remains a recursive process, it too can never hope to transcend the range of the denumerables, and so cannot hope to encompass the whole of the transdenumerable range of descriptive facts about a thing. (Moreover, even within the denumerable realm, our attempt at deductive systematization runs into difficulties: as is known from Kurt Goedel's work, one cannot even hope to systematize—by any recursive, axiomatic process—all of the inherently denumerable truths of arithmetic. It is one of the deepest lessons of modern mathematics that we cannot take the stance that if there is a fact of the matter in this domain, then we can encompass it within the deductive means at our disposal.)

Cognitive Dynamics

But there is yet another way of substantiating our point. For the preceding considerations related to the limits of knowledge that can be rationalized on a fixed and given conceptual basis—a full-formed, developed language. But, in real life, languages are never full-formed and a conceptual basis is never "fixed and given." Even with such familiar things as birds, trees, and clouds, we are involved in a constant reconceptualization in the course of progress in genetics, evolutionary theory, and thermodynamics. Our conceptions of things always present a moving rather than a fixed object of scrutiny and this historical dimension must also be reckoned with.

Any adequate theory of inquiry must recognize that the ongoing process of information acquisition at issue in science is a process of conceptual innovation, which always leaves certain facts about things wholly outside the cognitive range of the inquirers of any particular period. Caesar did not know—and in the then extant state of the cognitive arts could not have known—that his sword contained tungsten and carbon. There will always be facts (or plausible candidate facts) about a thing that we do not know because we cannot even conceive of them in the prevailing order of things. To grasp such a fact means taking a perspective of consideration that as yet we simply do not have, since the state of knowledge (or purported knowledge) is not yet advanced to a point at which such a consideration is feasible. Any adequate worldview must recognize that the ongoing progress of scientific inquiry is s process of conceptual innovation that always leaves various facts about the things of this world wholly outside the cognitive range of the inquirers of any particular period.

The language of emergence can perhaps be deployed usefully to make the point. But what is at issue is not an emergence of the features of things, but an emergence in our knowledge about them. Blood circulated in the human body well before Harvey; substances containing uranium were radioactive before Becquerel. The emergence at issue relates to our cognitive mechanisms of conceptualization, not to the objects of our consideration in and of themselves. Real-world objects must be conceived of as antecedent to any cognitive interaction—as being there right along, "pregiven" as Edmund Husserl put it. Any cognitive changes or innovations are to be conceptualized as something that occurs on our side of the cognitive transaction, and not on the side of the objects with which we deal.[16]

The prospect of substantive change can never be eliminated in this domain. The properties of any real are literally open-ended: we can always discover more of them. Even if we were (surely mistakenly) to view the world as inherently finitistic—espousing a Keynesian principle of "limited variety" to the effect that nature can be portrayed descriptively with the materials of a finite taxonomic scheme—there will still be no a priori guarantee that the progress of science will not lead *ad indefinitum* to changes of mind regarding this finite register of descriptive materials. And this conforms exactly to our expectation in these matters. For where the real things of the world are concerned, we not only expect to learn more about them in the course of scientific inquiry, *we expect to have to change our minds about their nature and modes of comportment*. Be the items at issue elm trees, or volcanoes, or quarks, we have every expectation that in the course of future scientific progress people will come to think about their origin and their properties differently from the way we do at this juncture.

This cognitive opacity of real things means that we are not—and will never be—in a position to evade or abolish the contrast between "things as we think them to be" and "things as they actually and truly are." Their susceptibility to further elaborative detail—and to changes of mind regarding this further detail—is built into our very conception of a "real thing." To be a real thing is to be something regarding which we can always, in principle, acquire more and possibly discordant information. This view of the situation is supported rather than impeded once we abandon the naive cumulativist/preservationist view of knowledge acquisition for the view that new discoveries need not supplement but can displace old ones. We realize that people will come to think differently about things from the way we do—even when thoroughly familiar things are at issue—recognizing that scientific progress generally entails fundamental changes of mind about how things work in the world.

In view of the cognitive opacity of real things, we must never pretend to a cognitive monopoly or cognitive finality. This recognition of incomplete information is inherent in the very nature of our conception of a "real thing." It is a crucial facet of our epistemic stance toward the Real world to recognize that every part and parcel of it has features lying beyond our present cognitive reach—at any "present" whatsoever.

Much the same story holds when our concern is not with physical things, but with types of such things. To say that something is copper or magnetic is to say more than that it has the properties we think copper or magnetic things have, and to say more than that it meets our test conditions for being copper (or being magnetic). It is to say that this thing is copper or magnetic. And this is an issue regarding which we are prepared at least to contemplate the prospect that we have got it wrong.

Certainly, it is imaginable that natural science will come to a stop, not in the trivial sense of a cessation of intelligent life, but in Charles Sanders Peirce's more interesting sense of eventually reaching a condition after which even indefinitely ongoing effort at inquiry will not—and indeed actually cannot—produce any significant change. Such a position is, in theory, possible. But we can never know—be it in practice or in principle—that it is actually realized. We can never establish that science has attained such an omega-condition of final completion: the possibility of further change lying "just around the corner" can never be ruled out finally and decisively. Thus, we have no alternative but to presume that our science is still imperfect and incomplete, that no matter how far we have pushed our inquiries in any direction, regions of terra incognita yet lie beyond.

Conceptual Basis of Realism as a Postulate

The skeptical tendency of these remarks is of a very restricted sort, however. It is not a fact-skepticism but a concept-skepticism. For to tell the truth is one thing, to tell the whole truth another. And the former is certainly possible without the latter. The fact that our knowledge of the world is incomplete clearly does not mean that it is incorrect.

But in this regard it is important to distinguish between a true *statement* and a true *conception*. The informative incompleteness of a statement does not preclude its truth. If the only thing I know about you is that you dislike peaches, so be it—I am still in possession of a fact about you and the statement at issue is a genuine truth. However, having a true *conception* of something is far more demanding. On the positive side it requires that we have all of the important facts about you straight—that we know all of the *important* truths that are relevant here. And on the negate side it means that the things that we accept about you—important or not—are all true.

Accordingly, while immediate information about something is perfectly compatible with knowing a true fact about it, it is not compatible—or not likely to be compatible with knowing a true conception of it, seeing that the things we do not know may well include some important details.

The fact is that as long as our information about something is incomplete we can never know that our conception of it is correct. For to know this would require that none of the unknown facts about it are important, that what we do not know does not matter substantially for the correctness of our conception—its adequacy overall. And this sort of knowledge about what we do not know is in principle unavailable.

The concept-skepticism at issue here is closely bound up with a realism.

Metaphysical realism is the doctrine that the world exists in a way that is substantially independent of the thinking beings it contains that can inquire into it, and that its nature—its having whatever characteristics it does actually have—is also comparably knowledge-transcending. In saying of something that it is "a real thing," an object existing as part of the world's furniture, we commit ourselves to various (obviously interrelated) points:

1. *Self-subsistence.* Being a "something" (an entity or process) with its own unity of being. Having an enduring identity of its own.
2. *Physicality or reality.* Existing within the causal order of things. Having a place on the world's physical scene as a participant of some sort.
3. *Publicity or multilateral accessibility.* Admitting universality of access. Being something that different investigators proceeding from different points of departure can get hold of.
4. *Autonomy or independence.* Being independent of mind. Being something that observers find and learn about rather than create make up in the course of their cognitive endeavors.

In natural science we try to get at the objective matters of fact regarding physical reality in ways that are accessible to all observers alike. (The "repeatability of experiments" is crucial.) And the salient factor enters in with that fourth and final issue—autonomy. The very idea of a thing so functions in our conceptual scheme that real things are thought of as having an identity, a nature, and a mode of comportment wholly indifferent to and independent of the cognitive state-of-the-art regarding them—and potentially even very different from our own current conceptions of the matter.

The conception of a thing that underlies our discourse about the things of this world inherently involves tentatively and fallibilism—the implicit recognition that our own personal or even communal conception of particular things may, in general, be wrong, and is in any case inadequate. At the back of our thought about things there is always a certain wary skepticism that recognizes the possibility of error. The objectivity of real existents projects beyond the reaches of our subjectively conditioned information. There is wisdom in Hamlet's dictum: "There are more things in heaven and on earth, Horatio . . ." The limits of our knowledge may be the limits of *our* world, but they are not the limits of *the* world. We do and must recognize the limitations of our cognition.

The fundamental idea of realism is that the existence and nature of the world are matters distinct from anyone's thinking about it: that—minds themselves and their works aside—the world is what it is without any reference to our cognitive endeavors and that the constituents of nature are themselves *impervious,* as it were, to the state of our knowledge or belief regarding them. As one expositor puts it: "Even if there were no human thought, even if there were no human beings, whatever there is other than human thought (and what depends on that, causally or logically) would still be just what it actually is."[17] Such a realism is based on a commitment to the notion that human inquiry addresses itself to what really and truly is the condition of things whose existence and character are altogether independent of our cognitive activities. Reality is not subordinate to the operations of the human mind; on the contrary, man's mind and its dealings are but a minuscule part of reality. The nature of things reaches beyond what we—or any other sort of finite being—happen to know or think. We cannot justifiably equate reality with what in known to us, nor equate reality with what is expressible in our language. And what is true here for our sort of mind is true for any other sort of finite mind as well. It is inherent in our conception of physical reality that any physically realizable sort of cognizing being can only know a part or aspect of the real.

This "objectivity" in the sense of mind-transcendence is pivotal for realism. A fact is objective in this mode through obtaining independently of whatever thinkers may think about relevant issues, so that changes merely in what is thought by the world's intelligences would leave it unaffected. With objective facts (unlike those which are merely a matter of intersubjective agreement), what thinkers think is never something determinative: they are thought-invariant or thought-insensitive.[18]

Realism accordingly has two indispensable and inseparable constituents—the one existential and ontological, and the other cognitive and epistemic. The former maintains that there indeed is a real world—a realm of thought-transcendent objective physical reality. The latter maintains that we can to some extent secure adequate descriptive information about this mind-independent realm. This second contention obviously presupposes the first. But how can that first, ontological thesis be secured?

Metaphysical realism is clearly not an inductive inference secured through the scientific systematization of our observations. Rather it represents a regulative presupposition that makes science possible in the first place. If we did not assume from the very outset that our sensations somehow relate to an extramental reality, we could clearly make no use of them to draw any inference whatever about "the real world." The realm of mind-independent reality is something we cannot *discover*—we do not learn that it exists as a result of inquiry and investigation. How could we ever learn from our observations that our mental experience is itself largely the causal product of the machinations of a mind-independent matrix, that all those phenomenal appearances are causally rooted in a physical reality? This is obviously something we have to suppose from the very outset. What is at issue is, all too clearly, a *precondition* for empirical inquiry—a presupposition for the usability of observational data as sources of objective information. That experience is indeed objective, that what we take to be evidence *is* evidence, that our sensations yield information about an order of existence outside the experiential realm itself, and that this experience constitutes not just a mere phenomenon but an appearance of something extramental belonging to an objectively self-subsisting order, all this is something that we must always *presuppose* in using experiential data as "evidence" for how things stand in the world.

The fact is that we do not learn or discover that there is a mind-independent physical reality; we have no alternative but to *presume or postulate* it. Realism represents a postulation made on *functional* (rather than *evidential*) grounds: we endorse it in order to be in a position to learn by experience at all. As Kant clearly saw, objective experience is possible only if the existence of such a real, objective world is *presupposed* from the outset rather than being seen as a matter of *ex post facto* discovery about the nature of things.[19]

To be sure, once we have made a start by accepting an objective reality and its concomitant causal aspect, more or less by sheer postulation, then principles of inductive systematization, of explanatory economy, and of common cause consilience can work wonders in exploiting the phenomena of experience to provide the basis for plausible claims about the nature of the real. But we indispensably need that initial existential presupposition to make a start. Without a commitment from the very outset to a reality to serve as ground and object of our experience, its cognitive import will be lost. Only on this basis can we proceed evidentially with the exploration of the interpersonally public and objective domain of a physical world-order that we share in common.

Of course, that second, descriptive (epistemic) component of realism stands on a very different footing. Unlike its *existence*, reality's *nature* is something about which we can only make warranted claims through examining it.

Substantive information must come through inquiry—through evidential validation. Once we are willing to credit our observational data with objectivity, and thus with evidential bearing, then we can, of course, make use of them to inform ourselves as to the nature of the real. But that initial presumption has to be there from the start.

Let us examine this basic reality postulate somewhat more closely. Our standard conception of inquiry involves recognition of the following facts: (1) The world (the realm of physical existence) has a nature whose characterization in point of description, explanation, and prediction is the object of empirical inquiry. (2) The real nature of the world is in the main independent of the process of inquiry that the real world canalizes or conditions. (3) In virtue of these considerations, we can stake neither total nor final claims for our purported knowledge of reality. Our knowledge of the world must be presumed incomplete, incorrect, and imperfect, with the consequence that "our reality" must be considered to afford an inadequate characterization of "reality itself."

Our commitment to realism thus centers on a certain practical *modus operandi*, encapsulated in the precept: "Proceed in matters of inquiry and communication on the basis that you are dealing with an objective realm, existing quite independently of the doings and dealings of minds." Accordingly, we standardly operate on the basis of the "presumption of objectivity" reflected in the guiding precept: "Unless you have good reason to think otherwise (i.e., as long as *nihil obstat*), treat the materials of inquiry and communication as veridical—as representing the nature of the real." The ideal of objective reality is the focus of a family of effectively indispensable regulative principles—a functionally useful instrumentality that enables us to transact our cognitive business in the most satisfactory and effective way.

And so the foundations of objectivity are not provided by the findings of science. They precede and underlie science, which would itself not be possible without a precommitment to the capacity of our senses to warrant claims about an objective world order. Mind transcendence is not a *product* of inquiry; we must precommit ourselves to it to make inquiry as we understand it possible. It is a necessary (a priori) input into the cognitive project and not a contingent (a posteriori) output thereof. The objective bearing of experience is not something we can preestablish; it is something we must presuppose in the interest of honoring Peirce's pivotal injunction never to bar the path of inquiry.

What we learn from science is not *that* an unobservable order of physical existence causally undergirds nature as we observe it, but rather *what* these underlying structures are like. Science does not (cannot) teach us that the observable order is explicable in terms of underlying causes and that the phenomena of observation are signs or symptoms of this extra- and subphenomenal order of existence; we must acknowledge this prior to any venture in developing an empirical science. It is something we must accept a priori to hold of any world in which observation as we understand it can transpire. (After all, observations are, by their very nature, the results of interactions.) And what science does teach us (and metaphysics cannot) is what the descriptive character of this extraphenomenal order can reasonably be supposed to be in the light of our experience of it.

Hidden Depths: The Impetus to Realism

The fact that we do and should always think of real things as having hidden depths inaccessible to us finite knowers—that they are always cognitively opaque to us to some extent—has important ramifications that reach to the very heart of the theory of communication.

Any particular thing—the moon, for example—is such that two related but critically different versions can be contemplated:

1. the moon, the actual moon as it "really" is

and

2. the moon as somebody (you or I or the Babylonians) conceives of it.

The crucial fact to note in this connection is that it is virtually always the former item—the thing itself—that we *intend* to communicate or think (= self-communicate) about, the thing *as it is,* and not the thing *as somebody conceives of it*. Yet we cannot but recognize the justice of Kant's teaching that the "I think" (I maintain, assert, etc.) is an ever-present implicit accompaniment of every claim or contention that we make. This factor of attributability dogs our every assertion and opens up the unavoidable prospect of "getting it wrong."

However, this fundamental objectivity-intent—the determination to discuss "the moon itself" (the real moon) regardless of how untenable one's own *ideas* about it may eventually prove to be—is a basic precondition of the very possibility of communication. It is crucial to the communicative enterprise to take the egocentrism-avoiding stance of an epistemological Copernicanism that rejects all claims to a privileged status for *our own* conception of things. Such a conviction roots in the fact that we are prepared to "discount any misconceptions" (our own included) about things over a very wide-range indeed—that we are committed to the stance that factual disagreements as to the character of things are communicatively irrelevant within enormously broad limits.

We are able to say something about the (real) Sphinx thanks to our subscription to a fundamental communicative convention or "social contract": to the effect that we *intend* ("mean") to talk about it—the very thing itself as it "really" is—our own private conception of it notwithstanding. We arrive at the standard policy that prevails with respect to all communicative convention or "social contract" to the effect that we *intend* ("mean") to talk about it—the very thing itself as it "really" is—our own private conception of it notwithstanding. We arrive at the standard policy that prevails with respect to all communicative discourse of letting "the language we use," rather than whatever specific informative aims we may actually "have in mind" on particular occasions, be the decisive factor with regard to the things at issue in our discourse. When I speak about the Sphinx—even though I do so on the basis of my own conception of what is involved here—I will nevertheless be taken to be discussing "the *real* Sphinx" in virtue of the basic conventionalized intention at issue with regard to the operation of referring terms.

Communication requires not only common *concepts* but common *topics*—shared items of discussion, a common world of self-subsistently real, "an sich" objects basic to shared experience. The factor of objectivity reflects our basic commitment of a shared world as the common property of communicators. Such a commitment involves more than merely de facto intersubjective agreement. For such agreement is a matter of a posteriori discovery, while our view of the nature of things puts "the real world" on a necessary and a priori basis. This stance roots in the fundamental convention of a socially shared insistence on communicating—the commitment to an objective world of real things affording the crucially requisite common focus needed for any genuine communication.

Any pretensions to the predominance, let alone the correctness of our own potentially idiosyncratic conceptions about things, must be put aside in the context of communication. The fundamental intention to deal with the objective order of this "real world" is crucial. If our assertoric commitments did not transcend the information we ourselves have on hand, we would never be able to "get in touch" with others about a shared objective world. No claim is made for the *primacy* of our conceptions, or for the *correctness* of our conceptions, or even for the mere *agreement* of our conceptions with those of others. The fundamental intention to discuss "the thing itself" predominates and overrides any mere dealing with the thing as we ourselves conceive of it.

To be sure, someone might object:

> But surely we can get by on the basis of personal conceptions alone, without invoking the notion of "a thing itself." My conception of a thing is something I can convey to you, given enough time. Cannot communication proceed by correlating and matching personal conceptions, without appeal to the intermediation of "the thing itself."

But think here of the concrete practicalities. What is "enough time"? When is the match "sufficient" to underwrite out right identification? The cash value of our commitment to the thing itself is that it enables us to make this identification straightaway by imputation, by fiat on the basis of modest indicators, rather than on the basis of an appeal to the inductive weight of a body of evidence that is always bound to be problematic. Communication is something *we set out* to do, not something we ultimately discern, with the wisdom of eventual hindsight, to have accomplished retrospectively.

The objectifying imputation at issue here lies at the very heart of our cognitive stance that we live and operate in a world of real and objective things. This commitment to the idea of a shared real world is crucial for communication. Its status is a priori: its existence is not something we learn of through experience. As Kant clearly saw, objective experience is possible only if the existence of such a real, objective world is *presupposed* at the onset rather than seen as a matter of *ex post facto* discovery about the nature of things.

The information that we may have about a thing—be it real or presumptive information—is always just that, namely, information that *we* lay claim to. We cannot but recognize that it is person-relative and in general person-differentiated. Our attempts at communication and inquiry are thus undergirded by an information-transcending stance—the stance that we communally inhabit a shared world of objectively existing things—a world of "real things" among which we live and into which we inquire but about that we do and must presume ourselves to have only imperfect information at any and every particular stage of the cognitive venture. This is not something we learn. The "facts of experience" can never reveal it to us. It is something we postulate or presuppose to be able to put experience to cognitive use. Its epistemic status is not that of an empirical discovery, but that of a presupposition that is a product of a transcendental argument for the very possibility of communication or inquiry as we standardly conceive of them.

And so, what is at issue here is not a matter of *discovery*, but one of *imputation*. The element of community, of identity of focus is not a matter of ex post facto learning from experience, but of an a priori predetermination inherent in our approach to language-use. We do not *infer* things as being real and objective from our phenomenal data, but establish our perception as authentic perception *of* genuine objects through the fact that these objects are given—or rather, *taken*—as real and objectively existing things from the first.[20] Objectivity is not deduced but imputed. We do, no doubt, *purport* our conceptions to be objectively correct, but whether this is indeed so is something we cannot tell with assurance until "all the returns are in"—that is, never. This fact renders it critically important *that* (and understandable *why*) conceptions are communicatively irrelevant. Our discourse *reflects* our conceptions and perhaps *conveys* them, but it is not in general substantively *about* them but rather about the things in which they actually or supposedly bear.

We thus reach an important conjuncture of ideas. The ontological independence of things—their objectivity and autonomy of the machinations of mind—is a crucial aspect of realism. And the fact that it lies at the very core of our conception of a real thing that such items project beyond the cognitive reach of mind betokens a conceptual-scheme fundamentally committed to objectivity. The only plausible sort of ontology is one that contemplates a realm of reality that outruns the range of knowledge (and indeed even of language), adopting the stance that character goes beyond the limits of characterization. It is a salient aspect of the mind-independent status of the objectively real that the features of something real always transcend what we know about it. Indeed, yet further or different facts concerning a real thing can always come to light, and all that we *do* say about it does not exhaust all that *can and should* be said about it. In this light, objectivity is crucial to realism and the cognitive inexhaustibility of things is a certain token of their objectivity.

As these deliberations indicates, authentic realism can only exist in a state of tension. The only reality worth having is one that is in some degree knowable. But it is the very limitation of our knowledge—our recognition that there is more to reality than what we do and can know or ever conjecture about it—that speaks for the mind-independence of the real. It is important to stress against the skeptic that the human mind is sufficiently well attuned reality that some knowledge of it is possible. But it is no less important to join with realists in stressing the independent character of reality acknowledging that reality has a depth and complexity of makeup that outruns the reach of the cognitive efforts of mind.

The Pragmatic Foundation of Realism as a Basis for Communication and Discourse

But what is it—brute necessity aside—that validates those communicative presuppositions and postulations of ours? The prime factor at work here is simply our commitment to utility. Given that the existence of an objective domain of impersonally real existence is not a *product* of but a *precondition* for empirical inquiry, its acceptance has to be validated in the manner appropriate for postulates and prejudgments of any sort—namely in terms of its ultimate utility. Bearing this pragmatic perspective in mind, let us take a closer look at this issue of utility and ask: What can this postulation of a mind-independent reality actually do for us?

The answer is straightforward. The assumption of a mind-independent reality is essential to the whole of our standard conceptual scheme relating to inquiry and communications. Without it, both the actual conduct and the rational legitimation of our communicative and investigative (evidential) practice would be destroyed. Nothing that we do in this cognitive domain would make sense if we did not subscribe to the conception of a mind-independent reality.

To begin with, we indispensably require the notion of reality to operate the classical concept of truth as "agreement with reality" (*adaequatio ad rem*). Once we abandon the concept of reality, the idea that in accepting a factual claim as true we become committed to how matters actually stand—"how it really is"—

would also go by the board. The very semantics of our discourse constrain its commitment to realism; we have no alternative but to regard as real those states of affairs claimed by the contentions we are prepared to accept. Once we put a contention forward by way of serious assertion, we must view as real the states of affairs it purports, and must see its claims as facts. We need the notion of reality to operate the conception of truth. A factual statement on the order of "There are pi mesons" is true if and only if the world is such that pi mesons exist within it. By virtue of their very nature as truths, true statements must state facts: they state what really is so, which is exactly what it is to "characterize reality." The conception of *truth* and of *reality* come together in the notion of *adaequatio ad rem*—the venerable principle that to speak truly is to say how matters stand in reality, in that things actually are as we have said them to be.

In the second place, the nihilistic denial that there is such a thing as reality would destroy once and for all the crucial Parmenidean divide between appearance and reality. And this would exact a fearful price from us: we would be reduced to talking only of what we (I, you, many of us) *think* to be so. The crucial contrast notion of the *real* truth would no longer be available: we would only be able to contrast our *putative* truths with those of others, but could no longer operate the classical distinction between the putative and the actual, between what people merely *think* to be so and what actually *is* so. We could not take the stance that, as the Aristotelian commentator Themistius put it, "that which exists does not conform to various opinions, but rather the correct opinions conform to that which exists."[21]

The third point is the issue of cognitive coordination. Communication and inquiry, as we actually carry them on, are predicated on the fundamental idea of a real world of objective things, existing and functioning "in themselves," without specific dependence on us and so equally accessible to others. Intersubjectively valid communication can only be based on common access to an objective order of things. The whole communicative project is predicated on a commitment to the idea that there is a realm of shared objects about which we as a community share questions and beliefs, and about which we ourselves as individuals presumably have only imperfect information that can be criticized and augmented by the efforts of others.

This points to a fourth important consideration. Only through reference to the real world as a *common object* and shared focus of our diverse and imperfect epistemic strivings are we able to effect communicative contact with one another. Inquiry and communication alike are geared to the conception of an objective world: a communally shared realm of things that exist strictly "on their own" comprising an enduring and independent realm within which and, more important, with reference to which inquiry proceeds. We could not proceed on the basis of the notion that inquiry estimates the character of the real if we were not prepared to presume or postulate a reality for these estimates to be estimates of. It would clearly be pointless to devise our characterizations of reality if we did not stand committed to the proposition that there is a reality to be characterized.

The fifth item is a recourse to mind-independent reality that makes possible a "realistic" view of our knowledge as potentially flawed. A rejection of this commitment to reality *an sich* (or to the actual truth about it) exacts an unacceptable price. For in abandoning this commitment we also lost those regulative contrasts that canalize and condition our view of the nature of inquiry (and indeed shape our conception of this process as it stands within the framework of our conceptual scheme). We could no longer assert: "What we have there is good enough as far as it goes, but it is presumably not 'the whole real truth' of the matter." The very idea of inquiry as we conceive it would have to be abandoned if the contract conceptions of "actual reality" and "the real truth" were no longer available. Without the conception of reality we could not think of our knowledge in the fallibilistic mode we actually use—as having provisional, tentative, improvable features that constitute a crucial part of the conceptual scheme within whose orbit we operate our concept of inquiry.

Reality (on the traditional metaphysicians' construction of the concept) is the condition of things answering to "the real truth"; it is the realm of what really is as it really is. The pivotal contrast is between "mere appearance" and "reality as such," between "our picture of reality" and "reality itself," between what actually is and what we merely think (believe, suppose, etc.) to be. And our allegiance to the conception of reality, and to this contrast that pivots on it, root in the fallibilistic recognition that at the level of the detailed specifics of scientific theory, anything we presently hold to be the case may well turn out otherwise—indeed, certainly will do so if past experience gives any auguries for the future.

Our commitment to the mind-independent reality of "the real world" stand together with our acknowledgment that, in principle, any or all of our *present* scientific ideas as to how things work in the world, at *any* present, may well prove to be untenable. Our conviction in a reality that lies beyond our imperfect understanding of it (in all the various senses of "lying beyond") roots in our sense of the imperfections of our scientific world-picture—its tentativity and potential fallibility. In abandoning our commitment to a mind-independent reality, we would lose the impetus of inquiry.

Sixth and finally, we need the conception of reality in order to operate the causal model of inquiry about the real world. Our standard picture of man's place in the scheme of things is predicated on the fundamental idea that there is a real world (however imperfectly our inquiry may characterize it) whose causal operations produce *inter alia* causal impacts on us, providing the basis of our world-picture. Reality is viewed as the causal source and basis of the appearances, the originator and determiner of the phenomena of our cognitively relevant experience. "The real world" is seen as causally operative both in serving as the external molder of thought and as constituting the ultimate arbiter of the adequacy of our theorizing. (Think here again of C. S. Peirce's "Harvard experiment.")

In summary, then, we need that postulate of an objective order of mind-independent reality for at least six important reasons.

1. To preserve the distinction between true and false with respect to factual matters and to operate the idea of truth as agreement with reality.
2. To preserve the distinction between appearance and reality, between our *picture* of reality and reality itself.
3. To serve as a basis for intersubjective communication.
4. To furnish the basis for a shared project of communal inquiry.
5. To provide for the fallibilistic view of human knowledge.
6. To sustain the causal mode of learning and inquiry and to serve as basis for the objectivity of experience.

The conception of a mind-independent reality accordingly plays a central and indispensable role in our thinking about communication and cognition. In both areas alike we seek to offer answers to our questions about how matters stand in this "objective realm" and the contrast between "the real" and its "merely phenomenal" appearances is crucial here. Moreover, this is also seen as the target and *telos* of the truth-estimation process at issue in inquiry, providing for a common focus in communication and communal inquiry. The "real world" thus constitutes the "object" of our cognitive endeavors in both senses of this term— the *objective* at which they are directed and the *purpose* for which they are exerted. And reality is seen as pivotal here, affording the existential matrix in which we move and have our being, and whose impact on us is the prime mover for our cognitive efforts. All of these facets of the concept of reality are integrated and unified in the classical doctrine of truth as it corresponds to fact (*adaequatio ad rem*), a doctrine that only makes sense in the setting of a commitment to mind-independent reality.

Accordingly, the justification for this fundamental presupposition of objectivity is not *evidential* at all; postulates are not based on evidence. Rather, it is *functional*. We need this postulate to operate our conceptual scheme. The justification of this postulate accordingly lies in its utility. We could not form our existing conceptions of truth, fact, inquiry, and communication without presupposing the independent reality of an external world. We simply could not think of experience and inquiry as we do. (What we have here is a "transcendental argument" of sorts from the character of our conceptual scheme to the acceptability of its inherent presuppositions.) The primary validation of that crucial objectivity postulate lies in its basic functional utility in relation to our cognitive aims.

It is worthwhile to explore the implications of this foundationalism more fully.

The commitment to an objective reality that lies behind the data that people secure is indispensably demanded by any step into that domain of the publicly accessible objects that is essential to communal inquiry and interpersonal communication about a shared world. We do—and must—adopt the standard policy that prevails with respect to all communicative discourse of letting the language we use, rather than whatever specific informative aims we may actually have in mind on particular occasions, be the decisive factor with regard to the things at issue in our discourse. For if we were to set up our own conception of things as somehow definitive and decisive, we would at once erect a barrier not only to further inquiry but also—no less important—to the prospect of successful communication with one another. Communication requires not only common *concepts* but common *topics,* interpersonally shared items of discussion, a common world constituted by the self-subsistently real objects basic to shared experience. The factor of objectivity reflects our basic commitment to a communally available world as the common property of communicators. Such a commitment involves more than merely de facto intersubjective agreement. Such agreement is a matter of a posteriori discovery, while our view of the nature of things puts "the real world" on a necessary and a priori basis. This stance roots in the fundamental convention of a shared social instance on communication. What links my discourse with that of my interlocutor is our common subscription to the governing presumption (a defensible presumption, to be sure) that we are both talking about the shared thing, our own possible misconceptions of it notwithstanding. This means that no matter how extensively we may change our minds about the *nature* of a thing or type of thing, we are still dealing with exactly the same thing or sort of thing. It assures reidentification across discordant theories and belief systems.

Our concept of a *real thing* is such that it provides a fixed point, a stable center around which communication revolves, an invariant focus of potentially diverse conceptions. What is to be determinative, decisive, definitive, and so on, of the things at issue in my discourse is not my conception, or yours, or indeed anyone's conception at all. The conventionalized intention discussed means that a coordination of conceptions is not decisive for the possibility of communication. Your statements about a thing may well convey something to me even if my conception of it is altogether different from yours. To communicate we need not take ourselves to share views of the world, but only take the stance that we share the world being discussed. This commitment to an objective reality that underlies the data at hand is indispensably demanded by any step into the domain of the publicly accessible objects essential to communal inquiry and interpersonal communication about a shared world. We could not establish communicative contact about a common objective item of discussion if our discourse were geared to the substance of our own idiosyncratic ideas and conceptions.

And so an important lesson emerges. The rationale of a commitment to ontological objectivity is in the final analysis functionally or pragmatically driven. Without a presuppositional commitment to objectivity, with its acceptance of a real world independent of ourselves that we share in common, interpersonal communication would become impracticable. Objectivity is an integral part of the sine qua non presuppositional basis of the project of meaningful communication. To reemphasize, if our own subjective conceptions of things were to be determinative, informative communication about a world of shared objects and processes would be rendered unachievable.

The Idealistic Aspect of Metaphysical Realism

Realism, then, is a position to which we are constrained not by the push of evidence but by the pull of purpose. Initially, at any rate, a commitment to realism is an *input* into our investigation of nature rather than an *output* thereof. At bottom, it does not represent a discovered fact, but a methodological presupposition of our praxis of inquiry; its status is not constitutive (fact-descriptive) but regulative (praxis-facilitating). Realism is not a factual discovery, but a practical postulate justified by its utility or serviceability in the context of our aims and purposes, seeing that if we did not *take* our experience to serve as an indication of facts about an objective order we would not be able to validate any objective claims whatsoever. (To be sure, what we can—and do—ultimately discover is that by taking this realistic stance we are able to develop a praxis of inquiry and communication that proves effective in the conduct of our affairs.)

The ontological thesis that there is a mind-independent physical reality to which our inquiries address themselves more or less adequately—and always imperfectly—is the key contention of realism. But on the telling of the presenting analysis, this basic thesis has the epistemic status of a presuppositional postulate that is initially validated by its pragmatic utility and ultimately retrovalidated by the satisfactory results of its implementation (in both practical and theoretical respects). Our commitment to realism is, on this account, initially not a product of our *inquiries* about the world, but rather reflects a facet of how we *conceive* the world. The sort of realism contemplated here is accordingly one that pivots on the fact that we *think* of being real in a certain sort of way, and that in fact the very conception of the real is something we employ because doing so merits our ends and purposes.

Now insofar as realism ultimately rests on a pragmatic basis, it is not based on considerations of independent substantiating evidence about how things actually stand in the world, but rather on considering, as a matter of practical reasoning, how we do (and must) think about the world within the context of the projects to which we stand committed. In this way, the commitment to a mind-independent reality plays an essentially utilitarian role as providing a functional requisite for our intellectual resources (specifically for our conceptual scheme in relation to communication and inquiry). Realism thus harks back to the salient contention of classical idealism that values and purposes play a pivotal role in our understanding of the nature of things. And we return also to the characteristic theme of idealism—the active role of the knower not only in the constituting but also in the constitution of what is known.

To be sure, this sort of idealism is not substantive but methodological. It is not a rejection of real objects that exist independently of mind and as such are causally responsible for our objective experience; quite the reverse, it is designed to facilitate their acceptance. But it insists that the justificatory *rationale* for this acceptance lies in a framework of mind-supplied purpose. For our commitment to a mind-independent reality is seen to arise not *from* experience but *for* it—for the sake of putting us into a position to exploit our experience as a basis for validating inquiry and communication with respect to the objectively real.

"Reality as such" is no doubt independent of our beliefs and desires, but what can alone concern us is reality as we view it. And the only view of reality that is available to us is one that is devised by us under the aegis of principles of acceptability that we subscribe to because doing so serves our purposes.

A position of this sort is in business as a realism all right. But seeing that it pivots on the character of our concepts and their *modus operandi,* it transpires that the business premises it occupies are actually mortgaged to idealism. The fact that objectivity is the fruit of communicative purpose allows idealism to infiltrate into the realist's domain.

And the idealism at issue cuts deeper yet. No doubt, we are firmly and irrevocably committed to the idea that there is a physical realm out there that all scientific inquirers inhabit and examine alike. We hold to a single, uniform physical reality, insisting that all investigations exist within and investigate *it:* this one single shared realism, this one single manifold of physical objects and laws. But this very idea of a single, uniform, domain of physical object and laws represents just exactly that—*an idea of ours.* And the idea is itself a matter of how we find it convenient and efficient to think about things: it is no more—though also no less—than the projection of a theory devised to sort the needs and conveniences of our intellectual situation.

This approach endorses an object-level realism that rests on a presuppositional idealism at the justificatory infralevel. We arrive, paradoxical as it may seem, at a realism that is founded, initially at least, on a fundamentally idealistic basis—a realism whose ultimate *justificatory basis* is ideal.

SCIENCE AND REALITY

How is our scientific knowledge related to our everyday knowledge of things? What is the relationship between nature as science sees it and as it figures in the experience and discourse of everyday life?

Much of the thought of recent epistemologists takes its starting point in the famous two-tables discussion of the British physicist-astronomer Sir Arthur Eddington, who contrasts the solid table of everyday experience with the physicists' table of a manifold of electromagnetic oscillations in mainly empty space. He then maintains that the later is the real table as it exists in nature, and that (1) is merely an appearance, a delusion—a mirage, as it were—existing in people's minds. Our ordinary life view of the world is a matter of mental (rather than optical) illusion.

In principle one could, the same, also take the opposite line. It is the table of ordinary experience that is alone real (it would then be said). What that physicist has done is not to give us a different view of that table, but simply to abolish it. With all the physicists talk of alone and their spatial distribution and interaction there is nothing left of that table. It is abolished; it no longer figures on the agenda of concern; we have completely changed the subject. Science does not illustrate the world of everyday experience, it puts it aside in its concern with other matters—hypothetical and theoretical objects and entities that might explain experience but do not figure on it. Authentic reality is human reality—reality as we experience it. The objects of physical science are something altogether different.

The sensible course with respect to this seeming conflict between the world-picture of common life and the world-picture of natural science is to see the issue not in terms of opposition but in terms of collaboration.

To be sure the discourses of science and everyday life are different and employ different languages. The one is concerned with our observational and interpersonally communicative experience, the other is concerned with the taxonomy and modus operandi of the underlying causal process that makes the world go round. They are functionally different enterprises with purposes and concepts of their own. To be sure, they have interrelationships. Natural science can explain the underlying causality of common life presentation and thought. Common life phenomenology is needed not only to use but even to make the observational and experimental mechanisms by which alone science can do its work.

Reality is not uniform but functionally diversified. One and the same landscape can figure in the concerns of the agronomist, the landscape architect, and the military engineer. All of them look to different things and have different orientations of factual concern. They raise different issues and pose different questions and the stories they tell accordingly do not disagree or conflict they simply "talk past each other." The processes of "food preparation" by baking and cooking and so forth can be analyzed both in the language of chemistry and in that of gastronomy. But they deal with altogether different matters.

And so, the phenomenology of human experience and the causal commerce of natural processes involve different and noncompetitive issues. In speaking of sunrises and insects there is nothing we say or intent that is at variance with a Copernicanism that positions the sun at the center of the solar system. In both cases we are dealing with the selfsame world—the selfsame "nature." But we are dealing with it via the concept manifold of entirely different but uncompetitive enterprises. There is no conflict because we are dealing with different projects and different issues. These enterprises are functionally and purposively different. They involve different issues and raise different questions, and different questions have different—and generally nonconflicting—answers. The world descriptions of science and common life are accordingly not competitive and conflicting but complementary.

A crucial question yet remains. How close a relationship can we reasonably claim to exist between the answers we give to our factual questions at the level of scientific generality and precision and the reality they purport to depict?

Scientific realism is the doctrine that *science describes the real world*: that the world actually is as science takes it to be, and that its furnishings are as science envisages them to be.[22] If we want to know about the existence and the nature of heavy water or quarks, of man-eating mollusks or a luminiferous ether, we are referred to the natural sciences for the answers. On this realistic construction of scientific theorizing, the theoretical terms of natural science refer to real physical entities and describe their attributes and comportments. For example, the "electron spin" of atomic physics refers to a behavioral characteristic of a real, albeit unobservable, object—and electron. According to this currently fashionable theory, the declarations of science are—or will eventually become—factually true generalizations about the actual behavior of objects that exist in the world. Is this "convergent realism" a tenable position?

It is quite clear that it is not. There is clearly insufficient warrant for and little plausibility to the claim that the world indeed is as our science claims it to be—that we have got matters altogether right, so that *our* science is *correct* science and offers the definitive "last word" on the issues. We really cannot reasonably suppose that science as it now stands affords the real truth as regards its creatures-of-theory.

One of the clearest lessons of the history of science is that where scientific knowledge is concerned, further discovery does not just *supplement* but generally *emends* our prior information. Accordingly, we have little alternative but to take the humbling view that the incompleteness of our purported knowledge about the world entails its potential incorrectness as well. It is now a matter not simply of *gaps* in the structure of our knowledge, or errors of omission. There is no realistic alternative but to suppose that we face a situation of real *flaws* as well, of errors of commission. This aspect of the matter endows incompleteness with an import far graver than meets the eye on first view.

Realism equates the paraphernalia of natural science with the domain of what actually exists. But this equation would work only if science, as it stands, has actually "got it right." And this is something we are surely not inclined—and certainly no *entitled*—to claim. We must recognize that the deliverances of science are bound to a methodology of theoretical triangulations from the data that binds them inseparably to the "state of the art" of technological sophistication in data acquisition and handling.

The supposition that the theoretical commitments of our science actually describes the world is viable only if made *provisionally*, in the spirit of "doing the best we can now do, in the current state-of-the-art" and giving our best estimate of the matter. The step of reification is always to be taken qualifiedly, subject to a mental reservation of presumptive revisability. We do and must recognize that we cannot blithely equate *our* theories with *the* truth. We do and must realize that the declarations of science are inherently fallible and that we can only "accept" them with a certain tentativeness, subject to a clear realization that they may need to be corrected or even abandoned.

These considerations must inevitably constrain and condition our attitude toward the natural mechanisms envisaged in the science of the day. We certainly do not—or should not—want to reify (hypostasize) the "theoretical entities" of current science, to say flatly and unqualifiedly that the contrivances of *our* present-day science correctly depict the furniture of the real world. We do not—or at any rate, given the realities of the case, should not—want to adopt categorically the ontological implications of scientific theorizing in just exactly the state-of-the-art configurations presently in hand. Scientific fallibilism precludes the claim that what we purport to be scientific knowledge is in fact *real* knowledge, and accordingly blocks the path to a scientific realism that maintains that the furnishings of the real world are exactly as our science states them to be. Scientific theorizing is always inconclusive.

Convergent scientific realism of the Peircean type, which pivots on the assumption of an ultimately complete and correct scientific theory (let alone those stronger versions of such realism that hinge on our ability to arrive at recognizably true scientific theories), is in deep difficulty. For we have little choice but to deem science's grasps or "the real truth of things" as both tentative and imperfect.

According to one expositor, the scientific realist "maintains that if a theory has scientific merit, then we are thereby justified in concluding that ... the theoretical entities characterized by the theory really do exist."[23] But this sort of position encounters insuperable difficulties. Phlogiston, caloric, and the luminiferous ether all had scientific merit in their day, but this did not establish their existence. Why, then, should things be all that different with us? Why should *our* "scientific merit" now suddenly assure actual existence? What matters for real existence is clearly (and only) the issue of truth itself, and not the issue of what is *thought* to be true at some particular stage of scientific history. And here problems arise. For its changeability is a fact *about* science that is as inductively well-established as any theory of science itself. Science is not a static system but a dynamic process.

We must accordingly maintain a clear distinction between *our conception of reality* and *reality as it really is*. Given the equation,

Our (conception of) reality = the condition of things as seen from the standpoint of "our *putative* truth" (= the truth as we see it from the vantage point of the science of the day)

we realize full well that there is little justification for holding that our present-day science indeed describes reality and depicts the world as it really is. In our heart of hearts, then, our attitude toward our science is one of *guarded* affirmation. We realize that there is a decisive difference between what science *accomplishes* and what it *endeavors* to accomplish.

The world *that we describe* is one thing, the world, *as we describe it* is another, and they would coincide only if our descriptions were totally correct—something that we are certainly not in a position to claim. The world-as-known is a thing of our contrivance, and artifact we devise on our own terms. Even if the "data" uniquely determined a corresponding picture of reality, and did not underdetermine the theoretical constructions we base on them (as they always do), the fact remains that altered circumstances lead to altered bodies of "data." Our recognition of the fact that the world-picture of science is ever-changing blocks our taking the view that it is ever *correct*.

Accordingly, we cannot say that the world *is* such that the paraphernalia of our science actually exist as such—that is, exactly as our science characterizes them. Given the necessity of recognizing the claims of our science to be tentative and provisional, one cannot justifiably take the stance the it depicts reality. At best, one can say that it affords an *estimate* of it, an estimate that will presumably stand in need of eventual revision and whose creatures-of-theory may in the final analysis not be real at all. This feature of science must crucially constrain our attitude toward its deliverances. Depiction is in this regard a matter of intent rather than one of accomplishment. Correctness in the characterization of nature is achieved not by *our* science but only by *perfected* or *ideal* science—only by that (ineradicably hypothetical!) state of science in which the cognitive goals of the scientific enterprise are fully and definitively realized. There is no plausible alternative to the view that reality is depicted by *ideal* (or perfected or "complete") science, and not by the real science of the day. But, of course, it is this latter science that is the only one we have actually got—now or ever.

A viable scientific realism must therefore turn not on what *our* science takes the world to be like but on what *ideal or perfected* science takes the world to be like. The thesis that "science describes the real world" must be looked on as a matter of intent rather than as an accomplished fact, of aspiration rather than achievement, of the ideal rather than the real state of things. Scientific realism is a viable position only with respect to that idealized science which, as we full well realize, we do not now have—regardless of the "now" at issue. We cannot justifiably be scientific realists. Or rather, ironically, we can be so only in an idealistic manner—namely, with respect to an "ideal science" that we can never actually claim to possess.

The posture of scientific realism—at any rate, or a duly qualified sort—is nevertheless built into the very goal-structure of science. The characteristic task of science, the definitive mission of the enterprise, is to respond to our basic interest in getting the best answers we can to our questions about the world. On the traditional view of the matter, its question-resolving concern is the *raison d'être* of the project—to celebrate any final victories. It is thus useful to draw a clear distinction between a *realism of intent* and a *realism of achievement*. We are certainly not in a position to claim that science as we have it achieves a characterization of reality. Still, science remains unabashedly realistic in *intent* or *aspiration*. Its *aim* is unquestionably to answer our questions about the world *correctly* and to describe the world "as it actually is." The orientation of science is factual and objective: it is concerned with establishing the *true* facts about the *real* world. The theories of physics purport to describe the actual operation of real entities; those Nobel prizes awarded for discovering the electron, the neutron, the pi meson, and antiproton, the quark, and so on, we intended to recognize an enlargement of our understanding of nature, not to reward the contriving of plausible fictions or the devising of clever ways of relating observations.

The language of science is descriptively committal. At the semantical level of the content of its assertions, science makes firm claims as to how things stand in the world. A realism of intent or aspiration is built into science because of the genesis of its questions. The factually descriptive status of science is ultimately grounded in just this erotetic continuity of its issues with those of "prescientific" everyday life. We begin at the prescientific level of the paradigmatic realities of our prosaic everyday-life experience—the things, occurrences, and processes of our everyday world. The very reason for being of our scientific paraphernalia is to resolve our questions about this real world of our everyday-life experience. Given that the teleology of the scientific enterprise roots in the "real world" that provides the stage of our being and action, we are committed *within its framework*, to take the realistic view of its mechanisms. Natural science does not address itself to some world-abstracted realm of its own. Its concern is with this familiar "real world" of ours in which we live and breathe and have our being—however differently science may characterize it. While science may fall short in performance, nevertheless in aspiration and endeavor it is unequivocally committed to the project of modeling "the real world," for in this way alone could it realize its constituting mandate of answering our questions as to how things work in the world.

Scientific realism skates along a thin border between patent falsity and triviality. Viewed as the doctrine that science *indeed describes* reality, it is utterly untenable; but viewed as the doctrine that science *seeks to describe* reality, it is virtually a truism. For there is no way of sidestepping the conditional thesis:

> If a scientific theory regarding heavy water or electrons of quarks or whatever is correct—if it were indeed to be true—*then* its subject

materials would exist in the manner the theory envisages and would have the properties the theory attributes to them: the theory, that is, would afford descriptively correct information about the world.

But this conditional relationship reflects what is, in the final analysis, less a profound fact about the nature of science than a near truism about the nature of truth as *adequation ad rem*. The fact remains that "our reality"—reality as we conceive it to be—goes no further than to represent our best estimate of what reality is like.

When we look to *what science declares*, to the aggregate content and substance of its declarations, we see that these declarations are realistic in intent, that they *purport* to describe the world as it really is. But when we look to *how science makes its declarations* and note the tentativity and provisionally with which they are offered and accepted, we recognize that this realism is on an abridged and qualified sort—that we are not prepared to claim that this is how matters actually stand in the real world. At the level of generality and precision at issue in the themes of natural science, we are not now—or ever—entitled to lay claim to the scientific truth as such but only to the scientific truth as we and our contemporaries see it. Realism prevails with respect to the *language* of science (i.e., the asserted content of its declarations); but it should be abandoned with respect to the *status* of science (i.e., the ultimate tenability or correctness of these assertions). What science says is descriptively committal in making claims regarding "the real world," but the tone of voice in which it proffers these claims is (or should be) provisional and tentative.

The resultant position is one not of skepticism but of realism—in two senses: (1) it is realistic about our capabilities of recognizing that here, as elsewhere, we are dealing with the efforts of an imperfect creature to do the best it can in the circumstances; and (2) it recognizes the mind-transcendent reality of a "real world" that our own best efforts in the cognitive sphere can only manage to domesticate rather imperfectly. We do, and always must, recognize that no matter how far we manage to extend the frontiers of natural science, there is more to be done. Within a setting of vast complexity, reality outruns our cognitive reach; there is more to this complex world of ours than lies—now or ever—within our ken.[24]

Notes

NOTES TO INTRODUCTION

1. William James, "The Sentiment of Rationality," in *The Will to Believe and Other Essays in Popular Philosophy* (New York and London: Longmans Green & Co., 1897), pp. 78–79.

NOTES TO CHAPTER 1

1. This view harks back to the Theaetetus of Plato's dialogue of that name, who held that knowledge is a matter of true belief equipped with a suitable rationale (*logos*), Plato Theaetetus 201C–210D. For modern discussions see Edmund Gettier, "Is Justified True Belief Knowledge?" *Analysis*, vol. 24 (1963), pp. 121–23. The problem at issue has generated a very substantial literature. The following publications give a good indication of its scope and nature: William P. Alston, *Epistemic Justification* (Ithaca, NY: Cornell University Press, 1989); Roderick Chisholm, *Theory of Knowledge* (Englewood Cliffs, NJ: Prentice-Hall, 1966; 2nd ed. 1977; 3rd ed. 1989); Edward Craig, *Knowledge and the State of Nature* (Oxford: Clarendon Press, 1990); Alvin Goldman, "A Causal Theory of Knowing," *Journal of Philosophy* 64 (1967), pp. 357–72; Alvin Goldman, *Epistemology and Cognition* (Cambridge, MA: Harvard University Press, 1986); Keith Lehrer, *Theory of Knowledge* (Boulder, CO: Westview Press, 1990); Paul K. Moser, *Knowledge and Evidence* (Cambridge: Cambridge University Press, 1989); Alvin Plantinga, *Warrant and Proper Function* (Oxford: Oxford University Press, 1993); John Pollock, *Contemporary Theories of Knowledge* (Totowa, NJ: Rowman and Littlefield, 1986); M. D. Roth and L. Galis (eds), *Knowing: Essays in the Analysis of Knowledge* (New York: Random House, 1970); Robert K. Shope, *The Analysis of Knowing* (Princeton: Princeton University Press, 1983); Ernest Sosa, *Knowledge in Perspective: Selected Essays in Epistemology* (Cambridge: Cambridge University Press, 1991), Part 1; and Linda Trinkhaus Zagzebski, *Virtues of the Mind: An Inquiry into the Nature of Virtue and the Ethical Foundations of Knowledge* (Cambridge: Cambridge University Press, 1996).

2. This is why it makes no sense to say: "You know that p but I don't."

3. The practical aspect of presumption and its justification will be examined more fully in chapter 5.

4. Acceptance is not quite the same as belief. For belief is episodic and psychological, whereas acceptance is nontransitory and epistemological; moreover acceptance is a matter of choice while belief is involuntary.

5. Indeed we cannot even say the more guarded "x thinks he knows that p, but does not accept it." On the other hand, knowledge-avoiding locutions like "x says that p, but he really knows better" are unproblematic.

6. We are not concerned here with "the language as she is spoke," but with the *careful* usage of *conscientious* speakers—with what used to be called "correct" usage in the old normative days of grammatical theory.

7. For example, D. M. Armstrong, *Belief, Truth and Knowledge* (Cambridge, 1973), p. 189.

8. It is worth distinguishing between objective grounding ("there are good grounds for accepting that p"), which reflects an altogether impersonal aspect of the epistemic situation, and subjective grounding ("x has good grounds for accepting that p"). For the person whose knowledge is incomplete may well have good grounds for accepting something whose grounding in itself (taken as a whole) is objectively insufficient. Thus someone can have good grounds for p when (unbeknownst to him) decisive grounds for not-p in fact exist. However, with the conclusive grounding at issue in knowledge the distinction between the objective and the subjective disappears. One can only have *conclusive* grounds when whatever grounds that one has are in themselves conclusive.

9. G. E. Moore, *Some Main Problems of Philosophy* (London: Allen and Unwyn, 1953).

10. Actually, it is immaterial for these principles whether or not the same knower is at issue in the two cases. Not only must *a person's* knowledge be consistent, but the whole body of what is known (by someone or other) as well.

11. This tactic is, of course, of no use to a skeptic.

Notes to Chapter 2

1. On this issue see the author's *Scientific Progress* (Oxford: Basil Blackwell, 1978). As the discussion there makes clear, what is at issue is not *just* an induction from the history of science; there are also reasons of fundamental general principle why the scientific investigations of the future are pretty much foreordained to destabilize the findings that result from the investigations of the past.

2. On these and related issues see Roy A. Sorensen, *Blindspots* (Oxford, 1988).

3. For substantiation see R. G. Winkler and S. Makridakis, *Journal of the Royal Statistical Society*, 146 (1983): 150–57.

4. On the Preface Paradox see A. N. Prior, "On a Family of Paradoxes," *Notre Dame Journal of Formal Logic*, 2 (1961): 26–32; D. C. Makinson, "The Paradox of the Preface," *Analysis*, 25 (1965): 205–7.

5. For relevant considerations see also the author's *The Limits of Science* (Berkeley and Los Angeles, 1985).

6. "Pour juger des apparences que nous recevons des subjects, il nous faudroit un instrument judicatoire; por vérificer cet instrument, il nous fault de la démonstration; pour vérifier la démonstration, un instrument: nous voilà au rouet." *Essaies*, bk. II, ch. 12 ("An Apologie of Raymond Sebond"); p. 544 of the Modern Library edition of *The Essays of Montaigne* (New York, 1933). Francis Bacon, with the characteristic shrewdness of a lawyer, even managed to turn the *diallelus* into a dialectical weapon against his methodological opponents: "no judgement can be rightly formed either of my method, or of the discoveries to which it leads, by means of . . . the reasoning which is now in use, since one cannot postulate due jurisdiction for a tribunal which is itself on trial." (*Novum Organon*, bk I, sect. 33).

7. *Outlines of Pyrrhonism*, bk., II, sect. 20 (tr. R. G. Bury); compare bk. I, sects. 114–17. See also the article on Pyrro in Bayle's *Dictionary*.

8. See however the author's *Philosophical Standardism* (Pittsburgh, 1994) for the elaboration of a version of this approach.

9. For example in the work of Richard Sylvan. And see also Nicholas Rescher and Robert Brandom, *The Logic of Inconsistency* (Oxford, 1979).

10. Effectively subject to the rule that when there is a conflict the members of K^+ must give way to those of K+. On those matters see the author's *Plausible Reasoning* (Assen-Amsterdam, 1976).

11. See, for example, N. Rescher and R. Brandom *The Logic of Inconsistency* (Oxford, 1979).

12. On this issue see the author's *Scientific Realism* (Dordrecht, 1987).

13. For further details see the author's *The Limits of Science* (Berkeley and Los Angeles, 1984).

14. A. Conan Doyle, "The Sign of Four" (1890).

15. Fallibilism as a philosophical doctrine is principally due to C. S. Peirce, though the position has its precursors in the sceptical tradition from the days of the Academic sceptics of Greek antiquity onward. For modern expositions of cognate positions see K. R. Popper *Conjectures and Refutations* (London: Publisher, 1963; 2nd ed. 1969), esp. chapter 10 "Truth Rationality and the Growth of Scientific Knowledge." See also Hans Albert, *Traktat über kritische Vernunft* (Tübingen: Mohr, 1975), and Robert Almeder, *Blind Realism* (Savage, MD : Rowman & Littlefield, 1992).

16. Larry Laudan, as cited in Ilkka Niiniluoto, "Scientific Progress," *Synthese*, vol. 45 (1980), p. 446. Rejection of the idea that science gets at the truth of things goes back to Karl Popper.

17. See the author's *Induction* (Oxford: Blackwell, 1980).

18. Michael E. Levin, "On Theory Change and Meaning Change," *Philosophy of Science*, vol. 46 (1979), p. 418.

Notes to Chapter 3

1. His clear awareness of this made it possible for the Platonic Academy to endorse skepticism throughout the middle phase of its development in classical antiquity.

2. Thus one acute thinker has written:

[W]hat are my grounds for thinking that I, in my own particular case, shall die. I am as certain of it in my innermost mind, as I am that 1 now live; but what is the distinct evidence on which I allow myself to be certain? How would I tell it in a court of justice? How should I fare under a cross-examination upon the grounds of my certitude? Demonstration of course I cannot have of a future event.... [J. H. Newman, *A Grammar of Assent* [London and New York: Longmans, Green, 1913], chap. 8, pt. 2, sect. I.)

3. Bertrand Russell, *Problems of Philosophy* (Oxford: Oxford University Press, 1912), p. 35.

4. It must be realized that the "certainty" at issue in these discussions is not the subjective psychological stare of a *feeling* of certainly at issue in locutions like "I *feel* certain that p." Rather it is a matter of the objective epistemic circumstances, and the relevant locutions are of the impersonal character of "It is certain that p". This is crucial to the skeptic's case, "And once we have noticed this distinction, we are forced to allow what we are certain of very much less than we should have said otherwise." (H. A. Pritchard, *Knowledge and Perception* [Oxford, 1930], p. 97.)

5. As Keith Lehrer has put it:

[I]t should also be clear why it is that ordinary men commonly, though incorrectly, believe that they know for certain that some contingent statements are true. They believe that there is no chance whatever that they are wrong in thinking some contingent statements are true and thus feel sure they know for certain that those statements are true. One reason they feel sure is that they have not reflected upon the ubiquity of ... change in all human thought. Once these matters are brought into focus. We may reasonably conclude that no man knows for certain that any contingent statement is true. ("Scepticism and Conceptual Change," in R. M. Chisholm and R. J. Swartz [eds.], *Empir-*

ical Knowledge [Englewood Cliffs, NJ: Prentice Hall, 1973], pp. 47–58 [see p. 53].) In a similar vein, L. S. Carrier has recently argued, in effect, that since belief concerning material objects can in theory turn out to be mistaken, no one ever knows that he knows such a belief to be true. See his "Scepticism Made Certain," *The Journal of Philosophy*, vol. 71 (1974), pp. 140–50.

6. For skepticism and its critique see Robert Audi, *Belief, Justification, and Knowledge* (Belmont, CA: Wadsworth, 1988); A. I., Goldman, *Epistemology and Cognition* (Cambridge, MA: Harvard University Press, 1986); Peter Klein, *Certainty: A Refutation of Scepticism* (Minneapolis: University of Minnesota Press, 1081); Keith Lehrer, *Knowledge* (Oxford: Oxford University Press, 1974); Robert Nozick, *Philosophical Explanations* (Cambridge, MA: Harvard University Press, 1981); and Peter Unger, *Ignorance: A Case for Scepticism* (Oxford: Clarendon Press, 1975) as well as the author's *Scepticism* (Oxford: Basil Blackwell, 1980), which gives extensive references to the literature.

7. For examples, we need not—in the usual course of things—be in a position to rule out the imaginative skeptic's recourse to uncannily real dreams, deceitful demons, and so on.

8. The "problem of knowledge" that figures in much of the epistemological literature is thus a creation of those philosophers who endow our knowledge claims with a hyperbolic absoluteness never envisaged in or countenanced by out ordinary usage of knowledge-terminology. Having initially created difficulties by distorting our usage, philosophers are then at great pains to try to revalidate it. This whole project gives an aura of unrealism to much of epistemological discussion. For an interesting discussion of relevant issues see Oliver A. Johnson. *The Problem of Knowledge* (The Hague: Martinus Nijhoff, 1974).

9. On this issue see J. L. Austin on "Other Minds" (1946), reprinted in his *Philosophical Papers* (Oxford: Clarendon Press, 1961), pp. 44–84. To be sure, the operation of the distinction between realistic and hyperbolic possibilities of error will to some extent depend on "the state of 'knowledge' of the day." But this simply carries us back to the truism that what people reasonably accept as known is state-of-knowledge dependent and that the plausibly purported knowledge of one era may turn out to be the error of another.

10. Compare Norman Malcolm's right-minded complaint:

Lewis and Carnap . . . make the mistake of identifying . . . absolute certainly with "theoretical certainty." Lewis, for example, uses the expression "absolutely certain" and "theoretically certain" interchangeably. . . . In what circumstances, supposing that such circumstances cold exist, would it be "theoretically certain" that a given statement is true? The answer is clear from the context of their arguments. It would be "theoretically certain" that a given statement is true only if an *infinite* number of "tests" or "acts or verification" had been performed. It is, of course, a *contradiction* to say that an infinite number of "tests" or acts of any sort have been performed by anyone. It is not that it is merely impossible in practice for anyone to perform an infinite number of acts. It is impossible *in theory*. Therefore these philosophers *misuse* the expression "theoretically certain." What they call "theoretical certainly" cannot be attained even in theory. But this misusage of an expression is in itself of slight importance. What is very important is that they identify what they mean by "theoretically certain: with what is ordinarily meant by "absolutely certain" would be contradictory. ("The Verification Argument" in Max Black (ed.), *Philosophical Analysis* [Englewood Cliffs, NJ: Prentic Hall, 1950], p. 278–79).

11. Compare Harry G. Frankfurt, "Philosophical Certainty," *The Philosophical Review*, vol. 71 (1962), pp. 303–27. However, the philosopher (unlike the natural scientist) is simply not free to switch over to "technical terms." His task is to elucidate the concepts we actually work with and in replacing them by technical reconstructions he merely changes the subject.

12. On this issue one should consult J. L. Austin's analysis of "I know" as a *performative* utterance that extends a guarantee in that the man who makes this claim stakes his reputation and binds himself to others. See "Other Minds" (op. cit.)

13. Keith Lehrer, *Knowledge* (Oxford: Clarendon Press, 1973), p. 239.

14. As William James said: "[Someone] who says 'Better to go without belief forever than believe a lie!' merely shows his own preponderant private horror of becoming a dupe . . . But I can believe that worse things than being duped may happen to a many in this world" (*The Will to Believe* (New York/London and Bombay, Longmans, Green, and Co., 1897), pp. 18–19).

15. "What I mean is this: that my not having been on the moon is as sure a thing for me as any grounds I could give for it." (Ludwig Wittgenstein, *On Certainty* [Oxford: Blackwell, 1961], sect. 111; cf. also sect. 516.)

16. In its fully developed form the argument goes back to the Academic Skeptic Carneades (c. 213–c. 128 B.C.) who headed the Platonic Academy; but its essentials are present in Pyrrho (c. 360–c. 270 B.C.), founder of the Skeptical school.

17. The skeptic might also hold that factual *knowledge* is not needed for praxis, since—so he could argue—mere *probability* suffices as a "guide to life."

18. On Peirce's project on economy of research, see the author's *Peirce's Philosophy of Science* (Notre Dame and London, 1976), as well as C. F. Delaney, "Peirce on 'Simplicity' and the Conditions of the Possibility of Science," in L. J. Thro, ed., *History of Philosophy in the Making* (Washington, DC: University Press of America, 1982), pp. 177–94.

19. Fridtjof Nansen as quoted in Roland Huntford, *The Last Place on Earth* (New York: Atheneum, 1985), p. 200.

20. William James, "The Sentiment of Rationality," in *The Will to Believe and Other Essays in Popular Philosophy* (New York/London and Bombay, Longmans, Green, and Co., 1897), pp. 78–9.

21. C. S. Peirce, *Collected Papers*, vol. II (Cambridge, MA, Harvard University Press, 1931), sect. 2.112.

22. Peter Unger suggests a "reformation" of language in the interests of skepticism: We want linguistic institutions and practices where our (universal) ignorance will not enjoin silence" (*Ignorance* [Oxford, 1975], p. 271). But, he offers no concrete proposals along these lines, and this is very understandable. It is difficult to see how a language could be formed, taught, and above all *used* in which assertion played no role and declaration carried with it no claims to veracity. And even if (per impossible) one had such a "language," what would be the *point* of using it in communicative contexts?

23. Further discussion of some of this chapter's themes is presented in the author's *Scepticism* (Oxford: Basil Blackwell, 1980).

Notes to Chapter 4

1. As standardly used, the claim "I see a cat" is amphibious and moves in both regions, seeing that it *conjoins* the subjective "I take myself to be seeing a cat" with the objective "There is a cat there that I actually see."

2. See, for example, Ernest Sosa "Mythology of the Given," *History of Philosophy Quarterly*, vol. 14 (1997), pp. 275–96.

3. In theory we can certainly have $K(\exists f)(f$ is a causal factor of an epistemically appropriate sort) without having $(\exists f)K(f$ is a causal factor of an epistemically appropriate sort). It is just that in matters of concrete fact there is no epistemically effective way to get at the former without getting there via the latter.

4. Immanuel Kant, *Critique of Pure Reason*, A532–B560.

5. Donald Davidson, "A Coherence Theory of Truth and Knowledge," in Ernest La Pore (ed.), *Truth and Interpretation: Perspectives on the Philosophy of Donald Davidson* (Oxford: Blackwell, 1986), pp. 313–14.

6. Ibid., p. 310.

7. To rely on evolution in this way is not to turn epistemology into an empirical science of contingent fact. For the point here is not that "what is a good reason is" is contingent, but that "what a good reason" may be contingent and depend on the empirically determinable conditions. (That "something bad for you should be avoided" is conceptual and noncontingent, but that "smoking is bad for you and should therefore be avoided" is contingent.)

8. See, for example, P. M. Churchland, *Matter and Consciousness* (Cambridge, MA: MIT Press, 1984) and P. S. Churchland, *Neurophysiology: Towards a Unified Science for the Mind-Brain* (Cambridge, MA: MIT Press, 1986).

9. No recent writer has stressed more emphatically than F. A. Hayek the deep inherent rationality of historical processes in contrast to the shallower calculations of a calculating intelligence that restricts its view to the agenda of the recent day. See especially his book, *The Political Order of a Free People* (Chicago: University of Chicago Press, 1979), vol. 3 of "Law, Liberty, and Civilization."

10. The French school of sociology of knowledge has envisioned a competition among and ultimately rational selection of culturally diverse modes of procedure in accounting for the evolution of logical and scientific thought. Compare Louis Rougier, *Traité de la Connaisance* (Paris: Gauthier-Villars, 1955), esp. pp. 426–28.

11. C. S. Peirce, *Collected Papers*, vol. V (Cambridge, MA: Harvard University Press, 1934), sect. 5.366.

12. "*Im Anfang war die Tat*" (In the beginning was the act) as Goethe's *Faust* puts it.

13. The position maintained here is closely akin to John McDowell's argumentation against Hilary Putnam (as discussed in Putnam's *Pragmatism: An Open Question* [Oxford: Blackwell, 1995], pp. 67–78). McDowell holds that "Once we think of [obviously mental achievement such as] hearing and seeing as *accessing information from the environment*—something with full right to be regarded as a rational accomplishment— there is no reason to accept the dictum that a perception can only *cause* (and not *justify*) a verbalized thought." But my position was arrived at independently of McDowell. Already in my 1971 Oxford lectures I argued at length against "an essentially linear model of understanding" and insisted on this coordinate synthesis of the causal and the rational order in the cognitive sphere, insisting that "these two orders must be grasped together in their systematic unity." (On the character and interrelationship of the mental and physical orders of concepts see the author's *Conceptual Idealism* (Oxford: Basil Blackwell, 1973), especially, pp. 184 ff; the quote is from pp. 191–91). Based on lectures given in Oxford in 1971, these deliberations antedate Davidson's John Locke lectures (let alone McDowell's response to Davidson in *Mind and World* [Cambridge, MA: Harvard University Press, 1994]).

14. Examples of this sort illustrate why philosophers are so reluctant to identify *knowledge* with *true belief*.

15. The epistemological program this essay sketches out combines ideas set out in considerable detail in several of the author's publications: *The Coherence Theory of Truth* (Oxford: Oxford University Press, 1973), *Plausible Reasoning* (Assen: Van Gorcum, 1976), *Scepticism* (Oxford: Blackwell, 1980), *Human Knowledge in Idealistic Perspective* (Princeton: Princeton University Press, 1991), *A Useful Inheritance* (Pittsburgh: University of Pittsburgh Press, 1994). The discussion has benefited from constructive comments by Alexander Pruss.

NOTES TO CHAPTER 5

1. "Consequences drawn by the law or the magistrate from a known to an unknown fact." *Code Civil*, bk. III, pt. iii, sect, art. 1349.

2. Sir Courtenay Ilbert, "Evidence," *Encyclopedia Britannica*, 11th ed., vol. 10, pp. 11–21 (see p. 15).

3. Richard A. Epstein, "Pleadings and Presumptions," *The University of Chicago Law Review*, vol. 40 (1973/4), pp. 556–82 (see pp. 558–59). As Lalande's philosophical

dictionary puts it: "La présomption est proprement et d'une manière plus précise une anticipation sur ce qui n'est pas prouvé." (A. Lalande, *Vocabulaire de la philosophie*, 9th ed. [Paris: Presses Universitaires de France, 1962], s.v. "présomption.")

4. The modern philosophical literature on presumption is not extensive. Some years ago, when I wrote *Dialectics* (Albany: State University of New York Press, 1977) there was little apart from Roland Hall's "Presuming," *Philosophical Quarterly*, col. 11 (1961), pp. 10–22. A most useful overview of more recent development is Douglas N. Wallon, *Argumentation: Schemes for Presumptive Reasoning* (Mahwah, NJ: L. Erlbaum Assoc., 1996). I am also grateful to Sigmund Bonk for sending me his unpublished study "Vom Vorurteil zum Vorausurteil."

5. C. S. Peirce put the case for presumptions in a somewhat different way—as crucial to maintaining the line between sense and foolishness:

> There are minds to whom every prejudice, every presumption, seems unfair. It is easy to say what minds these are. They are those who never have known what it is to draw a well-grounded induction, and who imagine that other people's knowledge is as nebulous as their own. That all science rolls upon presumption (not of a formal hut of a real kind) is no argument with them, because they cannot imagine that there is anything solid in human knowledge. These are the people who waste their time and money upon perpetual motions and other such rubbish. (*Collected Papers*, vol. VI, ed. by C. Hartshorne and P. Weiss [Cambridge, MA: Harvard University Press, 1935], sect. 6.423; compare vol. II, sect. 2.776–7)

Peirce is very emphatic regarding the role of presumptions in scientific argumentation and adduces various examples, for example, that the laws of nature operate in the unknown parts of space and time as in the known or that the universe is inherently indifferent to human values and does not on its own workings manifest any inclination toward being benevolent, just, wise, and so forth. As the above quotation shows, Peirce saw one key aspect of presumption to revolve about considerations regarding the economics of inquiry—that is, as instruments of efficiency in saving time and money. On this sector of his thought, see the author's paper "Peirce and the Economy of Research," *Philosophy of Science*, vol. 43 (1976), pp. 71–98.

6. R. A. Epstein, "Pleading and Presumptions," pp. 558–59.

7. Richard Whately, *Elements of Rhetoric* (London and Oxford: Oxford University Press, 1828), pt. I, ch. III, sect. 2.

8. As I. Scheffler puts a similar point in the temporal context of a change of mind in the light of new information: "That a sentence may be given up at a later time does not mean that its present claim upon us may be blithely disregarded. The idea that once a statement is acknowledged as theoretically revisable, it can carry no cognitive weight at all, is no more plausible then the suggestion that a man losses his vote as soon as it is seen that the rules make it possible for him to be outvoted" (*Science and Subjectivity* [Indianapolis, IN: Bobbs-Merrill, 1967], p. 118).

9. The next chapter will examine the idea of presumption in greater detail.

10. Historically this goes back to the conception of "the reasonable" (*to eulongon*), as discussed in Greek antiquity by the Academic Skeptics. Carneades (c. 213–c. 128 B.C.) for one worked out a rather well-developed (nonprobabilistic) theory of plausibility as it relates to the deliverances of the senses, the testimony of witnesses, and so on. For a helpful discussion of the relevant issues, see Charlotte L. Stough, *Greek Skepticism* (Berkeley and Los Angeles: University of California Press, 1969).

11. For an interesting analysis of presumptive and plausibilistic reasoning in mathematics and (to a lesser extent) the natural sciences, see George Polya's books *Introduction and Analogy in Mathematics* (Princeton: Princeton University Press, 1954) and *Patterns of Plausible Inference* (Princeton: Princeton University Press, 1954). Polya regards the patterns of plausible argumentation as defining the "rules of admissibility in scientific discussion" on strict analogy with the legal case, regarding such rules as needed because it is plain that not *anything* can qualify for introduction as probatively relevant.

12. This view of plausibility in terms of general acceptance either by the *consensus genium* or by the experts ("the wise") was prominent in Aristotle's construction of the plausible (*endoxa*) in book I of the *Topics*.

13. All this, of course, does not deal with question of the status of this rule itself and of the nature of its own justification. It is important in the present context to stress the *regulative* role of plausibilistic considerations. This now becomes a matter of *epistemic policy* ("Given priority to contentions which treat like cases alike") and not a metaphysically laden contention regarding the ontology of nature (as with the—blatantly false—descriptive claim "Nature is uniform"). The plausibilistic theory of inductive reasoning sees uniformity as a *regulative principle of epistemic policy* in grounding our choices, not as a *constitutive principle* of ontology. As a "regulative principle of epistemic policy" its status is *methodological*—and thus its justification is in the final analysis pragmatic. See the author's *Methodological Pragmatism* (Oxford: Basil Blackwell, 1976).

14. See Gerdinand Gonseth, "La Notion du normal," *Dialectics*, vol. 3 (1947), pp. 243–52. More generally, on the principles of plausible reasoning in the natural sciences see Norwood R. Hanson, *Patterns of Discovery* (Cambridge, 1958), and his influential 1961 paper, "Is There a Logic of Discovery," in H. Feigl and G. Maxwell (eds.), *Current Issues in the Philosophy of Science*, Vol. I (New York: Holt, Rinehart and Winston, 1961). The work of Herbert A. Simon is an important development in this area: "Thinking by Computers" and "Scientific Discovery and the Psychology of Problem Solving," in R. G., Colodny (ed.), *Mind and Cosmos* (Pittsburgh: University of Pittsburgh Press, 1966).

15. All three of these principles of plausibility were operative in the protoconception of plausibility represented by Aristotle's theory of *endoxa* as developed in book I of the *Topics*. For this embraced: (1) general acceptance by men at large or by the experts ("the wise"), (2) theses similar to these (or opposed to what is contrary to them, and (3) theses that have become established in cognitive disciplines (*kata technas*).

16. For a closer study of the notion of plausibility and its function in rational argumentation see the author's *Plausible Reasoning* (Assen: Van Gorcum, 1976).

17. The lottery paradox was originally formulated by H. K. Kyburg, Jr., *Probability and the Logic of Rational Relief* (Middletown, CT: Wesleyan University Press, 1961).

For an analysis of its wider implications for inductive logic see R. Hilpinen, *Rules of Acceptances and Inductive Logic* (Amsterdam: North-Holland Publishing Company, 1968; *Acta Philosophica Fennica*, fasc. 22), pp. 39–49.

18. Cf. also I. Levi, *Gambling with Truth* (New York: Knopf, 1967), chaps. 2 and 6.

19. See Rudolf Carnap, *Logical Foundations of Probability* (Chicago, IL: University of Chicago Press, 1950; 2d ed., 1960), sec. 50, and P. A. Schilpp, ed., *The Philosophy of Rudolf Carnap* (La Salle, IL: Open Court, 1963): pp. 972–73. Carnap's position is followed by Richard Jeffrey. See his "Valuation and Acceptance of Scientific Hypotheses," *Philosophy of Science*, vol. 23 (1956), pp. 237–46.

20. For further details regarding the important role of presumptions in the philosophical theory of knowledge, see the author's *Methodological Pragmatism* (Oxford: Basil Blackwell, 1976).

21. On the strictly empirical side, it is difficult to exaggerate the extent to which our processes of thought, communication, and argumentation in everyday life are subject to established presumptions and accepted plausibilities. This is an area chat is only beginning to be subjected to sociological exploration. (See, for example, Harold Garfinkel, *Studies in Ethnomethodology* [Englewood Cliffs, NJ.: Prentice Hall, 1967].) The pioneer work in the field is chat of Alfred Schutz, Der sinnhafte Aufbau der sozialen Welt (Wien, 1932).

22. For further discussion of this issue and for references to the literature, see the author's book on *The Coherence Theory of Truth* (Oxford: Clarendon Press, 1973).

23. See the author's *Methodological Pragmatism* (Oxford: Basil Blackwell, 1976) for a fuller development of this line of thought.

24. This is why adherence to custom is a cardinal principle of cognitive as well as practical rationality. Cf. William James, "The Sentiment of Rationality," in *The Will to Believe and Other Essays in Popular Philosophy* (New York, London and Bombay: Longmans, Green, and Co., 1897).

25. See the author's *Methodological Pragmatism* (Oxford: Basil Blackwell, 1976) for a fuller development of this line of thought.

26. Further detail regarding presumptions and their epistemic role is given in the author's *Dialectic's* (Albany: State University of New York Press, 1977). The economic aspect of cognition is examined in his *Cognitive Economy* (Pittsburgh: University of Pittsburgh Press, 1989).

Notes to Chapter 6

1. On priority conflicts, see R. K. Merton, "Priorities in Scientific Discovery," *American Sociological Review*, vol. 22 (1957), pp. 635–59.

2. The founding of the Royal Society in London, chartered in 1666, was a small but significant step toward openness. For an interesting analysis of the historical

situation, see Jerome R. Ravetz, *Scientific Knowledge and Its Social Problems* (Oxford: Clarendon Press, 1971). Ravet's presentation makes it clear that a considerable transformation was involved because "a significant proportion of the great 'scientists' of that age [sixteenth to seventeenth century] were even more concerned for the protection of their intellectual property, than for an immediate realization of its value through the prestige resulting from publication" (p. 249).

3. The situation is one of the sort called prisoner's dilemma by game theorists. For a good account, see Morton D. Davis, *Game Theory* (New York: Basic Books, 1970), pp. 92–103. See also A. Rappoport and A. M. Chammah, *Prisoner's Dilemma: A Study in Conflict and Cooperation* (Ann Arbor: University of Michigan Press, 1965); Anatol Rapoport, "Escape from Paradox," *American Scientist* 217 (1967), pp. 50–56; and Richmond Campbell and Lanning Sowden (eds.), *Paradoxes of Rationality and Cooperation* (Vancouver, BC: University of British Columbia Press, 1985).

4. Compare H. M. Vollmer and D. L. Mills, eds., *Professionalization* (Englewood Cliffs: Prentice Hall, 1966). This credit, once earned, is generally safeguarded and maintained by institutional means: licensing procedures, training qualifications, professional societies, codes of professional practice, and the like.

5. See John Hardwig, "The Role of Trust in Knowledge," *The Journal of Philosophy*, vol. 88 (1991), pp. 693–708.

6. On this matter, see Thomas Sowell, *Knowledge and Decisions* (New York: Basic Books, 1980), especially the discussion of "Informal Relationships," on pp. 23–30.

7. The literature on theoretical issues of trust and cooperation in contexts of inquiry is virtually nonexistent. However regard to morality in general this is not so. See, for example, Robert Axelrod, *The Evolution of Cooperation* (New York: Basic Books, 1984); David Gauthier, *Morals by Agreement* (Oxford: Oxford University Press, 1986); Raimo Tuomela, *Cooperation* (Dordrecht: Kluwer, 2000); and Nicholas Rescher, *Fairness* (New Brunswick, NJ: Transaction Press, 2001).

8. For a penetrating study of these and similar issues, see John Sabini and Maury Silver, *Moralities of Everyday Life* (Oxford: Clarendon Press, 1982), especially chap. 4.

9. Specifically, Solomon Asch found that in certain situations of interactive estimation, "whereas the judgments were virtually free of error under control conditions, one-third of the minority estimates were distorted toward the majority." See his "Studies of Independence and Conformity: I. A Minority of One Against a Unanimous Majority," *Psychological Monographs: General and Applied*, no. 70 (1956).

10. Ibid, p. 69.

11. John Sabini and Maury Silver, *Moralities of Everyday Life*, pp. 84–85.

12. Harry Kalven, Jr., and Hans Zeisel, *The American Jury* (Chicago: University of Chicago Press, 1966).

13. Further discussion of some of this chapter's themes can be found in the author's *Cognitive Economy* (Pittsburgh: University of Pittsburgh Press, 1989).

NOTES TO CHAPTER 7

1. J. H. Lambert, *Fragment einer Systematologie* in J. Bernouilli (ed.), Johann Heinrich Lambert: Heinrich: *Philosophische Schriften* (2 vols; reprinted Hildesheim, 1967).

2. Think here of Aristotle's extensive concern with "priority" in the order of justification, and his requirement that in adequate explanations the premises must be "better known than and prior to" the conclusion.

3. Thus Lambert wrote: "Grundregel des Systems: Das vohrgehende soll das folgende *klar* machen, in Absicht auf den Verstand, gewiss in Absicht auf die Vernunft..." ("Theorie des Systems," in *Philosophische Schriften*, vol. II [op. cit.], p. 510.)

4. Compare David Hilbert's view that successful axiomatization affords a "Tieferlegung der Fundamente der einzelnen Wissensgebiete," which renders the held at once more intelligible and more secure. "Axiomatisches Denken" (1918), reprinted in David Hilbert, *Hilbertiana* (Darmstadt: Wissenschaftliche Buchgesellschaft, 1964).

5. R. M. Chisholm, *The Theory of Knowledge*, 2nd ed. (Englewood Cliffs: Prentice Hall, 1977).

6. See especially Hugo Dingler, *Das System* (Munich: E. Reinhardt, 1930), pp. 19–20.

7. Aristotle, *Posterior Analytics*, I, 3; 72b5–24 (tr. W. D. Ross; Oxford: Clarendon Press, 1975).

8. The theory is to be met with in Plato's *Theaetetus* (200 D ff.) where the element added to truth and belief is the existence of a rationale or account (logos). Since this factor is essentially discursive, the theory is there criticized as conflicting with a foundationalism that admits of basic elements.

9. C. I. Lewis, *An Analysis of Knowledge and Valuation* (Lasalle, IL: Open Court Press, 1946), p. 186.

10. See Franz Brentano, *Wahrheit und Evidenz* (ed. O. Kraus, Leipzig: F. Meiner, 1930). The most influential present-day advocate of an epistemological position along Brentano-esque lines is R. M. Chisholm (see his *Theory of Knowledge* [Englewood Cliffs, NJ: Prentice Hall, 1966]). Compare also Roderick Firth, "Ultimate Evidence." *The Journal of Philosophy*, vol. 53 (1956), pp. 732–38; reprinted in A. J. Swartz (ed.), *Perceiving, Sensing, and Knowing* (Garden City, NY, Anchor Books, 1965), pp. 486–96. For an informative general survey and critique of present-day foundationalism sec Anthony Quinton, "The Foundations of Knowledge" in B. Williams and A. Montefiore (eds.), *British Analytical Philosophy* (New York, Humanities Press, 1966), pp. 55–86.

11. Compare Herbert Simon, "The Architecture of Complexity," *General Systems*, vol. 10 (Ann Arbor: University of Michigan Press, 1965), pp. 63–76.

12. Compare R. M. Chisholm, *The Theory of Knowledge*, 2nd ed. (Englewood Cliffs: Prentice Hall, 1977).

13. Recognition of these and other weaknesses of foundationalism has in recent times not been spearheaded by coherentists—the idealistic advocates were ineffectual, the positivist advocates (i.e., Neurath and his sympathizers) were unsuccessful. The only effective opposition has centered about the refutationism of K. R. Popper's *Logik der Forschung* (Wien: J. Springer, 1935: tr. as *The Logic of Scientific Discovery*, New York: Basic Books, 1959).

14. Some of this chapter's themes are also touched on in the author's "Foundationalism, Coherentism, and the Idea of Cognitive Systematization," *The Journal of Philosophy*, vol. 71 (1974), pp. 695–708.

NOTES TO CHAPTER 8

1. This consideration more than any other differentiates coherentism from reliabilism and data from reliably confined propositions.

2. The formal mechanism of best-fit analysis is described more fully in the author's books *The Coherence Theory of Truth* (Oxford: Clarendon Press, 1973) and *Plausible Reasoning* (Assen: Van Gorcum, 1976).

3. The epistemic problem with data is thus not just some of them are *uncertain*, but that, in general being collectively consistent, we know a priori that some of them must be false.

4. I. Levi, *Gambling with Truth* (New York: Knopf, 1967), p. 28. "To accept H as evidence is not merely to accept *H* as true but to regard as pointless further evidence collection in order to check on *H*" (*ibid.*, p. 149).

5. See A. C. Ewing, *Idealism: A Critical Survey* (London: MacMillan, 1934), p. 238, as well as his later essay on "The Correspondence Theory of Truth," where he writes: "that coherence is the test of truth can only be made plausible if coherence is interpreted not as mere internal coherence but as coherence with our experience" (*Non-Linguistic Philosophy* (London, 1968), pp. 203–4). For an author of the earlier period see H. H. Joachim who writes: "Truth, we said, was the systematic coherence which characterized a significant whole. And we proceeded to identify a significant whole with 'an organized individual experience, self-filling and self-fulfilled'" (*The* Nature *of Truth* [Oxford: Clarendon Press, 1906], p. 78).

6. B. Russell, *The Problems of Philosophy* (London: H. Holt, 1912), p. 191. Or compare M. Schlick's formulation of this point: "Since no one dreams of holding the statements of a story book true and those of a text of physics false, the coherence view fails utterly. Something more, that is, must be added to coherence, namely, a principle in terms of which the compatibility is to be established [sc. as factual], and this would alone then be the actual criterion" (M. Schlick, "The Foundation of Knowledge," in A. J. Ayer [ed.], *Logical Positivism* [Glencoe, IL: Free Press, 1959], pp. 209–27 [see p. 216]).

7. A. R. White, "Coherence Theory of Truth," in P. Edward (ed.), *The Encyclopedia of Philosophy*, vol. 2 (1967), pp. 130–3 (see p. 131). One critic of the coherence theory elaborates this important point with demonstrative clarity as follows:

That in the end only one sufficiently comprehensive system of statements would be found consistent, is a suggestion which runs counter to obvious facts about the nature of consistency and of systems; probably it strikes us as plausible because we are such poor liars, and are fairly certain to become entangled in inconsistencies sooner or later, once we depart from the truth. A sufficiently magnificent liar, however, or one who was given time and patiently followed a few simple rules of logic, could eventually present us with any number of systems, as comprehensive as you please, and all of them including falsehoods. Insofar as it is possible to deal with any such notion as "the whole of the truth," it is the Leibnizian conception of an infinite plurality of possible worlds which is justified, and not the conception of the historical coherence theory that there is just one all-comprehensive system, uniquely determined to be true by its complete consistency. . . . Thus if we start with any empirical [i.e., contingent] belief or statement "P," we shall find that one or other of every pair of further empirical statements, "Q" and "not-Q," "R" and "not-R," and so on, can be conjoined with "P" to form a self-consistent set. And exactly the same will likewise be true of its contradictory "not-P." *Every empirical supposition, being a contingent statement, is contained in some self-consistent system which is as comprehensive as you please.* And as between the truth of any empirical belief or statement "P" and the falsity of it (the truth of "not-P") consistency with other possible beliefs or statements, or inclusion in comprehensive and self-consistent systems, provides no clue or basis of decision. (C. I. Lewis, *An Analysis of Knowledge and Valuation* (La Salle, IL: Open Court, 1962), pp. 340–41).

8. On coherentism see R. Audi, *Belief Justification and Knowledge* (Belmont: Wadsworth, 1988); J. Bender (ed.), *The Current State of the Coherence Theory* (Dordrecht: Kluwer, 1989); L. BonJour, *The Structure of Empirical Knowledge* (Cambridge, MA: Harvard University Press, 1985); R. M. Chisholm, *Theory of Knowledge*, 3rd ed. (Englewood Cliffs: Prentice-Hall, 1989); A. Goldman, *Empirical Knowledge* (Berkeley: University of California Press, 1988); G. Harman, *Thought* (Princeton: Princeton University Press, 1973); K. Lehrer, *Theory of Knowledge* (Boulder: Westview, 1990), W. G. Lycan, *Judgement and Justification* (New York: Cambridge University Press, 1988); J. Pollock, *Contemporary Theories of Knowledge* (Totowa: Rowman & Littlefield, 1986), W. V. O. Quine and Joseph Ullian, *The Web of Belief* (McGraw-Hill Higher Education, 1979); N. Rescher, *The Coherence Theory of Truth* (Oxford: Oxford University Press, 1973); J. Rosenberg, *One World and Our Knowledge of It* (Dordrecht: D. Reidel, 1980).

9. On this term compare Michael Polanyi, *Personal Knowledge; Towards a Postcritical Philosophy* (New York: Harper & Row, 1964).

10. The question "Why should pragmatic success of the applications of the products of a cognitive method count as an index of its *cognitive* adequacy?," though seemingly straightforward, in fact plumbs hidden metaphysical depths. The complex issues that arise here are examined at considerable length in the author's *Methodological Pragmatism* (Oxford, 1976), where other considerations relevant to the present discussion are also set out at greater length.

11. It is worth stressing an important aspect of the pivotal role of pragmatic efficacy in the quality-control of cognitive systematization. Most of the theoretical parameters of systematizing adequacy (unity, uniformity, cohesiveness, etc.) exert an impetus in the direction of simplicity (economy austerity). Their operation would never in itself induce us to move from a system in hand that is relatively simple to one that is more complex. But the pursuit of applicative adequacy can reinforce the operation of completeness and comprehensiveness in counteracting the simplicity-oriented tendency of those parameters.

12. On this fact-ladenness of the fundamental ideas by which our very conception of nature is itself framed see Chapter VI, "A Critique of Pure Analysis" of the author's *The Primacy of Practice* (Oxford: Basil Blackwell, 1973).

13. Some of the key themes of this chapter are treated in author's *The Coherence Theory of Truth* (Oxford: Clarendon Press, 1973) and *A Useful Inheritance*: (Savage: Rowman & Littlefield, 1990). And some of its key concepts are further developed elsewhere: plausibility—*Plausible Reasoning* (Assen; Van Gorcum, 1976); systematicity—*Cognitive Systematization* (Oxford: Blackwell, 1979); retrospective justification—*Methodological Pragmatism* (Oxford: Basil Blackwell, 1977); cognitive efficiency—*Cognitive Economy* (Pittsburgh: University of Pittsburgh Press, 1989).

14. Arthur Pap, *Elements of Analytic Philosophy* (New York: Macmillan Co., 1949), p. 356. Compare H. H. Joachim, *Logical Studies* (Oxford: Clarendon Press, 1948).

15. The definition versus criterion dichotomy was the starting point of the author's *The Coherence Theory of Truth* (Oxford, 1973). It also provided the pivot for the critique of the coherentism of Blanshard's *The Nature of Thought* in his contribution to R. A. Schilpp (ed.), *The Philosophy of Brand Blanshard* (LaSalle: Open Court, 1980). Several subsequent publications have kept the pot boiling, in particular Scott D. Palmer, "Blanchard, Rescher, and the Coherence Theory of Truth, "*Idealistic Studies*, vol. 12 (1982) pp. 211–30, and Robert Tad Lehe, "Coherence: Criterion and Nature of Truth," ibid., vol. 13 (1983), pp. 177–89.

NOTES TO CHAPTER 9

1. The many good anthologies on the subject of relativism include: Martin Hollis and Steven Lukes (eds.), *Rationality and Relativism* (Oxford: Clarendon Press, 1982); Michael Krausz and Jack W. Meiland (eds.), *Relativism: Cognitive and Moral* (Notre Dame, IN: University of Notre Dame Press, 1982); J. Margolis, M. Krausz, and R. M. Burian (eds.), *Rationality, Relativism, and the Human Sciences* (Dordrecht, Kluwer, 1986); Michael Krausz (ed.) *Relativism: Interpretation and Confrontation* (Notre Dame, IN: University of Notre Dame Press, 1989); Geoffrey Sayre McCord (ed.), *Essays on Moral Relativism* (Ithaca, NY.: Cornell University Press, 1994). See also: Herbert Speigelberg, *Antirelativismus* (Zürich, Leipzig: M. Niehans, 1935); Johannes Thyssen, *Der philosophische Relativismus* (Bonn: L. Röhrscheid, 1941); Nancy L. Gifford, *When in Rome: And Introduction to Relativism and Knowledge* (Albany: State University of New York Press, 1983); Peter Unger, *Philosophical Relativity* (Oxford: Basil Blackwell, 1984);

Joseph Margolis, *The Truth About Relativism* (Oxford: Oxford University Press, 1991); H. Siegel, *Relativism Refuted* (Dordrecht: Kluwer, 1987); Larry Laudan, *Science and Relativism* (Chicago: University of Chicago Press, 1990); Richard Rorty, *Objectivity, Relativism, and Truth* (Cambridge; New York: Cambridge University Press, 1991); J. F. Harris, *Against Relativism* (La Salle, IL: Open Court, 1992).

2. William James, *Pragmatism* (New York: Appleton, 1907), p. 171. The basic line of thought goes back to the ancient skeptics. Compare Sextus Empiricus, *Outlines of Pyrrhonism*, I, 54, 59–60, 97, et passim.

3. William James, *Talks to Teachers on Psychology* (New York: Henry Holt, 1899), p. 4.

4. On "alternative standards of rationality" see Peter Winch, "Understanding a Primitive Society," *American Philosophical Quarterly*, vol. 1 (1964), pp. 307–24.

5. For an interesting critique of cognitive relativism that is akin in spirit though different in orientation from that of the present section see Lenn E. Goodman, "Six Dogmas of Relativism," in Marcello Dascal (ed.), *Cultural Relativism and Philosophy* (Leiden: E. J. Brill, 1991), pp. 77–102.

6. To be sure, someone could convince me that my understanding of the implications of my standards is incomplete and lead me to an *internally* motivated revision of my rational proceedings, amending those standards from within with a view to greater systemic coherence.

7. "Reste von christlicher Theologie innerhalb der philosophischen Problematik" (Martin Heidegger, *Sein and Zeit* [Halle a. d. S.: Niemeyer, 1931], p. 230).

8. William James, "Pragmatism and Humanism," in F. H. Burkhardt et al. (eds.), *Pragmatism* (Lecture VIII), *The Works of William James* (Cambridge, MA: Harvard University Press, 1975), p. 115.

9. Further discussion of some of this chapter's themes are presented in the author's *Objectivity* (Notre Dame, IN: University of Notre Dame Press, 1997).

NOTES TO CHAPTER 10

1. Two recent anthologies offer informative discussions of objectivity. S. C. Brown (ed.), *Objectivity and Cultural Divergence* (Cambridge: Cambridge University Press, 1984), and Alan McGill, *Rethinking Objectivity* (Durham NC: Duke University Press, 1994). Recent books on the subject include Richard J. Benstein, *Beyond Objectivism and Relativism* (Philadelphia: University of Pennsylvania Press, 1983); Hilary Putnam, *Realism with a Human Face* (Cambridge, MA: Harvard University Press, 1990), Nicholas Rescher, *Objectivity* (Notre Dame, IN: University of Notre Dame Press, 1997); and Richard Rorty, *Objectivity, Relativism and Truth* (Cambridge: Cambridge University Press, 1991).

2. See Robert Brandom, "Knowledge and the Social Articulation of the Space of Reasons," *Philosophy and Phenomenological Research*, vol. 55 (1995), pp. 895–908.

3. This is why it seems mistaken to characterize objectivity a *fundamental* epistemic value as Brian Ellis does in his *Truth* and Objectivity (Oxford: Basil Blackwell, 1990; pp. 228–31). The value of objectivity is not fundamental but instrumental and lies in its auspicity to facilitate the achievement of other, ulterior goal.

4. For an elaborate development of this position see the author's *Rationality: A Philosophical Inquiry into the Nature and the Rationale of Reason* (Oxford: Clarendon Press, 1988).

5. On our duty toward the cultivation of rationality see also pp. 204–9 of the author's *Rationality* (Oxford: Clarendon Press, 1988).

6. See Friedrich Nietzsche, *The Will to Power*, tr. by Walter Kaufmann and R. J. Hollingdale (New York: Vintage Books, 1968), p. 481.

7. Further discussion of some themes of this chapter can be found in the author's *Objectivity* (Notre Dame, IN: University of Notre Dame Press, 1997).

NOTES TO CHAPTER 11

1. In some situations (e.g., negotiations, games, and warfare) the intelligent thing at the level of policy or strategy may be occasionally to do a "stupid" thing at the level of particular acts or tactics. It is sometimes advantageous "to keep your opponent guessing" by not being too predictable and occasionally doing something unexpected—even though this is a "stupid" thing to do in the circumstances—as part of a deeper cunning. Nothing said here about rational procedure is intended to contradict the prospect of such wise "foolishness."

2. Some recent anthologies on rationality are: Stanley I. Benn and G. W. Mortimore, *Rationality and the Social Sciences* (London: Routledge and Kegan Paul, 1976); K. I. Manktelow and D. E. Over, *Rationality* (London: Routledge, 1993); and Paul K. Moser, *Rationality in Action* (Cambridge: Cambridge University Press, 1990). Recent book on the topic include: Jonathan Bennett, *Rationality: An Essay Toward Analysis* (London: Routledge & K. Paul, 1964); Christopher Cherniak, *Minimal Rationality* (Cambridge, MA: MIT Press, 1986); Stephen L. Darwall, *Impartial Reason* (Ithaca, NY: Cornell University Press, 1983); Stephan Nathanson, *The Ideal of Rationality* (Atlantic Highlands, NJ: Humanities Press, 1985); David Pears, *Motivated Individuality* (Oxford: Clarendon Press, 1984); Frederick Schick, *Having Reasons* (Princeton: Princeton University Press, 1984); and Herbert Simon, *Reason in Human Affairs* (Stanford, CA: Stanford University Press, 1983).

3. Kurt Baier, *The Moral Point of View* (New York: Random House, 1965; abr. edn.) 160–1.

4. Compare René Descartes, *Discourse on Method*, sect. iii, maxim 2.

5. In his interesting book on *A Justification of Rationality* (Albany: State University of New York Press, 1976), John Kekes argues that "The justification of rationality is . . . [as] a device for problem-solving and it should be employed because

everybody has problems, because it is in everybody's interest to solve his problems, and because rationality is the most promising way of doing so" (p. 168). This traditionally pragmatic view is very close to our own position except that it pivots rationality's justification on effectiveness in problem solving, while our own position is somewhat more cautious. It does not contend that the course of reason actually is our best recourse in problem solving, but only that it is so as best we can (rationally) judge. The present argumentation thus brings the aspect of reason's self-reliance to the fore as a critical aspect of the justification of rationality, and accordingly is not a pure pragmatism.

6. This view has become axiomatic for the entire "sociology of knowledge."

7. Peter Winch, "Understanding a Primitive Society," *American Philosophical Quarterly*, vol. 1 (1964), pp. 307–24; reprinted in B. R. Wilson (ed.), *Rationality* (Oxford: Clarendon Press, 1970).

8. Lévy-Bruhl, *Primitive Mentality* (London, 1923; first pub. in French, Paris, 1921).

9. *Witchcraft, Oracles and Magic Among the Azandi* (Oxford: Clarendon Press, 1937); *Nuer Religion* (Oxford: Clarendon Press, 1956).

10. The relevant issues are interestingly treated in John Kekes's book, *A Justification of Rationality* (Albany,: State University of New York Press, 1976), pp. 137–49.

11. "Est ridiculum querere quae habere non possumus," as Cicero wisely observed (*Pro Archia*, iv. 8).

12. Further discussion of some of this chapter's themes is presented in the author's *Rationality* (Oxford: Clarendon Press, 1988).

Notes to Chapter 12

1. A homely fishing analogy of Eddington's is useful here. He saw the experimentalists as akin to a fisherman who trawls nature with the net of his equipment for detection and observation. Now suppose (says Eddington) that a fisherman trawls the seas using a net of two-inch mesh. Then fish of a smaller size will simply go uncaught, and those who analyze the catch will have an incomplete and distorted view of aquatic life. The situation in science is the same. Only by improving our observational means of trawling nature can such imperfections be mitigated. (See A. S. Eddington, *The Nature of the Physical World* [New York: Macmillan, 1928].)

2. D. A. Bromley et al. *Physics in Perspective, Student Edition* (Washington, DC: National Academy of Sciences, 1973); pp. 13, 16. See also Gerald Holton, "Models for Understanding the Growth and Excellence of Scientific Research," in Stephen R. Graubard and Gerald Holton, eds., *Excellence and Leadership in the Democracy* (New York: Free Press, 1962), p. 115.

3. Note, however, that an assumption of the finite dimensionality of the phase space of research-relevant physical parameters becomes crucial here. For if these were

limitless in number, one could always move on to the inexpensive exploitation of virgin territory.

4. "Looking back, one has the impression that the historical development of the physical description of the world consists of a succession of layers of knowledge of increasing generality and greater depth. Each layer has a well-defined field of validity; one has to pass beyond the limits of each to get to the next one, which will be characterized by more general and more encompassing laws and by discoveries constituting a deeper penetration into the structure of the Universe than the layers recognized before." (Edoardo Amaldi, "The Unity of Physics," *Physics Today*, vol. 261, no. 9 [September 1973], p. 24.) See also E. P. Wigner, "The Unreasonable Effectiveness of Mathematics in the Natural Sciences," *Communication on Pure and Applied Mathematics*, vol. 13 (1960), pp. 1–14; as well as his "The Limits of Science," *Proceedings of the American Philosophical Society*, vol. 93 (1949), pp. 521–26. Compare also chapter 8 of Henry Margenau, *The Nature of Physical Reality* (New York: Knopf, 1950).

5. D. A. Bromley et al., *Physics in Perspective, Student Edition* (Washington DC: National Academy of Sciences, 1973), p. 23.

6. Gerald Holton, "Models for Understanding the Growth and Excellence of Scientific Research," in Stephen R. Graubard and Gerald Holton (editors), *Excellence and Leadership in a Democracy* (op. cit.), p. 115.

7. Gunther S. Stent, *The Coming of the Golden Age* (Garden City: American Museum of Natural History, 1969), pp. 111–113. One notable exception to this view is that of the eminent Russian physicist Peter Kapitsa. After surveying various fundamental discoveries of the past he writes: "If we honestly extrapolate this curve we see it does not have any tendency toward saturation and that in the very near future many more such discoveries, which give us the possibility of increasing our control over nature and put new strength in our hands, will be made. Subjectively it seems that we know all there is to know about nature. However, when we read the works of scientists of the Newtonian era we see that they felt precisely the same. We can, therefore, be sure that further discoveries must still be made" (The Future Problems of Science," in M. Goldsmith and A. Mackay [eds.], *The Science of Science* [London, 1964], pp. 102–13 [see pp. 105–6].) This reflects standard, "party-line" thinking in the U.S.S.R. See pp. 123–31 below, and also see footnote 33.

8. See the author's *Peirce's Philosophy of Science*, (Notre Dame and London: University of Notre Dame Press, 1978).

9. W. Stanley Jevons, *The Principles of Science*, 2nd ed. (London: Macmillan and Co., 1877), pp. 752–53.

10. The present critique of convergentism is thus very different from that of W. V. O. Quine. He argues that the idea of "convergence to a limit" is defined for numbers but not for theories, so that speaking of scientific change as issuing in a "convergence to a limit" is a misleading metaphor. "There is a faulty use of mathematical analogy in speaking of a limit of theories, since the notion of a limit depends on that of a 'nearer than,' which is defined for numbers and not for theories" *Word and Object* (Cambridge, MA: Technology Press of the Massachusetts Institute of Technology, 1960), p. 23. I am per-

fectly willing to apply the metaphor of substantial and insignificant differences to theories, but am concerned to deny that, as a matter of fact, the course of scientific theory-innovation must eventually descend to the level of trivialities.

11. Some of the themes of this chapter are also addressed in the author's *Scientific Progress* (Oxford, Blackwell, 1977), and in *Scientific Realism* (Dordrecht: D. Reidel, 1987).

12. This realization is something of which we can make no effective use: while we realize *that* many of our scientific beliefs are wrong, we have no way of telling *which* ones, and no way of telling *how* error has crept in. And, per contra, the same holds for truth. Cp. Robert Almeder, *Blind Realism* (Savage, MD: Rowman & Littlefield, 1992).

13. *Collected Papers*, ed. by C. Hartshorne et al., vol. I (Cambridge, MA: Harvard University Press, 1931), pp. 44–45 (sects. 108–9); ca. 1896.

14. Derek J. Price, *Science Since Babylon* (New Haven: Yale University Press, 1961), p. 137.

15. *The Structure of Scientific Revolutions*, 2nd ed. (Chicago: University of Chicago Press, 1970). See also I. Lakatos and A. Musgrave, eds., *Criticism and the Growth of Knowledge* (Cambridge: Cambridge University Press, 1970).

16. The prime exponent of this position is Paul Feyerabend. See his essays "Explanation, Reduction, and Empiricism," in Herbert Feigl and Grover Maxwell, eds., *Minnesota Studies in the Philosophy of Science*, Vol. III (Minneapolis: University of Minnesota Press, 1962); "Problems of Empiricism," in R. G. Colodny, ed., *Beyond the Edge of Certainty* (Englewood Cliffs, NJ, 1965), pp. 145–260 and "On the 'Meaning' of Scientific Terms," *Journal of Philosophy*, vol. 62 (1965), pp. 266–74.

17. Practical problems have a tendency to remain structurally invariant. The sending of messages is just that, whether horse-carried letters or laser beams arc used in transmitting the information.

18. This (essentially Baconian) idea that control over nature is the pivotal determinant of progress—in contrast to purely intellectual criteria (such as growing refinement, complication, or precision; let alone cumulation or proliferation)—has been mooted by several writers in response to Kuhn. See, for example, Paul M. Quay, "Progress as a Demarcation Criterion for the Science," *Philosophy of Science*, vol. 41 (1974), pp. 154–70 (especially p. 158 and also Friedrich Rapp, "Technological and Scientific Knowledge," in *Logic, Methodology, and Philosophy of Science: Proceedings in the Vth International Congress of DLMPS/IUHPS: London, Ontario, 1975* (Toronto: University of Toronto Press, 1976). The relevant issues are treated in depth in the present author's *Methodological Pragmatism* (Oxford: Blackwell, 1977).

19. For Hobbes's ideas in this region, see Hans Fiebig, *Erkenntnis und technische Erzeugung: Hobbes' operationale Philosophie der Wissenschaft* (Meisenheim am Glan: Forscher, 1973).

20. Further discussion of some of this chapter's themes can be found in chapter 7, "Cost Escalation in Empirical Inquiry" of the author's *Cognitive Economy* (Pittsburgh: University of Pittsburgh Press, 1989). *Scientific Progress* (Oxford: Basil Blackwell, 1978),

and *The Limits of Science* (Berkeley and Los Angeles: University of California Press, 1984) are also relevant.

Notes to Chapter 13

1. Hans Reichenbach, *Experience and Prediction* (Chicago: University of Chicago Press, 1938), p. 376.

2. *Science and Hypothesis* (New York: Dover Press, 1914), pp. 145–46.

3. Immanuel Kant was the first philosopher clearly to perceive and emphasize this crucial point: "But such a principle [of systematicity] does not prescribe any law for objects . . . , it is merely a subjective law for the orderly management of the possessions of our understanding, that by the comparison of its concepts it may reduce them to the smallest possible number; it does not justify us in demanding from the objects such uniformity as will minister to the convenience and extension of our understanding; and we may not, therefore, ascribe to the [methodological or *regulative*] maxim ['Systematize knowledge!'] any objective [or descriptively *constitutive*] validity." (CPuR, A306 = B362.) Compare also C. S. Peirce's idea that the systematicity of nature is a regulative matter of scientific attitude rather than a constitutive matter of scientific fact. See Charles Sanders Peirce, *Collected Papers*, vol. 7 (Cambridge, MA: Harvard University Press, 1958), sect. 7.134.

4. Herbert Spencer, *First Principles*, 7th ed. (London: Appleton's, 1889); see sects. 14–17 of part II, "The Law of Evolution."

5. On the process in general see John H. Holland, *Hidden Order: How Adaptation Builds Complexity* (Reading, MA: Addison Wesley, 1995). Regarding the specifically evolutionary aspect of the process see Robert N. Brandon, *Adaptation and Environment* (Princeton: Princeton University Press, 1990.)

6. On the issues of this paragraph compare Stuart Kaufmann, *At Home in the Universe: To Search for the Laws of Self-Organization and Complexity* (New York and Oxford: Oxford University Press, 1995).

7. An interesting illustration of the extent to which lessons in the school of better experience have accustomed us to expect complexity is provided by the contrast between the pairs: rudimentary/nonuanced; unsophisticated/sophisticated; plain/elaborate; simple/intricate. Note that in each case the second, complexity-reflective alternative has a distinctly more positive (or less negative) connotation than its opposite counterpart.

8. On this process see sections 1 of chapter 12.

9. See B. W. Petley, *The Fundamental Physical Constants and the Frontiers of Measurement* (Bristol: Hilger, 1985).

10. On the structure of dialectical reasoning see the author's *Dialectics* (Albany: State University of New York Press, 1977), and for the analogous role of such reasoning in philosophy see *The Strife of Systems* (Pittsburgh: University of Pittsburgh Press, 1985).

11. See John Dupré, *The Disorder of Things: Metaphysical Foundations of the Disunity of Science* (Cambridge, MA: Harvard University Press, 1993).

12. See Steven Weinberg, *Dreams of a Formal Theory* (New York: Pantheon, 1992). See also Edoardo Amaldi, "The Unity of Physics," *Physics Today*, vol. 261 (September, 1973), pp. 23–29. Compare also C. F. von Weizsäcker, "The Unity of Physics," in Ted Bastin (ed.) *Quantum Theory and Beyond* (Cambridge: Cambridge University Press, 1971).

13. The older figures are given in S. S. Visher, "Starred Scientists, 1903–1943," in *American Men of Science* (Baltimore, 1947). For many further details regarding the development of American science see the author's *Scientific Progress* (Oxford: Blackwell, 1978).

14. See, for example, Derek J. Price, *Little Science, Big Science* (New York: Columbia University Press, 1963), p. 11.

15. Cf. Derek J. Price, *Science Since Babylon*, 2nd ed. (New Haven, CT: Yale University Press, 1975); see in particular chapter 5, "Diseases of Science."

16. Data from *An International Survey of Book Production During the Last Decades* (Paris: UNESCO, 1985).

17. Raymond Ewell, "The Role of Research in Economic Growth," *Chemical and Engineering News*, vol. 33 (1955), pp. 2980–85.

18. It is worth noting for the sake of comparison that for more than a century now the *total* U.S. federal budget, its *nondefense subtotal*, and the aggregate budgets of all federal agencies concerned with the environmental sciences (Bureau of Mines, Weather Bureau, Army Map Service, etc.) have all grown at a uniform per annum rate of 9 percent. (See H. W. Menard, *Science: Growth and Change* [Cambridge, MA: Harvard University Press, 1971], p. 188.)

19. Data from William George, *The Scientist in Action* (New York: Arno Press, 1938).

20. "Impact of Large-Scale Science on the United State," *Science*, vol. 134 (1961; 21 July issue), pp. 161–64 (see p. 161). Weinberg further writes: "The other main contender [apart from space-exploration] for the position of Number One Event in the Scientific Olympics is high-energy physics. It, too, is wonderfully expensive (the Stanford linear accelerator is expected to cost 100×10^6), and we may expect to spend 400×10^6 per year on this area of research by 1970" (Ibid., p. 164).

21. There is thus no suggestion of cumulatively here. Clearly, knowledge, like populations, grows not just by additions (births) but also by abandonment (deaths). But with knowledge, in contrast to population, the concepts of displacement and replacement play a more crucial role. Knowledge never dies off without leaving progeny: in this domain you cannot eliminate something with nothing.

22. Note that a self-concatenated concept or fact is still a concept or fact, even as a self-mixed color is still a color.

23. In information theory, *entropy* is the measure of the information conveyed by a message, and is there measured by $k \log M$, where M is the number of structurally equivalent messages available via the available sorts of symbols. By extension, the $\log I$ measure

of the knowledge contained in a given body of information might accordingly be designated as the *enentropy*. Either way, the concept at issue measures informative actuality in relation to informative possibility. For there are two types of informative possibilities (1) structural/syntactical as dealt with in classical information theory, and (2) hermeneutic/semantical (i.e., genuinely meaning-oriented) as dealt with in the present theory.

24. Some writers have suggested that the subcategory of significant information included in an overall body of crude data of size I should be measured by I^k (for some suitably adjusted value $0 < k < 1$)—for example, by the "Rousseau's Law" standard of $\sqrt{I}$ (For details see Chapter VI of the author's *Scientific Progress* [Oxford: Blackwell, 1978].) Now since $\log I^k = k \log I$ which is proportional to $\log I$, the specification of *this* sort of quality level for information would again lead to a $K \approx \log I$ relationship.

25. See Stanislaw M. Ulam, *Adventures of a Mathematician* (New York: Scribner, 1976).

26. Max Planck, *Vorträge und Erinnerungen*, 5th ed. (Stuttgart: S. Hirzel, 1949), p. 376; italics added. Shrewd insights seldom go unanticipated, so it is not surprising that other theorists should be able to contest claims to Planck's priority here. C. S. Peirce is particularly noteworthy in this connection.

27. Henry Brooks Adams, *The Education of Henry Adams* (Boston: Houghton Mifflin, 1918).

28. To be sure, we are caught up here in the usual cyclic pattern of all hypothetico-deductive reasoning. In addition to explaining the various phenomena we have been canvasing that projected K/I relationship is in turn substantiated by them. This is not a vicious circularity but simply a matter of the systemic coherence that lies at the basis of inductive reasonings. Of course the crux is that there also be some predictive power, which is exactly what our discussion of deceleration is designed to exhibit.

29. Some of the themes of this chapter were also addressed in the author's *Scientific Progress* (Oxford: Blackwell, 1978). This book is also available in translation: German transl., *Wissenschaftlicher Fortschritt* (Berlin: De Gruyter, 1982); French transl., *Le Progrès scientifique* (Paris: Presses Universitaires de France, 1994).

30. Edward Gibbon, *Memoirs of My Life* (Harmondworth: Penguin Books, 1984), p. 63.

31. Gibbon's "law of learning" thus means that a body of experience that grows linearly over time yields a merely logarithmic growth in *cognitive* age. Thus a youngster of ten years has attained only one-eighth of his or her chronological expected life span but has already passed the halfway mark of his or her cognitive expected life span.

32. It might be asked: "Why should a mere accretion in scientific 'information'—in mere belief—be taken to constitute *progress*, seeing that those later beliefs are not necessarily *true* (even as the earlier one's were not)?" The answer is that they are in any case better *substantiated*—that they are "improvements" on the earlier one's by way of the elimination of shortcomings. For a detailed consideration is the relevant issues, see the author's *Scientific Realism* (Dordrecht: D. Reidel, 1987).

33. Derek J. de Solla Price, *Little Science, Big Science* (New York: Columbia University Press, 1963), pp. 6–8.

34. For the statistical situation in science see Derek J. de Solla Price, *Science Since Babylon* (New Haven, 1961; 2nd ed. 1975), See chapter 8, "Diseases of Science." Further detail is given in Price's *Little Science, Big Science* (op. cit.).

35. The data here are set out in the author's *Scientific Progress* (Oxford: Blackwell, 1978).

36. D. A. Bromley et al. *Physics in Perspective* (Washington, DC: National Academy of Science, 1973); pp. 16, 23. See also Gerald Holton, "Models for Understanding the Growth and Excellence of Scientific Research," in Stephen R. Graubard and Gerald Holton, eds., *Excellence and Leadership in a Democracy* (New York: Columbia University Press, 1962), p. 115.

37. To be sure, we are caught up here in the usual cyclic pattern of all hypothetico-deductive reasoning. In addition to explaining the various phenomena we have been canvasing that projected K/I relationship is in turn substantiated by them. This is not a vicious circularity but simply a matter of the systemic coherence that lies at the basis of inductive reasonings.

38. This relationship conveys an important lesson. For the question arises: Is the situation of a diminishing returns on scientific effort not incompatible with the fact (so decidedly emphasized in chapter 2) that natural science is potentially incompletable? To see that no incompatibility arises it suffices to recall that an ever decreasing series need not yield a *convergent* sum as indeed the series 1/2, 1/3, 1/4, ... indeed does not. And it is just this series that corresponds to the relationship $\frac{d}{dt} K(t) = \frac{1}{t}$.

39. Recall here the discussion of chapter 12. On these matters see also the author's *Priceless Knowledge?* (Savage, MD: Littlefield Adams, 1996).

40. This follows because:

$$\frac{d}{dt} K = c \log I \cong \frac{1}{I} \frac{d}{dt} I.$$

And when $I \cong t$ (as will be the case with I's linear growth over time in line with a steady-state resource investment) then:

$$\frac{d}{dt} K \cong \frac{1}{t}$$

41. See Henry Brooks Adams, *The Education of Henry Adams: An Autobiography* (Boston: Houghton Mifflin, 1918).

42. The situation with automobiles is analogous. Modern cars are simpler to operate (self-starting, self-shifting, power steering, etc.). But they are vastly more complex to manufacture, repair, maintain, etc.

43. Further discussion of some of this chapter's themes is presented in the author's *Scientific Progress* (Oxford: Blackwell, 1978), *Cognitive Economy* (Pittsburgh: University of Pittsburgh Press, 1989), and *The Limits of Science*, 2nd ed. (Pittsburgh: University of Pittsburgh Press, 1999).

NOTES TO CHAPTER 14

1. The author's *Cognitive Systematization* (Oxford: Blackwell, 1979) deals with further aspects of these matters.

2. Note that this is independent of the question Would we ever want to do so? Do we ever want to answer all those predictive questions about ourselves and our environment, or are we more comfortable in the condition in which "ignorance is bliss"?

3. One possible misunderstanding must be blocked at this point. To learn about nature, we must interact with it. And so, to determine some feature of an object, we may have to make some impact on it that would perturb its otherwise obtaining condition. (That indeterminacy principle of quantum theory affords a well-known reminder of this.) It should be clear that this matter of physical interaction for data acquisition is not contested in the ontological indifference thesis here at issue.

4. S. W. Hawking, "Is the End in Sight for Theoretical Physics?" *Physics Bulletin*, vol. 32 (1981), pp. 15–17.

5. As stated this question involves a bit of anthropomorphism in its use of "you." But this is so only for reasons of stylistic vivacity. That "you" is, of course, only shorthand for "computer number such-and-such."

6. This sentiment was abroad among physicists of the *fin de siècle* era of 1890–1900. (See Lawrence Badash, "The Completeness of Nineteenth-Century Science," *Isis*, vol. 63 [1972], pp. 48–58.) and such sentiments are coming back into fashion today. See Richard Feynmann, *The Character of Physical Law*, Cambridge, MA: MIT Press (1965), p. 172. See also Gunther Stent, *The Coming of the Golden Age*, Garden City, NY: Natural History Press, 1969); and S. W. Hawking, "Is the End in Sight for Theoretical Physics?" *Physics Bulletin*, vol. 32 (1981), pp. 15–17.

7. See Eber Jeffrey, "Nothing Left to Invent," *Journal of the Patent Office Society*, vol. 22 (July 1940), pp. 479–81.

8. For this inference could only be made if we could move from a thesis of the format $\sim(\exists r)(r \in S \ \& \ r \Rightarrow p)$ to one of the format $(\exists r)(r \in S \ \& \ r \Rightarrow \sim p)$, where "$\Rightarrow$" represents a grounding relationship of "furnishing a good reason" and p is, in this case, the particular thesis "S will at some point require drastic revision." That is, the inference would go through only if the lack (in S) of a good reason for p were itself to assure the existence (in S) of a good reason for $\sim p$. But the transition to this conclusion from the given premise would go through only if the former, antecedent fact itself constituted such a good reason that is, only if we have $\sim(\exists r)(r \in S \ \& \ r \Rightarrow p) \Rightarrow \sim p$. Thus, the inference would go through only if, by the contraposition, $p \Rightarrow (\exists r)(r \in S \ \& \ r \Rightarrow p)$. This thesis claims that the vary truth of p will itself be a good reason to hold that S affords a good reason for p—in sum, that S is complete.

9. One possible misunderstanding must be blocked at this point. To learn about nature, we must interact with it. And so, to determine some feature of an object, we may have to make some impact upon it that would perturb its otherwise obtaining condition. (That indeterminancy principle of quantum theory affords a well-known reminder of this.) It should be clear that this matter of physical interaction for data acquisition is not contested in the ontological indifference thesis here at issue.

10. Compare Chapter 10 of the author's *Limits of Science* (Pittsburgh: University of Pittsburgh Press, 1999).

11. Immanuel Kant, *Critique of Practical Reason*, p. 122 [Akad.].

12. See the author's *Peirce's Philosophy of Science* (Notre Dame, IN: University of Notre Dame Press, 1978).

13. Some further aspects of this chapter's themes are presented in the author's *Limits of Science* (Berkeley and Los Angeles: University of California Press, 1984; rev. ed. Pittsburgh: University of Pittsburgh Press, 2000).

Notes to Chapter 15

1. Erwin Schroedinger, *What Is Life?* (Cambridge: Cambridge University Press, 1945), p. 31.

2. Eugene P. Wigner, "The Unreasonable Effectiveness of Mathematics in the Natural Sciences," *Communications on Pure and Applied Mathematics*, vol. 13 (1960), pp. 1–14 (see p. 2).

3. Ibid, p. 14.

4. Albert Einstein, *Lettres à Maurice Solovine* (Paris: Gauthier-Villars, 1956), pp. 114–15.

5. K. R. Popper, *Objective Knowledge* (Oxford: Clarendon Press, 1972), p. 28.

6. Mary Hesse, *Revolutions and Reconstructions in the Philosophy of Science* (Bloomington, IN: University of Indiana Press, 1980), p. 154.

7. Conversations with Gerald Massey have helped in clarifying this part of the argument.

8. Galileo Galilei, *Dialogo II in Le Opere di Galileo Galilei* (Edizio Nazionale, vols. I–XX (Florence, 1890–1909), vol. VII, p. 298. (I owe this reference to Juergen Mittelstrass.) Kepler wrote, "Thus God himself was too kind to remain idle, and began to play the game of signatures, signing his likeness into the world. I therefore venture to think that all nature and all the graceful sky are symbolized in the art of geometry." (Quoted in Freeman Dyson, "Mathematics in the Physical Sciencs" in *The Mathematical Sciences*, ed., by the Committee on Support of Research in the Mathematical Sciences [Cambridge, MA: MIT Press, 1969], p. 99.)

9. To say that such a world must be *understandable* in such terms is not, of course, to say anything about how far intelligent beings will actually succeed in understanding it.

10. Just this approach is the salient feature of the Quinean program of "epistemology naturalized."

11. It is this unavoidable error-tolerant aspect of nature that blocks any prospect of a naive "it works, therefore it's true" pragmatism at the level of *theses*. To be sure, as regards large-scale *methods* from providing action-guiding theses, the situation is different. Here "it works *systematically*, therefore it is cogent (as a cognitive method—i.e., its deliverances are rationally credible)" is something else again. (See the author's *Methodological Pragmatism* [Oxford: Basil Blackwell, 1977].)

12. Further deliberations regarding some of this chapter's themes are presented in the authors' *Scientific Realism* (Dordrecht: D. Reidel, 1987).

Notes to Chapter 16

1. E. Purcell in A. G. W. Cameron (ed.), *Interstellar Communication: A Collection of Reprints and Original Contributions* (New York and Amsterdam: W. A. Benjamin, 1963), p. 142.

2. See Stephen H. Dole, *Habitable Planets for Man* (New York: Blaisdell, 1964; 2nd ed., New York: American Elsevier, 1970) and also Carl Sagan, *Cosmos* (New York: Random House, 1980).

3. See p. 29 of A. G. W. Cameron (ed.), *Interstellar Communication: A Collection of Reprints and Original Contributions* (New York and Amsterdam: W. A. Benjamin, 1963).

4. See Su-Shu Huang "Life Outside the Solar System," *Scientific American* 202, vol. 4 (April 1960), p. 55.

5. See Sir Arthur Eddington, *The Nature of the Physical World* (New York: Macmillan; Cambridge, Eng.: The University Press, 1928), p. 177.

6. See I. S. Shklovskii and Carl Sagan, *Intelligent Life in the Universe* (San Francisco, London, Amsterdam: Holden-Day, 1966), p. 411.

7. *Ibid.*, p. 359.

8. See A. G. W. Cameron (ed.), *Interstellar Communication: A Collection of Reprints and Original Contributions* (New York and Amsterdam: W. A. Benjamin, 1963), p. 190.

9. On this issue see George Gaylord Simpson, "The Nonprevalence of Humanoids," *Science*, vol. 143 (1964), pp. 769–775, chapter 13 of *This View of Life: The World of an Evolutionist* (New York: Harcourt Brace, 1964).

10. See A. G. W. Cameron (ed.), *Interstellar Communication: A Collection of Reprints and Original Contributions* (New York and Amsterdam: W. A. Benjamin, 1963), p. 75.

11. Compare A. G. W. Cameron (ed.), *Interstellar Communication,* p. 312.

12. Christiaan Huygens, *Cosmotheoros: The Celestial Worlds Discovered—New Conjectures Concerning the Planetary Worlds, Their Inhabitants and Productions* (London, 1698; reprinted London: F. Cass & Co., 1968), p. 359.

13. This chapter draws upon the author's essay "Extraterrestrial Science," *Philosophia Naturalis,* vol. 21 (1984), pp. 400–24.

Notes to Chapter 17

1. The thesis "I know that p is a known fact that I don't know" comes to:

$$Ki[(\exists x)Kxp \;\&\; \sim Kip])\; (\text{here } i = \text{oneself})$$

This thesis entails my knowing both $(\exists x)Kxp$ and $\sim Kip$. But the former circumstance entails Kip, and this engenders a contradiction.

2. In maintaining (i.e., claiming to know) $p\;\&\sim Kip$ we claim:

$$Ki(p\;\&\sim Kip).$$

But since $Kx(p\;\&\;q) \Rightarrow (Kxp\;\&\;Kxq)$ obtains, we obtain both Kip and $Ki(\sim Kip)$. But the latter of these entails $\sim Kip$. And so a manifest contradiction results.

3. Accordingly there is no problem about "t_o is a (particular) truth *you* don't know," although I could not then go on to claim modestly that "You know everything that I do." For the contentions $\sim Kyt_o$ and $(\forall t)(Kit \rightarrow Kyt)$ combine to yield $\sim Kit_o$ which conflicts with the claim Kit_o that I stake in characterizing $t°$ as a truth.

4. The progress of science offers innumerable illustrations of this phenomenon, as does the process of individual maturation:

> After three or thereabouts, the child begins asking himself and those around him questions, of which the most frequently noticed are the "why" questions. By studying what the child asks "why" about one can begin to see what kind of answers or solutions the child expects to receive. . . . A first general observations is that the child's whys bear witness to an intermediate precausality between the efficient cause and the final cause. Specifically, these questions seek reasons for phenomena which we see as fortuitous but which in the child arouse a need for a finalist explanations. "Why are there two Mount Salèves, a big one and a little one?" asked a six year-old boy. To which many of his contemporaries, when asked the same question, replied, "One for big trips and another for small trips." (Jean Piaget and B. Inhelder, *The Psychology of the Child,* trans. by H. Weaver [New York: Basic Books, 1969], pp. 109–10)

5. Immanuel Kant, *Prolegomena to Any Future Metaphysic* (1783), sect. 57; *Akad.,* p. 352.

6. Or perhaps alternatively: always after a certain time—at every stage subsequent to a certain juncture.

7. *Socrates dicens, se ispum dicere falsum, nihil dicit.* (Prantl, *Geschichte der Logik in Abenlande* [Leipzig: S. Hirzel, 1955], vol. IV, p. 139 n. 569.) It became a commonly endorsed doctrine in late medieval times that paradoxical statements do not preset propositions and for this reason cannot be classed as true or false. (See E. J. Ashworth, *Language and Logic in the Post-Medieval Period* (Dordrecht: Reidel, Yale University Press, 1974), p. 115 for later endorsements of this approach.) Thus later writers dismissed insolubles as not being propositions at all, but "imperfect assertions" (*orationes imperfectae*). (See E. J. Ashworth, *Language and Logic in the Post-Medieval Period* [Dordrecht: Reidel, 1974], p. 116.)

8. This work was published together with a famous prior (1872) lecture *On the Limits of Scientific Knowledge* as *Ueber Die Grenzen des Naturekennens: Die Sieben Walträtsel—Zwei Vorträge* (11th ed., Leipzig, 1916). The earlier lecture has appeared in English trans. "The Limits of Our Knowledge of Nature," *Popular Scientific Monthly*, vol. 5 (1874), pp. 17–32. For Reymond cf. Ernest Cassirer, *Determinism and Indeterminism in Modern Physics: Historical and Systematic Studies of the Problems of Causality* (New Haven: Yale University Press, 1956), part 1.

9. Bonn, 1889. Trans. by J. McCabe as *The Riddle of the Universe—at the Close of the Nineteenth Century* (New York and London, 1901). On Haeckel see the article by Rollo Handy in *The Encyclopedia of Philosophy* (ed. by Paul Edwards), vol. III (New York, 1967).

10. Haeckel, op. cit., pp. 365–66.

11. Charles Sanders Peirce, *Collected Papers*, ed. by C. Hartshorne et al., vol. VI (Cambridge, MA, 1929), sect. 6.556.

12. Of course these questions already exist—what lies in the future is not their existence but their presence on the agenda of active concern.

13. Some of this chapter's themes are also treated in the author's *Limits of Science*, 2nd ed. (Pittsburgh: University of Pittsburgh Press, 2000). And see also Timothy Williamson, *Knowledge and Its Limits* (Oxford: Oxford University Press, 2000).

NOTES TO CHAPTER 18

1. Michael Dummett, "Truth," *Proceedings of the Aristotelian Society*, vol. 59 (1958–59), p. 160.

2. C. S. Peirce, *Collected Papers*, ed. by C. Hartshorne and P. Weiss (Cambridge, MA: Harvard University Press, 1934), vol. V, sect. 5.64–67.

3. On the other hand, there is also the fact that we can control the content and the outcome of our dreams as little as those of conscious experience.

4. *Mediations*, No. VI: *Philosophical Works*, edited by E. S. Haldane and G. R. T. Ross, vol. I (Cambridge, MA: Cambridge University Press, 1911), pp. 187–89.

5. An informative of philosophical issues located in this general area, see Vincent Julian Fecher, *Error, Deception, and Incomplete Truth* (Rome: Officium Libri Catholici, 1975).

6. To be sure, abstract things, such as colors or numbers, will not have dispositional properties. For being divisible by four is not a disposition of sixteen. Plato got the matter right in Book VII of the *Republic*. In the realm of abstracta, such as those of mathematics, there are not genuine processes—and process is a requisite of dispositions. Of course, there may be dispositional truths in which numbers (or colors, etc.) figure that do not issue in any dispositional properties of these numbers (or colors, etc.) themselves—a truth, for example, such as my predilection for odd numbers. But if a truth (or supposed truth) does no more than to convey how someone thinks about a thing, then it does not indicate any property of the thing itself. In any case, however, the subsequent discussion will focus on *realia* in contrast to *fictionalia* and *concreta* in contrast to *abstracta*. (Fictional things, however, can have dispositions: Sherlock Holmes was addicted to cocaine, for example. Their difference from *realia* is dealt with below.)

7. This aspect of objectivity was justly stressed in the "Second Analogy" of Kant's *Critique of Pure Reason*, though his discussion rests on ideas already contemplated by Leibniz, *Philosophische Schriften*, edited by C. I. Gerhardt, vol. VII, pp. 319 ff.

8. See C. I. Lewis, *An Analysis of Knowledge and Valuation* (La Salle, IL: Open Court, 1962), pp. 180–81.

9. This also indicates why the dispute over mathematical realism (Platonism) has little bearing on the issue of physical realism. Mathematical entities are akin to fictional entities in this—that we can only say about them what we can extract by deductive means from what we have explicitly put into their defining characterization. These abstract entities do not have nongeneric properties since each is a "lowest species" unto itself.

10. Our position thus takes no issue with P. F. Strawson's precept that "facts are what statements (when true) state." ("Truth," *Proceedings of the Aristotelian Society*, Supplementary Vol. 24 (1950), pp. 129–56; see p. 136.) Difficulty would ensue only if an "only" were inserted.

11. But can any sense be made of the idea of *merely* possible (i.e., possible but nonactual) languages? Of course it can! Once we have a generalized conception (or definition) of a certain kind of thing—be it a language or a caterpillar—then we are inevitably in a position to suppose the theoretical prospect of items that meet these conditions are over and above those that in fact do so. The prospect of supposing the existence of certain "mere" possibilities cannot be denied—that, after all, is just what possibilities are all about.

12. Note, however, that if a Davidsonian translation argument to the effect that "if it's sayable at all, then, it's sayable in *our* language" were to succeed—which it does not—then the matter would stand on a very different footing. For it would then follow that any possible language can state no more than what can be stated in our own (actual) language. And then the realm of facts (i.e., what is (correctly) statable in some *possible* language) and of that of truths (i.e., what is (correctly) statable in some *actual* language) would necessarily coincide. Accordingly, our thesis that the range of facts is larger than that of truths hinges crucially on a failure of such a translation argument. (See Donald

Davidson, "The Very Idea of a Conceptual Scheme," *Proceedings and Addresses of the American Philosophical Association*, vol. 47 [1973–1974], pp. 5–20, and also the critique of his position in chapter II of the author's *Empirical Inquiry* [Totowa, NJ: Rowman & Littlefield, 1982].)

13. Compare Philip Hugly and Charles Sayward, "Can a Language Have Indenumerably Many Expressions?" *History and Philosophy of Logic*, vol. 4 (1983), pp. 73–82.

14. Compare F. H. Bradley's thesis: "Error is truth, it is partial truth, that is false only because partial and left incomplete," *Appearance and Reality* (Oxford: Clarendon Press, 1893), p. 169.

15. The author's *Empirical Inquiry* discusses further relevant issues.

16. One possible misunderstanding must be blocked at this point. To learn about nature, we must interact with it. And so, to determine a feature of an object, we may have to make some impact on it that would perturb its otherwise obtaining condition. (The indeterminacy principle of quantum mechanics affords a well-known reminder of this.) It should be clear that this matter of physical interaction for data-acquisition is not contested in the ontological indifference thesis here at issue.

17. William P. Alston, "Yes, Virginia, There Is a Real World," *Proceedings and Addresses of the American Philosophical Association*, 52 (1979), pp. 779–808 (see p. 779). Compare: "[T]he world is composed of particulars [individual existing things or processes] which have *intrinsic characteristics*—i.e., properties they have or relationships they enter into with other particulars independently of how anybody characterizes, conceptualizes, or conceives of them." Frederick Suppe, "Facts and Empirical Truth," *Canadian Journal of Philosophy*, col. 3 (1973), pp. 197–212 (see p. 200).

18. Compare deliberations about realism with those regarding objectivity in chapter 9 above. The same pragmatic rationale is at work in both contexts.

19. Kant held that we cannot experientially learn through perception about the objectivity of outer things, because we can only recognize our perceptions as *perceptions* (i.e., representations of outer things) if these outer things are supposed as such from the first (rather than being learned or inferred). As Kant summarizes in the "Refutation of Idealism": "Idealism assumed that the only immediate experience is inner experience, and that from it we can only *infer* outer things—and this, moreover, only in an untrustworthy manner. . . . But in the above proof it has been shown that outer experience is really immediate" (*Critique and Pure Reason*, B276).

20. The point too is Kantian in its orientation.

21. Maimonides, *The Guide for the Perplexed*, I, 71, 96a.

22. For some recent discussions of scientific realism, see Wilfred Sellars, *Science Perception and Reality* (London: Humanities Press, 1963); E. McKinnon, ed., *The Problem of Scientific Realism* (New York: Appleton-Century-Crofts, 1972); Rom Harré, *Principles of Scientific Thinking* (Chicago: University of Chicago Press, 1970); and Frederick Suppe, ed., *The Structure of Scientific Theories*, 2nd ed. (Urbana: University of Illinois Press, 1977).

23. Keith Lehrer, "Review of *Science, Perception, and Reality* by Wilfred Sellars," *The Journal of Philosophy*, vol. 63 (1966), p. 269.

24. Further discussion of some of this chapter's themes is presented in the author's *Limits of Science* (Pittsburgh, University of Pittsburgh Press, 2000).

Index of Names

Adams, Henry Brooks, 247, 254, 392n27, 393n41
Albert, Hans, 371n15
Almeder, Robert, 371n15, 389n12
Alston, William P., 369n1, 400n17
Amaldi, Edoardo, 388n4, 391n12
Anaximander of Miletus, 343
Archimedes, 115
Aristotle, xiii, 57, 113, 115–18, 122, 236, 327, 378n12, 381n2&7
Armstrong, D. M., 370n7
Asch, Solomon, 109, 380n9
Ashworth, E. J., 398m7
Audi, Robert, 373n6, 383n8
Austin, J. L., 373n9, 374n12
Axelrod, Robert, 380n7
Ayer, A. J., 118

Bacon, Francis, 214, 225, 371n6
Badash, Lawrence, 394n6, 395n9
Baier, Kurt, 193, 386n3
Ball, J. A., 307
Becquerel, Henri, 260, 346
Benn, Stanley I., 386n2
Bennett, Jonathan, 386n2
Benstein, Richard J., 385n1
Bernoulli, Daniel, 248
Bayle, Pierre, 371n7
Blanshard, Brand, 384n15
Bohr, Neils, 32
Boltzmann, Ludwig, 236
BonJour, Lawrence, 383n8
Bonk, Sigmund, 377n4
Bradbury, Ray, 308

Bradley, F. H., 133, 344, 400n14
Brandom, Robert, 175, 371n9&11, 385n2
Brandon, Robert N., 390n5
Brentano, Franz, 93, 118, 381n10
Bridgman, P. W., 25
Bromley, D. A., 387n2, 388n5, 393n36

Caesar, Julius, 64
Carnap, Rudolf, 24, 373n10, 379n19
Carneades, 374n16, 378n10
Carrier, L. S., 373n5
Cassirer, Ernest, 398n8
Chammah, A. M., 380n3
Cherniak, Christopher, 386n2
Chisholm, Roderick M., 93, 115, 118, 381n5, 381n10&12, 383n8
Churchland, P. M., 375n8
Churchland, P. S., 375n8
Cicero, 387n11
Comte, Auguste, 218, 326
Copernicus, 33
Craig, Edward, 369n1

Dalton, John, 32
Darwall, Stephen L., 386n2
Davidson, Donald, 68, 74, 375n5, 376n13, , 399n12, 400n12
Davis, Morton D., 380n3
De Bois-Reymond, Emil, 324, 325
Delaney, C. F., 374n18
Derrida, Jacques, 74
Descartes, René, 68, 85, 93, 117, 118, 155, 338, 386n4
Dewey, John, 25, 153

Dilthey, Wilhelm, 153
Dingler, Hugo, 116, 381n6
Dole, Stephen H., 300, 396n2
Doyle, A. Conan, 371n14
Dummett, Michael, 336, 398n1
Dupré, John, 391n11
Durkheim, Émile, 198
Dyson, Freeman, 395n8

Eddington, Arthur, 302, 362, 387n1, 396n5
Einstein, Albert, 280, 289, 311, 395n4
Ellis, Brian, 386n3
Epstein, Richard A., 376n3, 377n6
Euclid, 113
Evans-Prirchard, E. E., 199
Ewell, Raymond, 239, 391n17
Ewing, A. C., 381n5

Fecher, Vincent Julian, 399n5
Feyerabend, Paul K., 46, 389n16
Feynmann, Richard, 394n6
Fiebig, Hans, 389n19
Firth, Roderick, 381n10
Frankfurt, Harry G., 374n11

Galen, 156
Galileo, Galilei, 33, 286, 395n8
Garfinkel, Harold, 379n21
Gauthier, David, 380n7
George, William, 391n19
Gettier, Edmund, 3, 369n1
Gibbon, Edward, 64, 248, 392n30, 392n31
Gifford, Nancy L., 384n1
Goedel, Kurt, 319, 345
Goethe, Wolfgang von, 120
Goldman, A. I., 373n6, 383n8
Goldman, Alvin, 369n1
Gonseth, Gerdinand, 378n14
Goodman, Lenn E., 385n5

Haeckel, Ernest, 324, 325, 398n9, 398n10
Hall, Roland, 377n4
Handy, Rollo, 398n9

Hanson, Norwood R., 378n14
Hardwig, John, 380n5
Harman, Gilbert, 383n8
Harré, Rom, 400n22
Harris, J. F., 385n1
Hartshorne, Charles, 389n13
Harvey, William, 260, 346
Hawking, S. W., 394n4&6, 395n9
Hayek, F. A., 375n9
Hegel, G . F. W., 261, 344
Heidegger, Martin, 385n7
Heraclitus, 323
Hesse, Mary, 281, 395n6
Hilbert, David, 381n4
Hilpinen, Risto, 379n17
Hobbes, Thomas, 225, 389n19
Holland, John H., 390n5
Holton, Gerald, 387n2, 388n6, 393n36
Huang, Su-Shu, 396n4
Hugly, Philip, 400n12
Hume, David, 50–51, 188
Huntford, Roland, 369n1, 374n19
Husserl, Edmund, 260, 346
Huygens, Christiaan, 312, 397n12

Ilbert, Courtenay, 376n2
Inhelder, B., 397n4

James, William, xvii, 50–51, 55, 68, 153, 156, 168, 169, 174n20, 369n2, 374n14, 379n24, 385n2&3&8
Jeffrey, Eber, 394n7, 395n10
Jeffrey, Richard, 379n19
Jevons, W. Stanley, 218, 388n9
Joachim, H. H., 381n5, 384n14
Johnson, Samuel, 155
Johnson. Oliver A., 373n8

Kalven, Harry Jr., 380n12
Kant, Immanuel, 68, 180, 188, 271, 315, 321–323, 326, 350, 352, 354, 375n4, 390n3, 395n11, 397n5, 399n7, 400n19
Kapitsa, Peter, 388n7
Kaufmann, Stuart, 390n6
Kekes, John, 386n5, 387n10
Kepler, Johannes, 72, 142, 395n8

Index of Names

Klein, Peter, 373n6
Koyré, Alexandre, 311
Kuhn, Thomas, 224, 389n18
Kyburg, H. K. Jr., 378n17

Lalande, André, 376n3, 377n3
Lambert, J. H., 114, 381n1, 381n3
Laudan, Larry, 372n16, 385n1
Lehe, Robert Tad, 384n15
Lehrer, Keith, 43, 369n1, 372n5, 373n6, 374n13, 383n8, 401n23
Levi, Isaac, 379n18, 381n4
Levin, Michael E., 372n18
Lévy-Bruhl, Lucien, 199, 387n8
Lewis, C. I., 118, 381n9, 340, 373n10, 383n7, 399n8
Locke, John, 376n13
Lycan, W. G., 383n8

Maimonides, 400n21
Makinson, D. C., 371n4
Makridakis, S., 371n3
Malcolm, Norman, 373n10
Manktelow, K. I., 386n2
Margenau, Henry, 388n4
Margolis, Joseph, 385n1
Massey, Gerald, 395n7
Maxwell, James Clark, 57
McDowell, John, 376n13
McGill, Alan, 385n1
McTaggart, J. M. E., 319
Menard, H. W., 391n18
Mendeleev, D. I., 236
Merton. R. K., 379n1
Mill, John Stuart, 218
Mittelstrass, Juergen, 395n8
Montaigne, Michel de, 22
Moore, G. E., 11, 155, 370n9
Mortimore, G. W., 386n2
Moser, Paul K., 369n1, 386n2

Nansen, Fridtjof, xvii, 54, 369n1, 374n19
Nathanson, Stephan, 386n2
Newman, J. H., 372n2

Newton, Isaac, 33, 57, 102, 115, 214, 236, 238, 266, 280, 294
Nietzsche, Friedrich, 24, 184, 386n6
Niiniluoto, Ilkka, 372n16
Nozick, Robert, 373n6

Over, D. E., 386n2

Palmer, Scott D., 384n15
Pap, Arthur, 384n14
Pappus, 115
Pearman, J. P. T., 301, 302
Pears, David, 386n2
Peirce, Charles Sanders, 25, 52, 57, 71, 155, 218, 223, 238, 267, 270, 272, 282, 326, 329, 335, 336–39, 347, 358, 371n15, 374n18, 374n21, 376n11, 377n5, 390n3, 392n26, 398n2&11
Petley, B. W., 390n9
Piaget, Jean, 397n4
Planck, Max, 235, 247, 392n26
Plantinga, Alvin, 369n1
Plato, 369n1, 381n8, 399n6
Poincaré, Henri, 230
Polanyi, Michael, 383n9
Pollock, John, 369n1, 383n8
Polya, George, 378n11
Popper, Karl R., 25, 281, 371n15, 372n16, 382n13, 395n5
Price, Derek J. De Solla, 389n14, 391n14, 391n15, 393n33&34
Prior, A. N., 371n4
Pritchard, H. A., 372n4
Protagoras, 335
Pruss, Alexander, 376n15
Ptolemy, 33, 57, 115, 236
Putnam, Hilary, 376n13, 385n1
Pyrro, 371n7

Quay, Paul M., 389n18
Quine, W. V. O., 68, 383n8, 388n10
Quinton, Anthony, 381n10

Rapp, Friedrich, 389n18
Rappoport, Anatol, 380n3
Ravetz, Jerome R., 380n2

Reichenbach, Hans, 24, 390n1
Rorty, Richard, 385n1, 385n1
Rosenberg, Jay, 383n8
Rougier, Louis, 376n10
Royce, Josiah, 344
Russell, Bertrand, 4, 75, 139, 372n3, 381n6
Rutherford, Ernst, 32

Sabini, John, 380n8, 380n11
Sagan, Carl, 300, 304, 313, 396n6
Sartre, Jean Paul, 163
Sayward, Charles, 400n13
Scheffler, Israel, 377n8
Schick, Frederick, 386n2
Schiller, F. S. C., 153
Schlick, Moritz, 381n6
Schroedinger, Erwin, 280, 289, 395n1
Schutz, Alfred, 379n21
Sebond, Raymond, 371n6
Sellars, Wilfred, 400n22, 401n23
Sextus, Empiricus, 22, 24, 58, 385n2
Shakespeare, William, 287, 334
Shklovskii, I. S., 304, 313, 396n6
Shope, Robert K., 369n1
Siegel, Harvey, 385n1
Silver, Maury, 380n8&11
Simon, Herbert, 378n14, 381n11, 386n2
Simpson, George Gaylord, 396n9
Smith, Adam, 109
Solovine, Maurice, 280
Sorensen, Roy A., 371n2
Sosa, Ernest, 62–63, 78, 369n1, 375n2
Sowell, Thomas, 380n6
Speigelberg, Herbert, 384n1
Spencer, Herbert, 234, 235, 390n4

Spinoza, Baruch, 115, 280, 344
Stent, Gunther S., 217, 388n7, 394n6
Stephen, James, 83
Stough, Charlotte L., 378n10
Strawson, P. F., 399n10
Suppe, Frederick, 400n17
Sylvan, Richard, 371n9

Themistius, 356
Thyssen, Johannes, 384n1
Tuomela, Raimo, 380n7
Twain, Mark, 75

Ulam, Stanislaw M., 392n25
Ullian, Joseph, 383n8
Unger, Peter, 24–25, 373n6, 375n22, 384n1

Van Fraassen, Bas, 25
Visher, S. S., 391n13

Wallon, Douglas N., 377n4
Weinberg, Steven, 391n12
Weinberg, Alvin M., 240
Weizsaecker, C. F. von, 391n12
Whately, Richard, 85, 377n7
White, A. R., 381n7
Whitehead, A. N., 68
Wigner, Eugene P., 280, 388n4, 395n2
Williamson, Timothy, 398n13
Winch, Peter, 198, 385n4, 387n7
Winkley, R. G., 371n3
Wittgenstein, Ludwig, 25, 51

Zagzebski, Linda Trinkhaus, 369n1
Zeisel, Hans, 380n12

Made in the USA
Lexington, KY
28 March 2017